AN INTRODUCTION
TO COAL TECHNOLOGY

AN INTRODUCTION TO COAL TECHNOLOGY

N. BERKOWITZ

Fuel Sciences Division
Alberta Research Council
Edmonton, Canada

and

Department of Mineral Engineering
University of Alberta
Edmonton, Canada

ACADEMIC PRESS   New York   San Francisco   London   1979

A Subsidiary of Harcourt Brace Jovanovich, Publishers

ACADEMIC PRESS, INC.
111 Fifth Avenue, New York, New York 10003

United Kingdom Edition published by
ACADEMIC PRESS, INC. (LONDON) LTD.
24/28 Oval Road, London NW1 7DX

Library of Congress Cataloging in Publication Data

Berkowitz, Norbert, Date
    An introduction to coal technology.

    (Energy science and engineering)
    Includes bibliographical references.
    1. Coal. I. Title. II. Series.
TP325.B46   662'.62   78–19663
ISBN 0–12–091950–8

PRINTED IN THE UNITED STATES OF AMERICA

80 81 82   9 8 7 6 5 4 3 2

For Sheila,
and for Jonathan, Brian, and Cheryl

CONTENTS

Contents

PREFACE

Since the late 1960s, and especially since the "oil crisis" of 1973, coal—which had for several decades played a steadily diminishing role in the energy economies of industrialized countries and was, indeed, often viewed as passé—has once more moved onto center stage of the energy scene. Projections of long-term availability and costs of alternative fossil fuels, and better appreciation of technical options for interchanging gas, oil, and coal, have not only led to a consensus that coal will again become an increasingly important component of future energy supplies, but also persuaded many that it will become a *preeminent* primary energy resource before the end of the century.

This book has been written with a view to assisting individuals who, in such circumstances, may wish to gain some technical familiarity with coal. It is not intended for specialists who in the course of their work routinely rely on the vast professional literature that a century of coal research has created. It addresses itself, rather, to scientists and engineers who are presently active in other fields, but who might now want to bring coal within the orbit of their interests, and to advanced students of chemical and mineral engineering who are contemplating careers in coal-related endeavors.

What I have attempted to provide is, in other words, an overview that tells what coal is, how it came into being, what its principal physical and chemical properties are, and how it is handled or processed for particular end uses. For each topic I have tried to provide sufficient information to enable an interested reader, when so inclined, to attack the more specialized literature.

In presenting such diverse material, I have followed a natural sequence by beginning with a brief account of the origin, formation, and distribution of coal (Chapter 1), and then summarizing its composition, classification, and most important properties (Chapters 2–7) before turning to beneficiation and handling (Chapters 8 and 9), combustion (Chapter 10), and various partial or complete conversion technologies (Chapters 11–14), and finally dealing with some aspects of pollution and pollution control (Chapter 15). This format effectively divides the book into two parts—the first (Chapters 1–7) centered on coal science and only the second (Chapters 8–15) concerned with technology *in sensu stricto*. But I hope that incursions into science and the at times fairly detailed discussions of coal chemistry—as, for example, in Section 5.7, which treats the molecular structure of coal—are not misunderstood and deemed to make my choice of title inappropriate. I believe that the scope, challenges, and limitations of coal technology cannot be fully appreciated without an understanding of coal compositions and properties; and since the complexity of coal does not make it easy to gain such understanding, I can only ask my readers' indulgence if I seem on occasion to have gone beyond what they might think sufficient for an overview.

A brief comment on my selection of material and bibliography may also be called for. As much as possible, I have confined myself to topics that bear more or less directly on the *technical* behavior of coal; and with respect to coal technologies I have sought to draw attention to processes that are either already of practical importance or that may reasonably be expected to gain such importance. (I have therefore included a short outline of tar processing, about which much can be learned from the experiences of the 1930s and 1940s, as well as summaries of coal gasification and liquefaction, whose potential still remains to be realized.) To facilitate follow-up reading, I have also, wherever possible, cited the most accessible literature—and, preferably, *English-language* literature. At first glance this certainly distorts the research and development scene, and I would therefore ask readers always to bear in mind that whatever convenience my choices offer them is at the expense of proper recognition of the outstanding contributions made in many other countries and reported in other languages.

ACKNOWLEDGMENTS

I should here like to acknowledge the assistance that I received from several colleagues at the Alberta Research Council and from a number of professional peers elsewhere whom I have the privilege of counting as friends. I am, in particular, deeply indebted to Professor Dr. M.-Th. Mackowsky for help with Section 2.1 (which deals with the very difficult subject of coal petrography) as well as for the microphotographs that are reproduced in Figs. 1.1.1 and 1.1.2, and to Dr. R. A. S. Brown, Dr. S. K. Chakrabartty, Dr. H. W. Habgood, and Mr. J. F. Fryer, who read and offered helpful comments on all or major parts of the text. I am also very grateful to Mr. F. Copeland and his staff in the Research Council's drafting office for undertaking the preparation of the numerous diagrams; to the Elsevier Publishing Company, John Wiley & Sons, Inc., IPC Business Press Ltd., NTIS, and Drs. M. and R. Teichmüller for permission to reproduce in the original or in slightly adapted form several figures that had previously appeared in their publications; and to my secretary, Mrs. Pearl Williams, for help in preparing the manuscript. Last, I want to acknowledge the encouragement and assistance given me by Academic Press, and to express my gratitute to my wife and children for their forbearance during many evenings and weekends when I devoted myself to writing rather than, as they had a right to expect, to them!

Yet, notwithstanding the help I received, what I present here is ultimately my responsibility, and I alone am to blame for any errors or shortcomings.

AN INTRODUCTION
TO COAL TECHNOLOGY

PART I

ORIGINS, FORMATION, AND PROPERTIES OF COAL

CHAPTER 1

ORIGINS AND FORMATION

Nature feeds on itself; and left undisturbed by man, even large accumulations of plant debris will in the end be completely destroyed. Sooner or later, through microbiological activity or, abiotically, through exposure to the elements, even the most resistant of plant tissues will be disrupted and their chemical constituents broken down to simpler compounds that can nourish succeeding floral generations. But occasionally an accident of nature intervenes to arrest decay and set the residual plant matter, in whatever state it may then be, on a different course. Through land subsidence or other changes in the local water regime, the debris is inundated and gradually covered by silts, which shield it against further degradation. And as it is buried under increasingly thick inorganic sediments, it is progressively compacted by overburden pressures and chemically altered by heat. Whenever this happens, coal forms.

The fact that coal is so widely distributed and occurs in the most diverse types of nonmarine strata makes it evident that termination of plant decay by flooding and subsequent burial was commonplace throughout a great part of geological history, and that coal-forming processes did not require particular kinds of debris. What went into the making of coal in any era was, quite simply, whatever plant life happened to flourish at the time; and massive coal deposits consequently began forming as soon as terrestrial vegetation became abundant. But the variety of this vegetation and the diverse conditions under which it accumulated and decayed had a profound effect on the kinds of coal that ultimately developed from it. For practical purposes, it is well to recognize that "coal" is merely a *generic* term and that the solids to which it is applied are often more dissimilar than alike.

1.1 Diagenesis and Metamorphism

The sharp dichotomy between the processes that acted on plant debris before and after its burial divides coal formation into two stages, the first

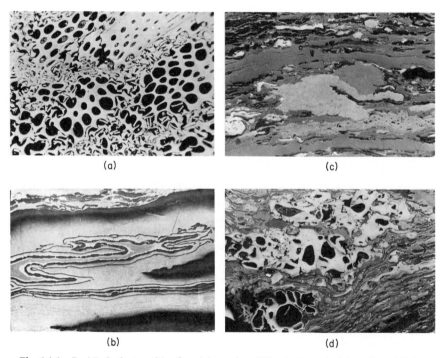

(a)   (c)

(b)   (d)

Fig. 1.1.1   Residual plant entities found in coal; × 290, photographed by reflected light. (a) Fusinite, showing variously preserved cell structure of "carbonized" woody tissues. (b) Cutinite, the remains of leaf cuticles. (c) Macrinite, unspecified detrital matter mostly derived from plant debris. (d) Sclerotinite, the remains of fungal sclerotia and mycelia. (Courtesy M.-Th. Mackowsky.)

characterized by extensive biochemical reactions and the second by abiotic thermal alteration of the organic mass.

Information about the biochemical or so-called *diagenetic* stage comes mainly from studies of plant remains that can be identified in thin sections or polished surfaces of coal under a microscope (Figs. 1.1.1 and 1.1.2); and these remains show that the conditions under which plant matter collected and decayed were generally quite as varied as in contemporary wilderness domains. In every geological period, major accumulations formed from forest vegetation as well as from hydrophilic species that thrived in swamps or shallow lakes and were then attacked by aerobic as well as by anaerobic microorganisms. Different types of environment coexisted in close spatial relationship or changed into each other with passage of time. Sometimes, debris drifted from its original (autochthonous) site to another (allochthonous) basin and then became intimately mixed with disparate "foreign" plant material and mineral matter en route. And wherever deposited, it

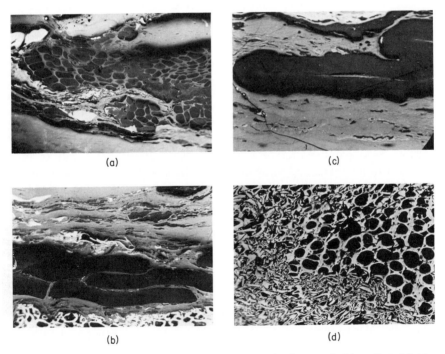

(a)

(c)

(b)

(d)

Fig. 1.1.2 Residual plant entities found in coal; ×290, photographed by reflected light. (a) Resin body, in tellinite derived from wood, bark, and cortical tissues of plants. (b) Sporinite, derived from fungal and other spores. (c) A macrospore embedded in tellinite. (d) Fusinite, showing partly crushed cell structures known as "Bogenstruktur." (Courtesy M.-Th. Mackowsky.)

often became a substrate for new growth, which, depending on circumstances, could also be comprised of a quite different assemblage of species.

Because of this diversity, individual diagenetic histories vary widely and are beyond the scope of this survey. But it is worth tracing their broad outlines, which can be inferred from what is now known about decay of modern debris under different environmental conditions.

All multicelled plants, regardless of their origin, are ultimately composed of

carbohydrates, i.e., mono-, di-, and polysaccharides, the latter mainly represented by starches, celluloses, and variously constituted "compound" celluloses;

glycosides, i.e., complexes of monosaccharides and hydroxylated aromatic or aliphatic compounds;

proteins, i.e., high-molecular-weight polypeptides, each carrying a specific sequence of amino acids;

fats, *waxes*, and *resins*, principally derived from terpenes or primary oxidation products of terpenes; and

a broad array of *alkaloids*, *purines*, *chitins*, *enzymes*, and *pigments*, the most important of the latter being chlorophyll and carotenoids.

In addition, woody tissues contain lignin, which is closely associated with hemicellulose[1] and diffused throughout the cell walls of such tissues. The structure of lignin is believed to be based on phenyl propane but is otherwise still uncertain and may, in fact, be variable. (Significant differences exist, for example, between lignins from conifers and angiosperms.)

How dead plant matter or individual "specialized" plant entities responded to microbial and/or abiotic attack depended therefore, in the first instance, on their composition. Plant fats, waxes, and resins generally resisted microbial degradation and, even in strongly aerobic environments, usually only underwent oxygen-promoted polymerization. Some pigments also survived for long periods without far-reaching alteration: Chlorophyll, for instance, commonly did so after intramolecularly rearranging to a stable porphyrin. However, except in completely stagnant media, in which *all* plant components were substantially preserved, cellulosic matter was fairly quickly broken down to simple sugars. Lignins were progressively oxidized to complex, variously structured "humic acids" and then to water-soluble benzenoid derivatives. Glycosides were hydrolyzed to sugars and a variety of aglycons, notably sapogenins and derivatives of hydroquinone. And proteins were "denatured" by random scission of the polypeptide chain and

---

[1] Because of this intimate association, some authorities speak of lignocellulose and regard this as one of the "compound" celluloses.

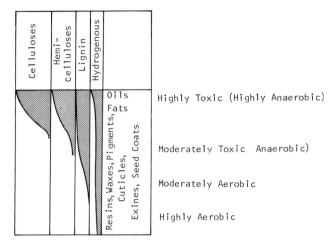

Fig. 1.1.3 Decay of plant components in different environments. (After White [1].)

yielded ill-defined slimes as well as free amino acids. Where not prematurely interrupted, these processes gradually converted the original plant matter into an undifferentiated "humus" in which many of the degradation products interacted with each other to further alter the residue.

But how fast decay proceeded, by what mechanisms it was sustained, and what kinds of residue it created depended also on factors that characterized the environment of the particular debris.

Provided local temperature fluctuations were not so great as to impede microbial activity, *rates* of decay generally increased with temperature; and decay processes therefore always proceeded appreciably faster in warm climates than in cool ones. And where the geological environment could neutralize acid decay products by infusion of dissolved alkaline mineral matter, it also maintained a higher level of bacterial activity than was otherwise the case (Note A^2). However, decay *mechanisms* were principally determined by the accessibility of the debris to oxygen. Figure 1.1.3, adapted from White [1], illustrates this. In substantially dry and fully exposed (aerobic) accumulations, decay proceeded entirely by "dry rotting," i.e., by abiotic oxidation and fungal hydrolysis of the debris which preferentially destroyed cellulosic matter, and little true humus accrued. Instead, the surviving residue consisted mostly of humic acids and their oxidation products, and of "fusinized" woody fragments whose texture resembled that of charcoal (see Section 2.1). On the other hand, in strongly anaerobic (swampy) environments, the dominant decay processes were bacterial hy-

---

[2] These notes can be found at the end of each chapter to which they refer.

drolysis and reduction (Note B) of plant matter, which tended to yield slimes and gels that penetrated, and thereby preserved, other plant entities (Note C). And between these two archetypal extremes stood a spectrum of *humid* environments in which fungal and bacterial attack on the plant matter, variously reinforced by abiotic oxidation, created correspondingly varied humic masses.

While all diagenetic processes thus tended to homogenize the organic mass, the compositions of the mixtures that they produced depended on site-specific factors and on the time interval between the onset and termination of decay. The only common feature of different situations was that all created increasingly aromatic and acidic residues as time went by. When decay was finally arrested, carbon contents (calculated on dry material) had usually increased from ~ 40 to 45% in the "fresh" plant debris to $>60\%$.

The nature of the decay processes implicitly defines the mechanisms that eventually terminated decay and brought the diagenetic stage to a close. Microbial activity may here and there be assumed to have ended when the decaying mass became too acidic for fungi and bacteria or when it was inundated to depths at which it became wholly stagnant. But since termination of microbial activity by excessive acidity would not, of itself, also have terminated *abiotic oxidation*, and destruction of the residue would therefore, albeit more slowly than before, have continued as long as it remained open to the air, total arrest of decay could only have come with flooding to stagnancy levels and subsequent siltation.

In the second or *metamorphic* stage of development, which began when siltation had built up an appreciable burden of inorganic sediments over the primordial coal mass, further (entirely abiotic) alteration proceeded mainly by compaction, dehydration, and a series of "stripping" and condensation reactions (Note D). These processes, which are terminated only when the coal is mined and which, if not so interrupted, would eventually result in formation of anthracites, further homogenized the mass and steadily enriched it in carbon, first at the expense of its oxygen content and later through abstraction of hydrogen (mainly in the form of methane).

It is now generally recognized that the principal impulse for this chemical "maturation" was provided by heat. Pressure alone only modified certain physical properties, though it may, on occasion, also have facilitated chemical reactions by maintaining the coal mass under high compression. But heat was variously imparted from at least three different sources and had correspondingly different effects on the rate of metamorphic development (Note E).

The most common pattern, so usual as to be termed "normal metamorphism," was sustained by heat flow from the earth's interior and can be directly associated with geothermal gradients. Because of the heterogeneity of the surficial layers of the earth crust, these gradients, which average 14°C per kilometer of depth, vary from one locale to another, ranging from less than 10°C to over 30°C per kilometer of depth, and geographically separated, but otherwise comparable coal masses are consequently often found to have attained quite different degrees of maturity. However, since rates of chemical change are always very sensitive to temperature changes and will, in many instances, double with each 10°C rise, even modestly greater than surface temperatures could over long periods of time induce quite extensive metamorphism. Geological studies have shown that burial to depths of 1000 to 2000 m sufficed to raise carbon contents from less than 60% to as much as 80 to 85%.

Not uncommonly, though, metamorphic change was accelerated or driven beyond "normal" (geothermal) levels by more localized heat sources, which subjected the coal mass to substantially higher temperatures. Particularly important in causing such "regional metamorphism" were incursions of molten magma into coal-bearing strata and the intense tectonic disturbances associated with formation of mountains (orogeny).

Magmatic incursions, especially when they took the form of sill-like intrusions, provided a heat source which, as a rule, simply superimposed itself on the local geothermal gradient.[3] Classic examples of this effect are afforded by the Stirling coal field of Scotland, where a vertical succession of six seams overlies a whinstone sill [3], and by Tertiary Sumatran coals which have been regionally metamorphosed by an extensive laccolith [4]. In both instances, the coals have undergone much greater metamorphic change than could have taken place had they been exposed only to normal geothermal gradients; and in the Stirling field, the diminishing heat effect of the sill can be traced by ascending the strata from the lowest seam, some 65 m above the sill, to the highest, some 300 m above it.

A similar enhancement of "normal" metamorphic change was brought about by tectonic disturbances during uplift of the earth's crust, and these do, in fact, seem to have played a particularly important role in coal development: "Mature" coals (with more than 85% carbon) are rarely found elsewhere than in heavily folded and/or faulted strata. But whether the incremental metamorphic changes attributable to tectonism were caused solely by frictional heat or whether the shear stresses that generated such heat contributed at least indirectly to them is still an open question. Because

---

[3] Where a magma intersected a coal mass or otherwise came into sufficiently close contact with it to overheat the coal, it caused thermal decomposition rather than the type of maturation associated with metamorphism and produced a so-called "natural coke" or "cinder coal" with properties quite unlike those of coal (see Chapter 6). The term "contact metamorphism," which is conventionally used to describe this phenomena, is, strictly speaking, a misnomer in the context of coal metamorphism.

anthracite, the most fully developed coal type, generally occurs only in mountain terrain and anthracitic solids can only be "synthesized" in the laboratory by subjecting less mature coals to temperatures of 500 or 550°C, Roberts [5] has argued that such or even higher temperatures must have been generated during orogeny. However, this conclusion is difficult to reconcile with the properties of anthracite;[4] and for the time being, it seems more appropriate to suppose that how great a role pressure played in coal metamorphism depended on how it affected the immediately surrounding strata. Where it rendered these strata less permeable and therefore made it more difficult for gaseous maturation products to escape from the coal, it undoubtedly tended to retard metamorphic change [6, 7]. On the other hand, where tectonism caused extensive fracturing or brecciation of the surrounding strata and consequently facilitated escape of gaseous matter, it almost certainly accelerated metamorphism.

In passing, it might be noted that there are also instances where coal has been metamorphosed by radioactive emissions rather than by heat. Good examples have been found [8] in a suite of Jurassic and Triassic coals and coalified woods from sandstones near uranium mines in Utah and Colorado. In these cases, uranium appears to have entered the coal some time after the late Cretaceous, probably as an aqueous solution of a complex uranyl carbonate, and to have been adsorbed as the uranyl ion UO_2^{2+}, which was then reduced to uranite (UO_2). However, the changes caused in this manner are radically different from those associated with normal or regional metamorphism. The elemental compositions of the altered coals suggest that alpha particles emitted during radioactive decay progressively dehydrogenated the coal substance, and the coals consequently show several abnormalities, among them unusually high density.

Proceeding from local geological histories and geothermal gradients, several investigators (e.g., Karweil [9]; Francis [10]; M. and R. Teichmüller [11]; Lopatin and Bostick [12]) have made estimates of the rates at which coal subject to normal metamorphism will mature (see Figs. 1.1.4 and 1.1.5; Table 1.1.1). All show that these rates are markedly temperature sensitive and that geological time and temperature are therefore interchangeable. However, as in many other sequential reaction series, e.g., certain condensation and polymerization reactions, successive steps in the metamorphic progression become increasingly more difficult and require increasingly more energy. Each step, therefore, demands a characteristic minimum temperature, and uninterrupted metamorphic development requires that the coal

---

[4] Aside from involving a circular argument, the conclusion is based on short-lived experiments that disregard the interchangeability of time and temperature (see below). Anthracite formation at 500°C or higher is, in any event, inconsistent with the fact that anthracites begin to suffer thermal decomposition at much lower temperatures (see Section 6.1).

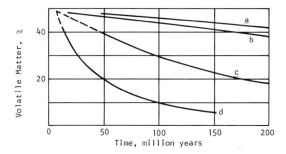

Fig. 1.1.4 Metamorphic development of coal: relationships between temperature, time, and coal rank as expressed by volatile matter contents. Curve a, 50°C ($\sim$1000- to 1300-meter depth); curve b, 60°C ($\sim$1200- to 1700-meter depth); curve c, 100°C ($\sim$2200- to 3000-meter depth); curve d, 150°C ($\sim$3500- to 4700-meter depth). (After Karweil [9]; by permission of M. and R. Teichmüller.)

mass is exposed to progressively higher temperatures. This is illustrated by Francis's estimates of the minimum temperatures and corresponding time intervals needed for various transitions (Table 1.1.2): While a "young" coal mass can, in principle, mature from 60 to 68% carbon at a temperature as low as 20°C (and will then take some 20 million years to complete the change), further development, to 75% carbon, cannot occur unless the temperature rises to at least 30°C. If the minimum temperatures are exceeded, the time intervals would, of course, be correspondingly shortened.

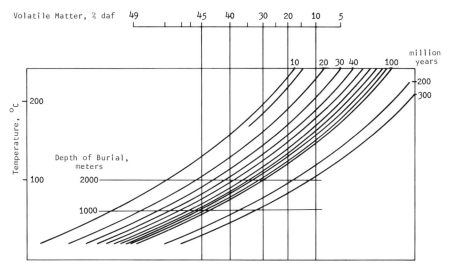

Fig. 1.1.5 Metamorphic development of coal; relationships between temperature, time, and coal rank as expressed by volatile matter contents. (After Karweil [9]; by permission of M. and R. Teichmüller.)

Table 1.1.1

Time–temperature relationship for development of anthracite[a]

Formation	Age (10^6 years)	Average formation temperature (°C)
Upper Carboniferous, Oklahoma	270	100
Jurassic, Texas	160	175
Pliocene/Pleistocene, Salton Geothermal Fields, California	<2	290–300

[a] After Bostick; cited by Lopatin and Bostick [12].

Table 1.1.2

Time–temperature relationship for metamorphic changes[a]

Carbon content (%) From	To	Minimum temperature (°C)	Minimum time (10^6 years)
	60	10	67
60	68	20	20
68	75	30	8.5
75	80	40	4
80	84	50	2.5
84	87	60	1.6
87	89	70	1
89	91	80	0.9
91	92	90	0.75
92	93	100	0.6
93	94.7	150	0.5

[a] After Francis [10].

Because Francis's estimates are generalized from studies of Indian and Australian coals, they are illustrative rather than definitive, and their accord with more specific geological studies is sometimes tenuous. For example, while Table 1.1.2 shows the final step (to 94.7% carbon) as requiring a minimum temperature of 150°C, Table 1.1.1 places the formation temperature of Oklahoma Upper Carboniferous anthracite (with similar carbon content) at 100°C. Despite their shortcomings, however, they express important basic aspects of coal metamorphism and offer a broad explanation for the existence of geologically old, yet relatively "immature," coals.

1.2 The Coal "Series"

Since metamorphic development, or *coalification*,[5] is, in chemical terms, synonymous with progressive enrichment of the coal substance in organically bound carbon, all coals, regardless of their origin, age, or type, can be arranged in an ascending order of carbon contents to form an apparently continuous coal "series." In North American terminology (Note F) this is written as

peat → lignite → subbituminous coal → bituminous coal → anthracite

and is, in effect, a rudimentary classification from which the more elaborate systems discussed in Section 3.3 have evolved. Conceptually, the endpoint of the series is graphite, which in some respects resembles anthracite and is in fact believed to be directly related to it but which is only very rarely reached (Note G).

Appearances to the contrary notwithstanding, the series does not, in itself, imply a close *genetic* connection between its members; and while it is fully consistent with the notion that lignites in time become subbituminous coals and that the latter are eventually transformed into bituminous coals, it must not be taken to indicate that such a progression is inevitable. There is at least a possibility that coalification of Late Mesozoic and Tertiary plant debris (which is now mostly found matured to lignite and subbituminous coal) may have proceeded along pathways that precluded further development into bituminous coals similar to those formed from older debris.

However, even without prior expansion into a more comprehensive classification scheme, the coal series provides a useful framework for ranking the different kinds of coal and for broadly assessing their suitability for particular end-uses. (As shown later, especially in Chapters 4–6, the covariation of physical and chemical properties with carbon contents is generally so systematic that behavior can be predicted with considerable confidence even when nothing beyond proximate analyses (see Section 2.2) or standings in the coal series are known.)

Rank, as applied to coal, carries the same meaning as degree (or extent) of maturation, and is therefore a qualitative measure of carbon contents. Lignites and subbituminous coals are customarily referred to as being low rank, while bituminous coals and anthracites are classed as high rank. (In European terminology, lignites and subbituminous coals are also called *soft* coals, while bituminous coals and anthracites are termed *hard* coals; see Section 3.3.) But in this connection it should be emphasized that rank

---

[5] This term, equivalent to the German *Inkohlung*, is commonly used to denote the gradual transition from chemically younger to older members of the coal series and has the same meaning as *increase in rank*.

is not synonymous with *grade* (which carries implications about quality), and that low-rank coals, although not as suitable for some uses as more mature members of the series, are often superior to them in other applications.

1.3 Major Coal-Forming Epochs

Since Devonian times, when terrestrial plant life first appeared on earth, diastrophic processes and repeated global changes in sea levels have continuously altered the areal extent as well as the form, topography, and climates of continental land masses; and regional geological histories therefore show periods particularly favorable for coal development alternating with less favorable ones.

Isolated small coal deposits, such as those found on Bear Island and a number of other Arctic islands, began forming in the Late Devonian from algae and plankton accumulations in open waters near the shore line of the Tethys Sea. This ocean then separated two Northern Hemisphere continents, Laurentia in the west and Angara in the east, from each other and from the huge Southern Hemisphere Gondwanaland. But massive and widespread formation of coal commenced only in the Carboniferous.

As the Tethys Sea began its slow retreat toward the present-day limits of the Mediterranean during Lower Carboniferous (or Mississippian) times, coal began being deposited in Pennsylvania and Virginia, in New Brunswick and Nova Scotia (Canada), in Scotland and Northumberland (Britain), and in Maine, Brittany, and Basse-Loire (France). And by Upper Carboniferous times (equivalent to the Permo-Carboniferous of North American and Australian terminology), coal development had reached into the interior Appalachian and Gulf regions of the United States, into Wales and the British Midlands, and into central and eastern Europe. For the first time large accumulations were then also forming in South Africa, northern India, northeast China, and Australia.

How closely shrinkage of the Tethys Sea throughout the Carboniferous period governed deposition of coal in the resultant swamplands is shown by lateral transitions in contemporaneous strata. For instance: while the first coal masses were laid down in swamps and shallow open water near the northern shore line of the Tethys Sea (in Scotland and Northumberland), great thicknesses of marine limestone, indicative of deep sea conditions, were deposited further south in England and Wales. However, the sea often advanced again over previously vacated land and repeated this cycle several times; and where this happened, distinctive *coal measures,* i.e., vertical successions of coal seams with intercalated inorganic sediments which reflect the depth and duration of flooding, came into being.

As in later ages, the vegetation that contributed the bulk of the organic mass depended on the local environment. In shallow open water along shore lines, the source materials were algae, plankton, and, to a lesser extent, primitive Bryophyta (mosses and liverworts). In swamplands they were mainly comprised of Pteridophyta, represented by ferns, horsetails, club mosses, and tree ferns. And on dry lands the major contributions were made by Spermatophyta, notably gymnosperms such as Cordaitales (the fore-runners of modern conifers) and Pteridospermeae (the forerunners of modern cycads). But plant remains in Carboniferous coals and in strata associated with them also show that each of these vegetative types was repre-sented by many different species, and that there were marked differences between the floras that contributed to coals in the Northern and Southern Hemispheres. In the Northern Hemisphere a relatively unchanging plant assemblage persisted throughout the Carboniferous. The most abundant club mosses were various species of *Lepidodendron* and *Sigillaria*. Horsetails were mostly represented by *Calamites*. Cordaitales included *Cordaites* (which resembled the modern tamarack) as well as several species resembling the kauri pine (*Agathis*). And the dominant Pteridospermeae were *Neuropteris*, *Alethopteris*, and *Pecopteris*. In contrast, the Permo-Carboniferous flora of the Southern Hemisphere was mainly comprised of glossopterids.

In the subsequent Permian and Triassic periods, climatic changes in the Northern Hemisphere created widespread desert conditions that generally precluded accumulation of peat, and coal formation was confined to a few scattered areas, notably Pennsylvania and Ohio, the Saar region of West Germany, central France, and western Poland. But more favorable condi-tions prevailed in the Southern Hemisphere where coal development con-tinued substantially as during the Permo-Carboniferous.

Similar disparities between north and south continued throughout most of the Jurassic period, during which major coal accumulations formed only in the trans-Ural regions of Russia, in Chungking Province of China, and in New Zealand and Australia. However, in the Northern Hemisphere a shallow Carboniferous sea, which bisected the North American continent, now began to shrink and thereby set the stage for subsequent deposition of massive coal deposits in western Canada and the western United States.

Toward the end of the Mesozoic, fragmentation of Gondwanaland and topographic changes in the Americas made the Cretaceous almost as pro-ductive of coal as the Carboniferous was some 150–200 million years earlier. Aside from the immense deposits laid down in a broad strip parallel to, and extending on either side of, the Rocky Mountain chain (Note H), a number of large coal accumulations in Mexico, Chile, Hokkaido (northern Japan),

Assam (northern India), Nigeria, Queensland (Australia), and New Zealand are all of Cretaceous origin. Because of their relatively recent origins, these coals, except where altered by regional metamorphism, have only reached subbituminous rank, and this sets them (as well as all later ones) superficially apart from older, more mature coals. Also noteworthy, however, is that Cretaceous coals developed from a distinctly different, more advanced flora. Pteridophyta (i.e., club mosses, ferns, horsetails, and glossopterids) and Gymnospermeae, which contributed so heavily to coal in the Carboniferous and Permo-Carboniferous, had by the late Mesozoic declined to a position of minor importance, with only conifers remaining widely distributed; and a dominant role was played by Angiospermae, which, because of their highly specialized structures, were usually much more susceptible to microbial degradation and "homogenization" during decay. As a result, Cretaceous and post-Cretaceous coals contain far fewer identifiable tissue remains than older ones.

With minor interruptions, conditions favorable for accumulation of coal were also widespread throughout most of the Tertiary era, particularly during the Eocene and Oligocene periods. Dating from these periods are the extensive lignite deposits of Saskatchewan (Canada), the Dakotas and Wyoming, Asiatic Russia, West Germany, Pakistan, northern India, Borneo, Sumatra, Manchuria, and the Arctic regions (Alaska and Canada's Northwest Territories).

Massive peat deposits, now generally found matured to lignites, continued to accumulate during the Miocene period, which, in particular, brought into being the unique 300-m-thick brown coal occurrence in the Latrobe Valley of Victoria (Australia) and coals in Central America (Venezuela and Mexico), southeastern Europe (southern Germany and the Volga region of Russia), and Asia (notably northern China).

The youngest coals are Pliocene lignites in Alaska, southeastern Europe, and southern Nigeria. However, extensive forest peats (in Alaska, Florida, and Wisconsin), moorland peats (in various regions of Europe and Siberia), and peat swamps (in southeast Asia)—all accumulated in Quaternary times—make it evident that the process of coal formation continues wherever sufficient vegetation and suitable topographies present themselves.

1.4 The Distribution of Coal

The extent of coal resources in any particular region is usually expressed in terms of

(a) ultimate in-place resources, and
(b) measured or proved reserves.

Ultimate in-place resources include all tonnages known or suspected to exist in the region within a stipulated distance from the surface but sometimes exclude coal in seams of less than some arbitrarily defined thickness. (This exclusion reflects the view that coal in such seams does not warrant recovery or could not be economically extracted in any foreseeable circumstances.) Data for assessing these resources are obtained from outcrops, seismic surveys, and exploratory drill holes (including water wells and holes drilled in the search for oil, gas, or minerals). However, since "observation points" are almost always fairly widely spaced, broad assumptions about the continuity of coal-bearing strata between any two points are commonly necessary; and for this reason, estimates of ultimate resources are invariably subject to substantial upward *or* downward revisions as additional information becomes available. In contrast, *measured* or *proved reserves,* which usually represent only a small fraction of the ultimate resources, are comprised of tonnages that have been delineated by detailed exploration programs and can therefore be definitely counted upon. Figure 1.4.1 illustrates the qualitative meaning of terms commonly used in reference to coal resources.

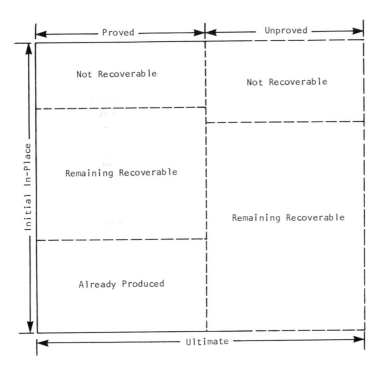

Fig. 1.4.1   Schematic representation of different categories of coal resources.

Table 1.4.1

Major coal resources of the world: all ranks, in 10^9 metric tons

	Estimated total	Percent of world total	Measured	Economically recoverable
North and Central America				
Canada	500	4.4	14	9
Mexico	12	0.1	5.3	0.7
United States	2925	27.1	364	182
South America				
Brazil	3.2		3.2	1.6
Chile	3.9		0.1	0.06
Colombia	5.3			
Venezuela	0.9			
Europe				
Britain	163	1.5	99	3.9
Belgium	1.8		0.5	0.3
Bulgaria	5.2		4.4	4.4
Czechoslovakia	21	0.2	13.8	2
France	31	0.3	1.4	0.5
German Democratic Republic	30	0.3	15	3
German Federal Republic	287	2.7	99.5	39.5
Greece	1.6		0.7	0.2
Hungary	5.7		3	0.7
Netherlands	3.7		3.7	1.8
Poland	60.6	0.6	38.9	3.8
Romania	1.4		0.7	0.3
Soviet Union (all)	5714	53	273	273
Spain	3.5		2.2	1.6
Yugoslavia	21.8	0.2	18	17
Asia				
Bangladesh	1.6		0.8	0.5
China, Peoples' Republic	1012	9.4	200	101
India	83	0.8	23	11.5
Indonesia	2.5		2.1	0.08
Japan	8.6	0.1	8.6	1
Korea, North	2		1	0.6
Korea, South	1.5		0.9	0.5
Turkey	7.3		2.9	0.1
Africa				
Mozambique	0.4		0.1	
Nigeria	0.5		0.5	0.3
Republic of South Africa	44.3	0.4	24.2	10.6
Rhodesia	6.6		1.7	1.7
Oceania				
Australia	199	1.8	74.7	24.3
New Zealand	1.1		0.4	0.2
World total	11,172		1301.3	697.7

Statistical information on coal resources often also provides some data on economically recoverable coal, i.e., tonnages that are expected to be extractable at acceptable cost by present or reasonably anticipated technology. These estimates generally assume that some 85–90% of all in-place coal amenable to surface mining can be recovered, and that the extraction rate for deeper coal will average 45–50%.

There is, however, still little uniformity in the criteria that different jurisdictions use for assessing and reporting their ultimate and proved reserves. There are wide divergencies with respect to what coal is excluded from the estimates on grounds of excessive depth or insufficient thickness, and equally wide differences in the types of data that are used to define measured or proved reserves. (Not infrequently such discrepancies are even observed in data reported by different investigators from the *same* jurisdiction.) Meaningful correlation of coal resource estimates from different sources is therefore beset by considerable difficulties.

Recent information on global coal resources, mainly drawn from the 1968 World Power Conference Survey of Energy Resources, the 1974 World Energy Conference Survey of Energy Resources, and World Coal (November 1975), is summarized in Table 1.4.1. In this tabulation, countries with estimated ultimate resources of less than 500 million tons have been omitted, and all data have been rounded to the nearest 0.1×10^9 tons. Blanks in the second column indicate less than 0.1% of the currently estimated world total. A notable feature is that three countries—the United States, the USSR, and the Peoples' Republic of China—are now believed to hold, together, 89.5% of that total.

Notes

A. Bacterial activity is markedly slowed when the pH of the medium falls below ∼5 and, as a rule, ceases altogether at pH ∼3. Fungal activity usually ceases at pH ∼2.

B. It should, however, be observed that since anaerobic bacteria use oxygen from plant matter for their own life processes, the products of bacterial synthesis, as distinct from the reduced debris itself, are *oxidation* products.

C. Later, during the second or metamorphic stage of coal formation, these slimes and gels formed the essentially structureless matrices in which preserved entities are now found embedded. In petrographic parlance, they can be identified with *collinite* and *tellinite* (see Section 2.1).

D. These reactions involve the elimination of peripheral substituent groups (such as —OH, —COOH, —OCH$_3$ and —CH$_3$) and, concurrently, the production of condensation products of increasing molecular weight.

E. Some recent evidence suggests that particularly rapid maturation may on occasion also have affected the *nature* of the chemical reactions involved in metamorphic development of coal [2].

F. Elsewhere, a slightly different nomenclature is occasionally used. Some British authorities, for example, refer to subbituminous coals as lignitous.

G. The reason for this is undoubtedly the very high energy required for transforming even extensively aromatized carbon, such as exists in high-molecular-weight polycondensed compounds, into graphitic carbon. Unless aided by specific catalysts that are not found in the natural state, conversion of tetravalent carbon into its graphitic form occurs only at temperatures in excess of 1200 to 1500°C.

H. The Rocky Mountain chain was elevated by the Laramide revolution toward the end of the Cretaceous period. The Andes and ancient Appalachians were raised at about the same time.

References

1. D. White, *Econ. Geol.* **28,** 556 (1933).
2. N. Berkowitz, J. F. Fryer, B. S. Ignasiak, and A. J. Szladow, *Fuel* **53,** 141 (1974).
3. W. Francis, "Coal," p. 590. Arnold, London, 1961.
4. C. A. Seyler, *Proc. S. Wales Inst. Eng.* **63** (2), 213 (1948).
5. J. Roberts, *Coke Gas* **18,** 25 (1956).
6. H. Jüntgen, and J. Karweil, *Erdöl Kohle* **19,** 251, 339 (1966).
7. G. Huck and K. Patteisky, *Fortschr. Geol. Rheinl. Westfalen.* **12,** 551 (1964).
8. S. Ergun, W. F. Donaldson, and I. A. Breger, *Proc. Int. Conf. Coal Sci., 3rd, Valkenburg* (1959).
9. J. Karweil, *Z. Dtsch. Geol. Ges.* **107,** 132 (1956); *Brennst. Chem.* **47,** 161 (1966).
10. W. Francis, "Coal," pp. 572–575. Arnold, London, 1961.
11. M. Teichmüller, and R. Teichmüller, *in* "Coal Science," Adv. Chem. Ser. 55, p. 133. American Chemical Society, New York, 1966.
12. N. V. Lopatin, and N. H. Bostick, *in* "Nature of Organic Matter in Recent and Fossil Sediments." Nauka Press, Moscow, 1973.

CHAPTER 2

COMPOSITIONS

The diversity of source materials from which coal formed and the successive changes wrought by diagenesis and metamorphism made coal so heterogeneous that statements about its composition or structure have little meaning unless they are related to *scale*. How coal is perceived to be constituted depends, in other words, not only on its rank, but also—and more importantly—on whether one considers its macroscopic, microscopic, or submicroscopic aspects.

2.1 Physical Composition

A dominant feature of hand specimens of bituminous coals is their banded appearance, which is due to an irregular alternation of variously thick glossy, lustrous, dull, and fibrous layers. In early literature this caused coals to be designated as *bright*, *predominantly bright*, or *dull* (see, e.g., Muck [1] and Fayol [2]).

Stopes [3], Thiessen [4], and Winter [5], among others, have published detailed descriptions of four recognizably distinct *banded components*. According to Stopes, whose terminology was later universally adopted and incorporated into a systematic petrographic nomenclature,

(a) *vitrain* presents itself in narrow, brilliantly black, glossy or vitreous bands, fractures conchoidally or into small cubes, and appears structureless to the unaided eye;

(b) *clarain* occurs as black, often horizontally striated, layers or extended lenticular masses, has a silky luster, and presents a glossy surface when freshly broken;

(c) *durain*, which also occurs in layers or as lenticular masses, is a dull gray-black material, possesses a tight granular texture, and breaks to display fine-grained or matte surfaces; and

(d) *fusain*, resembling charcoal, is usually found in small lenses that parallel the bedding plane of the coal, is intrinsically soft and friable,[1] and breaks readily into fibrous strands or to a powder.

Table 2.1.1 relates Stopes's designation of these banded components, now known as *lithotypes*, to terms that others have used to identify them.

However, far from being basic "building blocks" of bituminous coals, lithotypes are themselves complex aggregates of discrete microscopic entities and primarily reflect diverse depositional environments. The irregularly layered arrangement of lithotypes in large pieces of coal has been attributed to fluctuating conditions in the primordial peat bog, to differential settling of heavier plant components in humic gels, and to syneresis.[2]

In immature coals, i.e., lignites and subbituminous coals, banding is as a rule poorly expressed, because such coals, almost invariably of post-Jurassic age, formed mostly from angiosperms, which decayed more completely and consequently produced a more homogeneous humus than vegetation from which geologically older coals developed [6]. But under a microscope even the youngest coals are found to be comprised of optically differentiated entities that correspond to those in bituminous coals.

Regardless of rank, the physical composition of coal is therefore such as to make coal the organic analog of a conglomerate, with lithotypes

---

[1] Not infrequently, however, fusain is found to be intimately impregnated and substantially reinforced by mineral matter.

[2] A term first used by T. Graham in 1861, syneresis denotes the exudation of liquid from a quiescent gel. Examples are afforded by the behavior of silicic acid gels and gelatin or agar gels. In the case of a heterogeneous humic gel of the type from which coals formed, such a process could easily cause partial separation of vitrain precursors.

Table 2.1.1

Designations of "banded components" (lithotypes) in bituminous coal[a]

Stopes [3], Britain	Thiessen [4], U.S.A.	Muck [1], Germany	Winter [5], Germany	Fayol [2], France[b]
Vitrain	Anthraxylon[c]	Glanzkohle	Vitrit	Lames claires (Charbon brilliant)
Clarain	Attritus	Halbglanzkohle	Clarit	Houille folaire (Charbon semi-brilliant)
Durain	Splint	Mattkohle	Durit	Houille grenue (Charbon mat)
Fusain	—	Faserkohle	Fusit	Fusain (Charbon fibreux)

[a] Only Stopes' terminology is now used. Winter's designations, originally intended as German equivalents, have later come to be applied to *micro*lithotypes (see Table 2.1.3).

[b] The parenthetical terms show later French usage, now obsolete.

[c] From the Greek *anthrax* (= coal) and *xylon* (= wood); the term was intended to indicate that this component formed from woody tissues, as distinct from such other plant "ingredients" as leaf cuticles, fungal spores, and resin bodies, which also contributed to coal (see Table 2.1.2).

representing different rocks and so-called macerals (see below) being the counterparts of inorganic minerals.

The broad outlines of this complex makeup began to emerge in the 1920s from the work of, among others, Potonié [7], Duparque [8], and Hickling [9]. But definitive identification of the various microscopic constituents and progress toward formulation of a uniform comprehensive nomenclature for them became possible only after World War II, as improved experimental methods were developed and coal petrographers established channels for closer international collaboration (Note A). As now codified by the International Committee for Coal Petrography (see Mackowsky [10, 11]), the ultimate microscopic constituents of coal are a series of *macerals*, which are characterized by their appearance, chemical composition, and optical properties, and which can in most cases be traced to specific components of the plant debris from which the coal formed. These macerals present themselves in three *maceral groups*, and the groups, in turn, occur in a variety of associations termed *microlithotype* groups.

Table 2.1.2 lists the individual macerals and maceral groups now recognized by the ICCP and indicates the plant components from which they derive; Table 2.1.3 identifies the microlithotype groups, and Fig. 2.1.1 shows the composition of the macroscopic lithotypes in terms of the microlithotype groups. (It is helpful to note that all maceral and maceral group names end in *-inite*, microlithotypes in *-ite*, and lithotypes in *-ain*.) The sys-

Table 2.1.2

Coal macerals and maceral groups recognized by the International Committee for Coal Petrography[a]

Maceral group	Symbol	Maceral	Composed of or derived from
Vitrinite	V	Collinite	Humic gels
		Tellinite	Wood, bark, and cortical tissue
		Vitrodetrinite[b]	
Exinite	E	Sporinite	Fungal and other spores
		Cutinite	Leaf cuticles
		Resinite	Resin bodies and waxes
		Alginite	Algal remains
		Liptodetrinite[b]	
Inertinite	I	Micrinite	Unspecified detrital matter, $<10\ \mu$m
		Macrinite[c]	Similar, but $10–100\ \mu$m grains
		Semifusinite ⎱	
		Fusinite ⎰	"Carbonized" woody tissues
		Sclerotinite	Fungal sclerotia and mycelia
		Inertodetrinite[b]	

[a] After Mackowsky [10].

[b] These terms are applied to small entities that, because of their reflectivity, must be assigned to this maceral group, but that cannot be unequivocally identified with any particular maceral within the group. Thus, vitrodetrinite is used to designate a maceral when it is not possible to distinguish between collinite and tellinite, and liptodetrinite is used where, e.g., it is impossible to differentiate between sporinite and cutinite on morphological grounds.

[c] This is sometimes also referred to as *massive micrinite*.

tem specified by the American Society for Testing Materials (ASTM D 2796-72)[3] and used in North America is summarized in Table 2.1.4. As can be seen from Table 2.1.2, maceral and maceral group designations are generally indicative of the respective source materials (e.g., *sporinite* and *cutinite*) or of their appearance (e.g., *vitrinite* and *micrinite*). The exception is *inertinite*, a term borrowed from carbonization practice (see Section 11.2) and intended to convey that macerals of this group behave as inert infusible diluents when a coal containing them and others is pyrolyzed.[4] Vitrinite and exinite are, in this context, termed "reactive" components.

In high-volatile bituminous coals, the three maceral groups of the ICCP system have widely different chemical compositions (see Section 2.2), and individual macerals, e.g., sporinites, resinites, and cutinites, are often found to have retained the shapes of their respective progenitors so well that they

[3] ASTM Standards available from American Society for Testing and Materials, Philadelphia, Pennsylvania 19103.

[4] Further reference to this matter, which is one of several that lend coal petrography great practical importance, is made below and in Section 6.3.

Table 2.1.3

The microlithotype groups recognized by the International Committee for Coal Petrography[a]

Microlithotype group	Composition (by maceral groups)
Vitrite	>95% V, <5% E + I
Liptite	>95% E, <5% V + I
Inertite[b]	>95% I, <5% V + E
Clarite	>95% V + E, <5% I
Durite	>95% I + E, <5% V
Vitrinertite	>95% V + I, <5% E
Trimacerite	V + E + I, each >5%

[a] After Mackowsky [10].
[b] When the inertinite component (I) consists principally of fusinite, semifusinite, and sclerotinite, i.e., when micrinite and macrinite are almost or completely absent, this is often called *fusite*.

can be identified by their botanical features. Cady [12] has used the term *phyteral* to denote such fossilized but otherwise perfectly preserved entities. However, as the rank of the coal increases and carbon contents of vitrinite and exinite approach that of inertinite (92–94%), chemical and optical differences between macerals (or maceral groups) progressively narrow and eventually almost disappear. In coal with >89% carbon, it is usually no longer possible to differentiate between exinite and vitrinite; and in anthracites (with >93% carbon), even a distinction between vitrinite and the inertinite macerals semifusinite and sclerotinite is difficult [13, 14].

Paralleling this progressive loss of maceral identity, the optical densities of macerals also increase with rank toward a common value near ~95% carbon; and since this makes it difficult to prepare sufficiently translucent thin sections of mature coals, modern petrographic studies of coal generally involve inspection of polished surfaces by incident light. Suitable specimens can be prepared [15, 16; ASTM D 2797-72] by plane polishing small slabs or coarse (−20 mesh = −850 μm) particles embedded in, e.g., polyvinyl acetate or epoxy resins; and where desired, structural features can be brought into sharper contrast by relief polishing [15] or by etching the plane-polished surface with dilute acids [16] or potassium permanganate in dilute H_2SO_4 [17].

For petrographic analysis, use is made of the fact that the refractive index n and the absorptive index k of coal macerals vary systematically with coal rank, and that macerals can therefore be uniquely characterized by determining their reflectivity (or "reflectance") R. This parameter is defined by

$$R = [(n - n')^2 + n^2 k^2]/[(n + n')^2 + n^2 k^2]$$

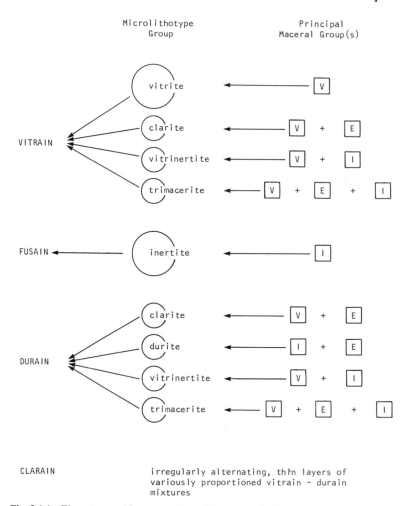

Fig. 2.1.1 The petrographic composition of (macroscopic) lithotypes in bituminous coal.

where n' is the refractive index of the medium (i.e., air or an oil; see below) in which R is determined. Such measurements were at first made [18, 19] with a Berek split-beam microphotometer [20] but have now been much simplified by using photomultipliers whose output is amplified and displayed by a galvanometer, recorder, or magnetic tape [21–23]. Incident light is a plane-polarized beam that is monochromatized by passage through filters with peak transmittance at 546 ± 5 nm (ASTM D 2798-72), and all measurements are made with the microscope lens immersed in a standard oil, e.g., Cargille's Type A or B (Note B).

Table 2.1.4

American maceral/maceral group system[a]

Macerals	Maceral groups[b]	
Vitrinite[c]	Xylenoid	Vitrinites in lignites and sub-bituminous coals
	Vitrinoid	Vitrinites in bituminous coals
	Anthrinoid	Vitrinites in anthracites
Exinite[d]	Exinoid	
Resinite[e]	Resinoid	
Fusinite[f]	Fusinoid	All ranks of coal
Semifusinite[g]	Semifusinoid	
Micrinite[h]	Micrinoid	

[a] Maceral definitions are taken from the fuller descriptions in ASTM D 2796-72.

[b] Unlike in the ICCP system, where it comprises a set of related macerals, a maceral *group* here contains the different, rank-dependent forms of a single maceral; e.g., exinites of different rank form the exinoid group.

[c] Produced by gradual alteration of plant cell substances; has lower reflectance than fusinite, but higher reflectance than exinite; comprises the bulk of most coals.

[d] Derived from cuticles, spore coats, and pollen exines; often (especially in low-rank coals) has the morphology of its source materials.

[e] Derived from resinous secretions and exudates of plant cells; in attrital coal material (= clarains) it occurs characteristically as discrete homogeneous bodies or clusters that, in a section, are usually round, oval, or rodlike.

[f] Derived from plant cell walls that were rapidly altered prior to or soon after incorporation into the enclosing sediment; has prominent cellular structure.

[g] A maceral transitional in optical properties between fusinite and vitrinite.

[h] A maceral similar in reflectance to fusinite but not exhibiting cellular structure; "granular" micrinite is composed of dispersed or aggregated ~ 1-μm particles; "massive" micrinite consists of optically homogeneous, dense > 10-μm particles.

For calibration, standard optical glass prisms, with accurately known refractive indices and reflectivities, are usually employed, but diamond as well as synthetic sapphires and garnets have also been used.

In practice, it is now customary to determine the "true" rank of the coal[5] by measuring the reflectivity of its *vitrinite* and to report the result as an average (percent) reflectance of at least 100 individual "point" readings.

[5] In this connection it should be noted that volatile matter and carbon contents of "whole" coal, which are also used to fix coal rank (see Section 1.2), are greatly influenced by petrographic composition and, hence, not only less certain indicators of the degree of metamorphism but also less reliable guides to industrial behavior.

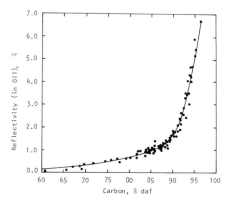

Fig. 2.1.2 The variation of vitrinite reflectivity with rank. The data points shown in this diagram have been compiled from several independent studies reported in the technical literature and cover European as well as North American coals.

For high-rank coals, in which some degree of optical anisotropy is commonly observed, this mean is computed from the maximum readings. Figure 2.1.2 shows how R_{max} (in oil) varies with carbon contents, and Fig. 2.1.3 illustrates the effects of anisotropy with a plot of R_{min} and R_{max} against average reflectance.

The petrographic composition of the coal is then determined with a point counter and reported as percentages of the three maceral groups (see Table 2.1.2) or, if the ASTM procedure (ASTM D 2799-72) is followed, as percentages of the six ASTM-recognized macerals (see Table 2.1.4).

Widely used for many years to characterize and correlate coal seams and to resolve questions about coal diagenesis and metamorphism (see Section 1.1), petrographic analysis has since the late 1940s also increasingly in-

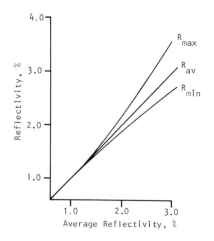

Fig. 2.1.3 The variation of optical anisotropy with rank. Anisotropy is here indicated by the difference between R_{max} and R_{min}, and rank is expressed in terms of the *average* reflectivity. Comparison with Fig. 2.1.2 shows that anisotropy becomes increasingly pronounced as carbon contents exceed 90%. (After De Vries *et al.* [23a].)

fluenced developments in coal preparation (see Chapter 8) and conversion technologies. For industrial purposes, it is particularly significant (a) that the microlithotypes coexisting in any one coal differ in hardness (or mechanical strength), and (b) that pronounced thermoplastic properties (see Section 6.3), which are essential for producing cokes from coal and, as a rule, characteristic of certain bituminous coals, are even in such coals only associated with vitrinites and exinites. This recognition has had an important impact on coal-blending practices in the carbonization industry (see Section 11.2), where petrographic data are used to select components of coke oven charges, and also prompted improvement of coal quality by selective crushing and sizing, i.e., by utilizing the fact that poorly caking (hard) durites and (friable) fusinites tend to accumulate in the larger and smallest size fractions, respectively. The Longwy–Burstlein (or "Sovaco") process [24] that was operated commercially in Belgium and a similar scheme in the Soviet Union [25] are examples of this type of coal preparation.

More recently, it has also been confirmed that inertinite macerals (such as fusinite and sclerotinite) are chemically much less reactive than other constituents of coal (see, e.g., Section 5.5). Petrographic analysis is therefore likely to become as important an adjunct of future coal conversion processes (in particular, coal liquefaction; see Section 13.2) as it now is of carbonization.

In this connection, however, it must be noted that *isolation* of macerals, while technically possible, is difficult. By comminuting coal to < 50 μm, suspending it in a liquid medium (such as carbon tetrachloride-toluene or aqueous $ZnCl_2$) adjusted to a specific gravity (s.g.) of ~ 1.25, and then centrifuging the suspension at 12,000 to 25,000 rpm, fractions containing up to 99% vitrinite, up to 97% exinite, and up to 95% inertinite have been obtained [14, 26, 27]. But such separation has so far only been used to provide material for laboratory studies.

2.2   Chemical Composition

For practical purposes the chemical composition of coal is always defined in terms of its "proximate" analysis, which affords a rough measure of the distribution of products obtained from it by destructive distillation, and by its "ultimate" or elemental composition. Neither offers significant information about coal structure, i.e., about how its constituent elements are molecularly combined (see Section 5.7), but both furnish data that in the light of long experience can be directly correlated with most facets of coal "behavior" (see, e.g., Chapters 4–6).

Proximate analysis determines (i) moisture contents, (ii) volatile matter

(VM) contents, (iii) ash, i.e., inorganic material left behind when all combustible substances have been burned off, and (iv) indirectly, the so-called fixed carbon contents (FC), which is defined by

$$\% \, FC = 100 - (\% \, H_2O + \% \, VM + \% \, ash)$$

Comprehensive reviews of the (largely arbitrary) experimental methods adopted for these purposes have been published by Abernethy and Walters [28] and Hattman and Ortuglio [29]. However, in most coal-producing and -consuming countries, national standard techniques are used; and through Technical Committee 27 (Solid Mineral Fuels) of the International Organization for Standardization (ISO), considerable progress has, since the late 1940s, also been made toward development of internationally acceptable procedures. For the greater part, the latter are based on standard methods formulated by the American Society for Testing Materials (ASTM), which are followed in Canada as well as in the United States, by the British Standards Institution (BSI), by the German Normenausschuss (DIN), and by Poland's Standards Committee (PN). All are subject to periodic review.

Moisture contents [ASTM D 3173-73; 30] are conventionally determined by measuring the percentage weight loss of a comminuted (<60 mesh) $\frac{1}{2}$- or 1-gm sample after heating for 1 hour at $107 \pm 4°C$ in vacuo or in an inert protective atmosphere (usually purified nitrogen). An alternative procedure, which is particularly appropriate for use with easily oxidizable high-moisture, low-rank coals, and therefore also permitted in several national standards (BS 1016; DIN 51718; PN 57, ISO 1015),[6] involves distilling the coal with a water-immiscible liquid of suitable boiling point (e.g., toluene or xylene). Other methods include

(i)   extraction of coal with reagent solutions that undergo corresponding changes in concentration or density (e.g., Japanese Patent 1100, 1954; Mielecki [31]), and

(ii)   electrical techniques that involve measurement of capacitances, resistances, or dielectric constants.

The latter are mainly used for quality control in coal preparation plants where great accuracy is not necessary (Note C).

The forms of moisture encompassed by these methods are

(a)   "bulk" water (sometimes also termed "superficial moisture"), which is present in large cracks and capillaries and possesses the normal vapor pressure of water, and

---

[6] All foreign standards are available in the United States only from the American National Standards Institute, New York, New York 10018.

(b) physically adsorbed water, which is held in small pores (see Section 4.1) and has a vapor pressure corresponding to its adsorbed state.

However, since most coals, particularly those of low rank, may also contain significant quantities of chemisorbed water and, in addition, begin to generate water by thermal decomposition at temperatures as low as 150–160°C (see Chapter 6), care must be taken that the conditions under which moisture contents are determined do not encourage *formation* of water. This is especially important when conventional heating is used on lignites and subbituminous coals: Even small amounts of oxygen in the protective atmosphere can in such cases form water by progressively abstracting hydrogen from the coal.

To understand what analytically determined moisture contents mean, a distinction must be made between *natural bed moisture* contents, also referred to as *capacity moisture*, and *as received* moisture. The former represents the amount of water that a coal will hold when fully saturated at nearly 100% relative humidity (as in an undisturbed coal seam) and thus reflects the total pore volume of the coal accessible to moisture (see Section 4.1). Figure 2.2.1 shows how this quantity varies with rank. The bed moisture content of mined coal that has lost some moisture through exposure to the atmosphere can be closely estimated by rewetting it, allowing it to drain, and then equilibrating it at $30 \pm 0.2°C$ in a desiccator over a fully saturated aqueous potassium sulfate solution, which maintains a relative humidity of 97 to

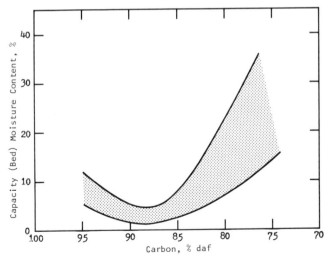

Fig. 2.2.1 Generalized variation of capacity (or bed) moisture contents with rank.

98% (ASTM D 1412-74; ISO 1018; Krumin [32]). In contrast, *as received* moisture contents relate to the amounts of water in samples as submitted for analysis, and these may be substantially smaller than bed moistures (if the samples have been allowed to partly dry) or greater (if they contain a significant quantity of excess surface water). The most often reported quantity is the *air-dried* moisture content, which, as the term implies, is measured on coal samples that have been allowed to equilibrate under "normal" laboratory conditions (20 ± 5°C and 30–60% relative humidity).

Volatile matter, which varies qualitatively and quantitatively with the rank and petrographic composition[7] of the coal, is comprised of a wide spectrum of hydrocarbons, carbon monoxide, carbon dioxide, and "chemically combined" water. Except for small amounts of methane and carbon monoxide, which may be chemisorbed on the coal, all these substances form by thermal decomposition of the coal (see Section 6.2).

Volatile matter contents are equated with the weight loss of a dried, comminuted sample (usually 1 gm, <60 mesh) when heated at a specified temperature for a specified time. (ASTM standards, e.g., ASTM D 3175-77, prescribe 7 min at 950 ± 20°C, but others, e.g., ISO 1171, call for temperatures between 875 and 1050°C and heating periods between 3 and 20 min.) But test results depend, among other matters, on the rate and duration of heating, and it is therefore extremely important to follow the prescribed standard procedures closely. Great care must also be exercised to ensure that the coal is not oxidized during the test and that rapid discharge of volatile matter (particularly between 350 and 500°C) does not eject coal particles from the crucible. Several laboratories have consequently adopted slightly different procedures for high- and low-rank coals. Where ASTM mèthods are used, high-rank coals, defined as coals with less than 10% air-dried moisture, are usually weighed into platinum crucibles, covered with well-fitting lids, and then placed in a muffle furnace (at 950 ± 20°C) for 7 min. Low-rank coals with >10% air-dried moisture are first brought to 600 ± 50°C at 100°C/min and then heated at 950 ± 20°C for an additional 6 min.

Ash contents are routinely determined from the residues left behind when weighed (1–2 gm) test samples are completely incinerated in air at 725 ± 25°C (ASTM D 3174), or at slightly higher temperatures in some European countries. Caution is, however, sometimes necessary, since retention of sulfur in the ash can adversely affect the accuracy of the determination through reaction of sulfur with alkaline oxides derived from carbonate minerals. When analyzing coals with high pyrite concentrations, it is therefore common practice to initiate incineration at the lowest possible tempera-

---

[7] In macerals isolated from a single coal, volatile matter contents always decrease in the order (a) exinite, (b) vitrinite, (c) inertinite.

ture in order to decompose pyrite *before* decomposing carbonates. According to Rees and Selvig [33], retention of sulfur can also be minimized by ashing the coal in thin layers.

It should be noted that ash contents determined by proximate analysis are neither qualitatively nor quantitatively identical with *mineral* matter in coal. (The differences and typical compositions of coal mineral matter and ash are discussed in Section 2.4).

The so-called *fuel ratio*, which is occasionally reported with proximate analysis of a coal, is the ratio of "fixed" carbon to volatile matter.

Ultimate or *elemental* analysis encompasses quantitative determination of *carbon, hydrogen, nitrogen,* (organic) *sulfur,* and *oxygen,* which make up the coal substance, and is usually performed with classic oxidation, decomposition, and/or reduction methods. In most cases, however, the "raw" analytical data require some correction. For example, measured carbon contents may include carbon from mineral carbonates. Sulfur determinations must take into account contributions from pyrite, marcasite, and sulfates. And unless the coal is carefully dried before analysis, some hydrogen may originate in residual water.

Subject to such subsequent corrections as may be necessary, *carbon* and *hydrogen* are usually determined by adaptations of the conventional Liebig method, i.e., by burning a 0.2–0.5 gm sample in pure dry oxygen at 800 to 900°C and completely converting the combustion products to CO_2 and H_2O by passing them through heated cupric oxide. The exit stream is then led over hot lead chromate and silver gauze, which remove all oxides of sulfur and chlorine, and finally sent through absorbers which collect CO_2 and H_2O. Carbon and hydrogen contents are calculated from the weight increases of the absorbents. Alternative procedures, increasingly widely used since the 1940s and commonly also permitted by various national standards, employ semi-micromethods developed by Pregl for organic compounds (see Ingram [34]). These allow accurate analyses of 50-mg samples and are claimed to be less time-consuming than the macromethod.

For *nitrogen,* the favored method is the Kjeldahl–Gunning procedure in which the coal (1 gm) is digested with concentrated sulfuric acid (30 ml) and potassium sulfate (8–10 gm) in the presence of a suitable catalyst (e.g., mercury, a mercury or selenium salt, cobaltic oxide, or perchloric acid). The cooled solution is then made alkaline, and liberated ammonia distilled into a standard boric or sulfuric acid solution from which nitrogen (as NH_3) is obtained by back-titration. In a semimicroversion of this procedure [35], 0.1 gm of coal is digested for 30 min with 3 ml sulfuric acid and a $32:5:1$ $K_2SO_4:HgSO_4:Se$ catalyst mixture. NH_3 is then freed by adding an aqueous hydrogen sulfide/sodium sulfide solution and steam-distilled into boric acid.

Other methods include modifications of the classic Dumas method by Unterzaucher [36] and Radmacher and Lange [37; DIN 51722], as well as a gasification procedure that has been developed by Mantel and Schreiber [38; DIN 15722] from the very early work of Varrentrapp and Will [39].

Methods for quantitative estimation of total (organic and inorganic) *sulfur* are based on combustion of sulfur-bearing compounds to sulfate ions, which can be determined gravimetrically or volumetrically.

Of the three alternatives allowed in various national standards (ASTM D 3177) the simplest and most generally useful [40] is the Eschka procedure, which in its modern version entails incinerating the coal sample with a 1:2 mixture of sodium carbonate and calcined magnesium carbonate in air at $800 \pm 25°C$. Resultant sodium sulfate is then extracted with an acid or alkaline solution and precipitated as barium sulfate. In the alternative bomb combustion method (ASTM D 3177), the test sample is decomposed with sodium peroxide or burned in oxygen at 2- to 3-MPa pressure. And in the so-called high-temperature combustion procedure (Beet and Belcher [41]; DIN 51724; BS 1016, Pt. 6; ISO 351), samples are heated in oxygen at 1250 to 1350°C in the presence of aluminum oxide, ferric phosphate, or kaolin (which assist decomposition), oxides of sulfur are absorbed in a standard hydrogen peroxide solution, and the resultant acidity in that solution (due to SO_3) is titrated against 0.05 N aqueous sodium borate.

In all cases, the proportion of *organically* bound sulfur in the coal is obtained by separately determining inorganic (sulfate and pyritic) sulfur by the conventional Powell and Parr method (Fieldner and Selvig [42]; BS 1016, Pt. 6) and subtracting this from the total sulfur content.

Oxygen, the only other element in the organic coal substance, is usually reported by difference, i.e., as

$$\% \text{ oxygen} = 100 - (\% \text{ C} + \% \text{ H} + \% \text{ N} + \% \text{ S}_{org})$$

but then reflects the accumulated errors of carbon, hydrogen, nitrogen, and sulfur determinations. Since the 1930s, several procedures for direct oxygen analysis, mainly involving oxidation [43, 44] or reduction of the coal [45–47], have therefore been developed. The most widely used method is due to Schütze [48] and Unterzaucher [49] and is based on pyrolysis of the test sample in nitrogen at $\sim 1100°C$. Volatile material thus released is then passed over carbon at 1100 to 1200°C, and the resultant CO is estimated. A recent study of this technique has been published by Abernethy and Gibson [50].

In the past few years, it has also become practical to determine oxygen contents by neutron activation, exemplified by the fast neutron reaction

$$^{16}O + n \rightarrow {}^{16}N + p$$

Oxygen concentrations in a test sample are then obtained by counting β or γ radiation from ^{16}N in the sample and comparing the result with the count for a specimen of known oxygen content [51, 52]. This technique has the advantage of being fast and very accurate. But if only *organic* oxygen is to be determined, the sample must be carefully demineralized before being tested; and because the half-life of ^{16}N is short (~ 7.3 sec), the irradiated sample must be rapidly transferred to the counting cell.

The *calorific value* Q, which is a complex function of elemental composition and determined as part of the proximate or ultimate analysis of the coal, is obtained by combusting a weighed sample, usually 1–2 gm, under oxygen (at ~ 25 MPa pressure) in a bomb calorimeter. For *adiabatic* measurement (see, e.g., ASTM D 2015-73; ISO 1928-76), the temperature of the calorimeter jacket is continuously adjusted to approximate that of the calorimeter itself, while in the *isothermal* procedure (see, e.g., ASTM D 3286-73), the jacket temperature is held constant and a correction for heat transfer from the calorimeter is applied. In either form of measurement, the recorded value of Q is the so-called *gross* heat of combustion (since all water generated during the test remains in liquid form). The corresponding *net* value, which is more useful for practical purposes because water would normally be allowed to evaporate, is obtained by subtracting 1030 Btu per lb (~ 2.395 kJ/gm) of water.

Where no measured calorific value is available, a close estimate can be made by using the Dulong formula

$$Q = 144.4(\% \, C) + 610.2(\% \, H) - 65.9(\% \, O) + 0.39(\% \, O)^2$$

or the Dulong–Berthelot formula

$$Q = 81,370 + 345[\% \, H - (\% \, O + \% \, N - 1)/8] + 22.2(\% \, S)$$

in which $\% \, C$, $\% \, H$, $\% \, O$, $\% \, N$, and $\% \, S$ denote the carbon, hydrogen, oxygen, nitrogen, and (organic) sulfur contents of the coal, all calculated on a dry, ash-free coal basis (see below). Thus calculated, Q (also on a dry, ash-free basis) is in Btu per lb *or* calories per gm; and if derived by means of the Dulong–Berthelot formula, it is usually within $\pm 2\%$ of the measured calorific value.

Since moisture and mineral matter (or ash) are, from some points of view, extraneous to the coal substance, analytical data can be expressed on several different bases in order to reflect

(a)   the composition of as received, air-dried, or fully water-saturated coal (at *capacity* moisture), or

(b)   the composition of *dry; dry*, ash-free (daf); or *dry*, mineral-matter-free (dmmf) coal.

To convert data for moist coal to the dry coal basis, it is merely necessary to multiply each measured quantity other than moisture by $100/(100 - \% \text{ H}_2\text{O})$; e.g.,

$$\% \text{ carbon, dry} = 100 \times \% \text{ C}/(100 - \% \text{ H}_2\text{O})$$

Similarly, to convert data for moist coal to the dry, ash-free basis, each datum *other than moisture and ash* is multiplied by $100/[100 - (\% \text{ H}_2\text{O} + \% \text{ ash})]$; e.g.,

$$\% \text{ carbon, daf} = 100 \times \% \text{ C}/[100 - (\% \text{ H}_2\text{O} + \% \text{ ash})]$$

However, conversion to the dry, *mineral-matter*-free basis is, strictly speaking, only possible if the composition of the mineral matter, and hence the actual weight relationship between the original mineral matter and the ash produced from it, is known (see Section 2.4). Where this is not known, but ash contents are moderate (less than 10–15%), a reasonable approximation, which suffices for most practical purposes, can be obtained by multiplying the analytical data by $100/[100 - (\% \text{ H}_2\text{O} + 1.1 \times \% \text{ ash} + 0.1 \times \% \text{ S}_{\text{inorg}})]$.

Because of the pronounced heterogeneity of coal, analytical accuracy does, of course, always depend on test samples being truly *representative*

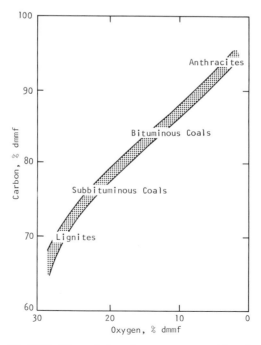

Fig. 2.2.2   The variation of oxygen contents with rank.

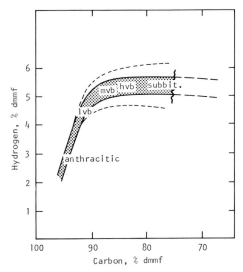

Fig. 2.2.3 The variation of hydrogen contents with rank (after Seyler). The dashed lines extend Seyler's coal band (see Section 3.3) to lignites, and the dotted lines indicate how the hydrogen contents of *per*hydrous and *sub*hydrous coals change with the degree of metamorphic development.

of the particular coal. Careful attention must therefore be given to sampling and preparation of the aliquots submitted for analysis, and national standards address themselves to this matter in considerable detail (see, e.g., ASTM D 346, D 2013, and D 2234; ISO 1988). Typical procedures call for successive reduction of the weight and average lump (or particle) size of the bulk samples (Note D) by a series of riffling steps, with the final 250 gm (or equivalent) of −60-mesh coal being repeatedly riffled to yield ∼ 50 gm for the laboratory.

Of the five elements that make up the organic substance of coal, nitrogen and (organic) sulfur are, however, always minor components. Nitrogen contents rarely exceed 1%; and organic sulfur, although occasionally ranging up to 4 or 5%, lies mostly between 0.2 and 1.5%. Neither nitrogen nor sulfur, moreover, is in any way systematically related to coal rank. For correlative purposes, it is therefore only possible to use carbon, hydrogen, and oxygen contents, and how these change with rank is schematically shown in Figs. 2.2.2 and 2.2.3. In both diagrams, concentrations refer to dry, mineral-matter-free coal.

Of special importance here is the near-constancy of hydrogen contents (at ∼ 5.4 ± 0.4%) up to 87–89% carbon and the fact that enrichment in carbon up to this point involves preferential abstraction of oxygen. Loss of hydrogen as well as of oxygen occurs only in the later stages of meta-

Table 2.2.1

Composition of macerals[a,b]

	Ash (%)	VM (%)	C (%)	H (%)	N (%)	S (%)	O (%)	Atomic H/C ratio
Coal no. 1								
Exinite	0.5	68.8	85.5	7.3	0.5	0.9	5.8	1.03
Vitrinite	1.3	36.1	83.5	5.1	0.8	0.9	9.7	0.73
Inertinite	3.8	22.5	86.8	3.9	0.6	0.7	8.0	0.54
Coal no. 2								
Exinite	0.6	59.8	87.4	6.7	0.6	0.5	4.8	0.93
Vitrinite	0.5	32.0	85.7	4.9	0.8	0.8	7.8	0.68
Inertinite	5.9	23.4	88.0	4.2	0.6	0.5	6.7	0.57
Coal no. 3								
Exinite	0.1	37.1	89.1	6.0	0.7	0.5	3.7	0.80
Vitrinite	1.6	28.4	88.4	5.1	0.8	1.0	4.7	0.69
Inertinite	1.1	19.2	89.6	4.3	0.6	0.5	5.0	0.58
Coal no. 4								
Exinite	1.9	22.6	89.3	4.9	1.5	0.6	3.7	0.66
Vitrinite	2.3	23.5	88.8	4.9	1.6	0.7	4.0	0.67
Inertinite	5.8	17.0	89.8	4.3	0.9	0.5	4.5	0.57

[a] After Teichmüller [53].

[b] Ash on dry coal basis; all other parameters on dry, ash-free basis. Original data rounded off to first decimal places, but H/C calculated (by author) from analytical data as reported.

morphic development. But in this connection it must be emphasized that while the band outlined in Fig. 2.2.3 defines the basic elemental compositions of *most* coals, it does not by any means encompass *all*. Especially in the region occupied by lignites and subbituminous coals, i.e., below $\sim 85\%$ carbon, many coals lie above or below the band, depending on whether they contain particularly high proportions of hydrogen-rich exinites or hydrogen-poor inertinites (Note E). Table 2.2.1, adapted from Teichmüller [53], illustrates how the compositions of these maceral groups differ from the compositions of the vitrinites with which they are associated.[8] How the carbon/hydrogen ratios of such *per*hydrous and *sub*hydrous coals tend to change during further naturation is indicated by the dotted lines in Fig. 2.2.3.

Volatile matter contents decrease with increasing carbon contents but are also proportional to the hydrogen contents of coal, and plots of carbon versus volatile matter consequently tend to show equally wide or even greater scatter.

[8] Since the H/C ratio has much to do with how a coal responds to pyrolysis (see Section 6.2) or to liquefaction (see Chapters 13 and 14), these data also imply that coals with closely similar average elemental compositions will not necessarily display correspondingly similar physical and chemical properties.

2.3 Relationship among Age, Location, and Rank of Coal

In undisturbed strata, in which increasing depth is synonymous with increasing geological age, there is also a direct (sometimes nearly linear) relationship between the rank of a coal and the depth at which it lies. This was first demonstrated by Hilt [54], who observed that volatile matter contents of coals in South Wales, western Germany, and the Pas de Calais region of northern France decreased regularly with increasing depth of cover, and was later found to be generally true [55–58]. Examples of what is now known as Hilt's rule, but variously expressed in terms of carbon contents, bed moisture, or volatile matter contents of "whole" coal, are illustrated in Figs. 2.3.1–2.3.3. (As shown in Section 2.2, these parameters are all interrelated and covariant with several others, including vitrinite reflectance; see Section 2.1.) In all cases, the rate of change with depth is a function of the local geothermal gradient.

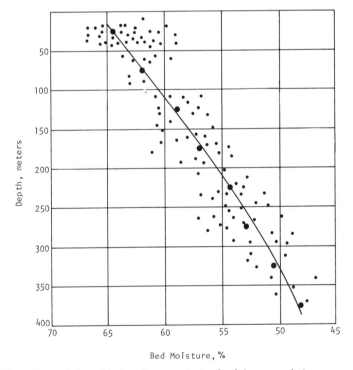

Fig. 2.3.1 The variation of bed moisture contents of soft brown coals (or unconsolidated lignites) with depth of burial; Cologne region. West Germany. Large points show statistically weighted averages for each 50-m depth interval. (After Kutzner [59]; by permission of M. and R. Teichmüller.)

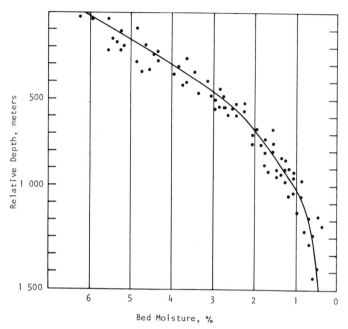

Fig. 2.3.2 The variation of bed moisture contents of high-rank coals with depth of burial; Saar region, West Germany. Data relate to 10 boreholes. (After Damberger *et al.* [60]; by permission of M. and R. Teichmüller.)

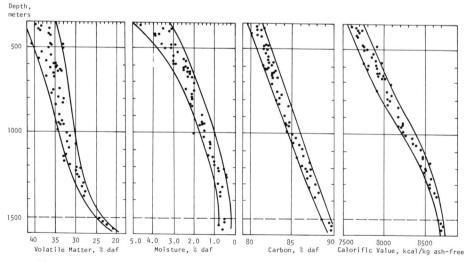

Fig. 2.3.3 The variation of rank indicators with depth of burial; Nordlicht Ost I borehole; West Germany. (Teichmüller and Teichmüller [62]; by permission of M. and R. Teichmüller.)

Because maceral distributions in coal seams intersected in any one locality may be very different, graphical representations of how rank parameters of "whole" coal vary with overburden thicknesses will usually show considerable scatter of data points; and for quantitative geochemical investigations, e.g., studies seeking to estimate rates of metamorphic change, it is therefore often preferable to use appropriate *vitrinite* characteristics as measures of rank. Figure 2.3.4 illustrates the much more clearly defined relationships that are then obtained. For quantitative work it is, however, also essential to bear in mind

(a)   that present overburden thicknesses may have been reduced by surface erosion (e.g., glaciation), and that the paleotemperatures to which the coal was exposed may therefore have been substantially higher than the temperatures calculated from present depths and geothermal gradients; and

(b)   since the extent of metamorphic change is as much a function of time as of temperature, that rank stages now observed in a seam succession may also reflect temperature fluctuations caused by periodic uplift and subsidence of the strata. (A history of such instabilities is exemplified by the 6-km-deep Münsterland-1 borehole in West Germany, where 21 coal

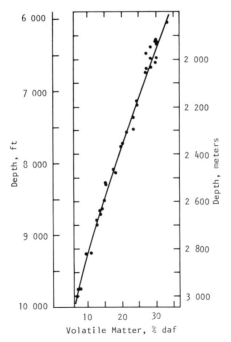

Fig. 2.3.4   Variation of volatile matter contents of some vitrinites with depth of burial. (Teichmüller and Teichmüller [62]; by permission of M. and R. Teichmüller.)

seams of Carboniferous age and nearly 80 younger ones were intersected (Note F). Lopatin [61] has here traced no fewer than 10 cycles of uplift and subsidence, with major upthrows in Permian and Cretaceous times reducing strata temperatures by as much as 25–30°C for periods up to 20 million years.)

"Anomalous" rank, inconsistent with what would be expected on the basis of present depth and geothermal gradients, may consequently be due to extensive surface erosion or strata uplift, or to both.

Maximum paleodepths, and hence the maximum temperatures attained by the coal, can be fairly closely estimated from a generalized correlation of vitrinite bed moisture contents and depths (see Fig. 2.3.3; Patteisky and Teichmüller [63]);[9] and such "reconstruction" has considerable practical importance in, inter alia, oil exploration programs (see below). However, detailed reconstruction of a history of strata uplift and subsidence, as in the case of the Münsterland-1 borehole, requires detailed geological information that is only rarely available; and in most cases, recourse must therefore be made to (less satisfactory) theoretical coalification models.

As would be expected from a general relationship between coal depth and rank, Hilt's rule applies also to gently tilting seams in which rank changes similar to those in a vertical seam succession can be traced. A particularly fine example is afforded by the Pittsburgh seam, which dips eastward from central Ohio to the Appalachian Mountains at an average rate of 5.5 m/km. Where tectonism has not seriously disturbed regional regularities, coal can consequently be used as a natural geological "thermometer." (In a more restricted sense, it can also be used as a thermometer in disturbed areas, but in such strata it has obviously much more limited value.)

The practical value of Hilt's rule lies mainly in its ability to predict the elemental compositions, and hence many properties, of the coals encountered or likely to be encountered in any particular location; and from the variation of coal composition with geothermal gradients and maximum burial depths, it has in turn proved possible to estimate the maximum temperatures at which transitions from one coal type to another occur. One such set of temperatures, which stands in reasonable agreement with estimates derived from the properties of coal resins in Tertiary lignites and from the conditions under which wood, lignin, and cellulose can be converted to "artificial coal" in the laboratory [64–66], is shown in Table 2.3.1.

There is also an important connection between the extent of metamorphic development of coal and the *occurrence of petroleum;* and where coal and

[9] In making such estimates, it is of course implicitly assumed that the local geothermal gradient itself has not significantly changed throughout the geological history of the formation. This is generally reasonable except where strata have been affected by magmatic intrusions.

Table 2.3.1

Maximum temperatures for major rank transitions[a]

Transition	Carbon contents[b] (%)	Maximum temperature (°C)
Peat		
Lignite	$51 \rightarrow 63$	200
Low-rank bituminous coal	$63 \rightarrow 79$	200–225
High-rank bituminous coal	$79 \rightarrow 85$	250–280
Anthracite	$85 \rightarrow 92$	280–350

[a] After Francis [67].
[b] Approximate values (interpolated by author).

petroleum coexist, coal can serve to indicate the stratigraphic limits beyond which no useful oil accumulations can be expected. This concept is illustrated by Fuller's observations [68] concerning the nature of petroleum in major US pools and the fixed carbon contents of coals associated with them (see Table 2.3.2), and has been discussed by Teichmüller [69].

Table 2.3.2

Relationship between fixed carbon in coal and the nature of associated petroleum accumulations[a]

Fixed carbon in coal (%)	Carbon contents[b] (%)	Petroleum quality
70	86–89	Petroleum and gas accumulations rare
65–70	83–89	No commercially significant petroleum or gas pools
60–65	79–83	Few commercial pools, but petroleum of excellent (light) quality; gas occurrences frequent, but small
55–60	75–79	Principal (light) petroleum and gas pools in the Appalachian region
50–55	71–75	Principal (medium) petroleum pools in Ohio, Indiana, and midcontinent states
50	71	Heavy petroleum accumulations in coastal plains and in Tertiary un-consolidated formations

[a] After Fuller [68].
[b] To facilitate comparison with Table 2.3.1, approximate corresponding carbon contents have been interpolated by the author.

A particularly important point established by Fuller's observations is that increasing temperatures affect coal and petroleum in opposite ways: While coal gradually loses its volatility, petroleum becomes progressively "lighter" as it cracks and, at sufficiently high temperatures, also yields "thermal" gas thereby. Recent confirmation of this has come from, inter alios, Hacquebard [70] who studied a series of Lower Cretaceous and Tertiary coals in order to determine the paleotemperatures at which oil and gas in 144 major Alberta (Canada) pools attained their present metamorphic status. The maximum burial depths of the coals were reconstructed from Patteisky and Teich-müller's correlation [63]; coal ranks were determined from vitrinite re-flectances; and subsurface contours connecting vitrinites of equal rank were drawn from time–temperature–rank relationships derived from Karweil's [71] and Bostick's [72] coalification models. Table 2.3.3 summarizes the principal results thus obtained. The "liquid window" (~ 65–$135°C$) shown in this table is virtually identical with similar windows previously reported by Landes [73], Tissot *et al.* [74], and others; but the paleotempera-tures that Hacquebard assigns to the coals are at least $100°C$ lower than the corresponding maximum transition temperatures given by Francis (see Table 2.3.1: 200–225°C for development to $\sim 80\%$ carbon, and 250–280°C for development to $\sim 88\%$ carbon). The reason for these discrepancies may lie in the fact that Francis's data make no allowance for interchangeability of time and temperature and also include coals that experienced relatively sudden, short-lived, high temperatures during abnormal regional meta-morphism.

Table 2.3.3

Relationship between metamorphic development of coal, oil, and natural gas[a]

Vitrinite reflectance R (%)	$\%C$[b]	Paleotemperature (°C)	Petroleum/gas
0.5–0.9	76–83	68–116	Almost 90% of all oil (with over 40% having reached current status at temperatures between 82 and 96°C)
>1.20	>88	>143	Petroleum absent
		106–177	"Thermal" gas (and petroleum up to 135°C)

[a] After Hacquebard [70].
[b] Approximate; estimated (by author) from R.

In disturbed strata, in which orogeny has often upset the original succession of strata by extensive thrust- and overfolding, variations of rank with depth (or distance) are rarely predictable; and there is, similarly, no direct relation between rank and geological age of the coal. In many instances, abnormally high temperatures generated by tectonism have, in fact, raised coal ranks far beyond the levels reached by normal metamorphism in the available time (since deposition). Cretaceous coals in western Canada and the western United States, which have only attained subbituminous rank in the plains, are thus found to contain up to 86% carbon in the Rocky Mountain foothills and up to 91% carbon in the mountains proper. In Sumatra (Indonesia), Tertiary coals with 77% carbon contain as much as 93% in equivalent mountain strata. And in Sarawak (Indonesia), Tertiary lignites with 69% carbon have been upgraded to coals with nearly 92% by orogeny. However, rank changes analogous to those found down-dip in gently tilting seams are often traceable in steeply pitching seams and along the limbs of syn- or anticlinal coal formations.

Magmatic sills or laccoliths, such as those referred to in Section 1.1, create situations intermediate between tectonically undisturbed and disturbed formations: Hilt's rule is followed (though with steeper than usual gradients), but because of the greater heat input, metamorphic development of the coal is much more advanced than its geological age would otherwise indicate. A classic example is the upgrading of Miocene-age Bohemian (Czechoslovakia) lignites by underlying volcanic strata, which still cause local geothermal gradients of 7 to 20°C per 100 m.

2.4 Mineral Matter in Coal

The term "mineral matter" is conventionally used to mean all forms of inorganic material associated with coal and includes optically identifiable mineral species (and mineral phases), as well as variously complexed metals and anions. Some of this material entered the coal as constituents of the parent vegetation and then depended on the particular plant type. (For example, whereas grasses will selectively absorb silica from the soil during active growth, lycopods will preferentially take up alumina. Ashes from lycopods have been found to contain 26–57% Al_2O_3 as compared with the usual 1–4% in other plant ashes.) But more important contributions were made

(a)   during the diagenetic stage of coal development, when inorganic matter was carried into the exposed decaying debris by wind and water;

(b)   during transport of debris from its original site to another location, when silts were sometimes so intimately mixed into it as to become colloidally dispersed,[10]

(c)   by deposition from percolating mineral waters during the metamorphic stage; and finally,

(d)   by various ion-exchange processes during either or both stages of development.

Mackowsky [75] has used the terms "syngenetic" and "epigenetic" to denote, respectively, mineral matter incorporated into coal during the diagenetic and metamorphic stages. These are technically more meaningful than the terms "inherent" and "adventitious" (or "extraneous"), which have often been used with similar intent, but do not in themselves differentiate between dissimilar associative forms of mineral matter.

Although all types of inorganic material in coal are important, either because of their deleterious effects in coal processing or because of their adverse environmental impact and potential health hazards [76; see also Section 15.1], the relative abundance of different inorganic constituents makes it appropriate to distinguish between major and minor constituents. The major constituents are

(1)   clay minerals (aluminosilicates), which occur mostly as illite, kaolinite, montmorillonite, and mixed illite–montmorillonite and commonly make up as much as 50% of the total mineral matter contents;

(2)   carbonate minerals, principally calcite ($CaCO_3$), siderite ($FeCO_3$), dolomite ($FeCO_3 \cdot MgCO_3$), and ankerite ($2CaCO_3 \cdot MgCO_3 \cdot FeCO_3$), but frequently also present as variously composed mixed carbonates of Ca, Mg, Mn, and Fe;

(3)   sulfides, which are mainly present as pyrite and marcasite (i.e., dimorphs of FeS_x, which only differ in crystal habit), but which have also been reported in the form of galena (PbS) and sphalerite (ZnS); and

(4)   silica, which is ubiquitous as quartz and usually accounts for up to 20% of all mineral matter.

Sulfates are relatively rare and, when present, exist mainly as variously hydrated iron sulfates ($FeSO_4 \cdot nH_2O$) or as a mixed Na·K·Fe sulfate (jarosite).

Minor inorganic constituents, mostly only present in ppm concentrations and therefore properly termed *trace* constituents, cover an even wider spectrum (see Table 2.4.1), but may be conveniently grouped according to their tendency to associate with minerals of a single class. According to

---

[10] Such colloidally dispersed clay minerals and other inorganic matter often amount to 15–20% of the coal and are typical of *allochthonous* coals.

Table 2.4.1

Minor and trace constituents of mineral matter

"Rare" elements in coal ash[a]		
Arsenic (8000; 500)	Germanium (11,000; 500)	Platinum (0.7; —)
Beryllium (1000; 300)	Gold (0.1; —)	Silver (10; 2)
Bismuth (200; 20)	Indium (2; —)	Thallium (5; —)
Boron (3000; 600)	Lead (1000; 100)	Tin (500; 200)
Cadmium (50; 5)	Lithium (500; —)	Yttrium (800; 100)
Cobalt (1500; 300)	Molybdenum (500; 200)	Zinc (10,000; 200)
Gallium (400; 100)	Nickel (8000; 700)	Zirconium (5000; —)

	Maximum concentrations of minor elements in ash of western US coals[b]			
Location	1.0–0.1	0.1–0.01	0.01–0.001	0.001–0.0001
(1)[c]	As, B, Ba, Be, Co, Mn, Mo, Ni, Pb, Sr, Ti, U, Zn, Zr	Cu, Cr, Ga, Ge, La, Li, Sc, V, Y	Sn	Ag, Yb
(2)[c]	B, Ba, Mo, P, Sr, Ti, U, Zr	Co, Cr, Cu, Ga, Ge, La, Mn, Ni, Pb, V, Zn	Sc, Sn, Y	Ag, Be, Yb
(3)[c]	B, Ba, Mn, Mo, Sr, Ti, Y	Co, Cr, Cu, Ni, Pb, V, Zr	Be, Ga, Ge, La, Sc, Sn, Yb, U	
(4)[c]	B, Ba, Mn, Ni, Sn, Sr, Ti	Co, Cr, Cu, Pb, V, Y, Zn, Zr	Be, Ga, Ge, Mo, Sc, Yb	

[a] Adapted from Goldschmidt [81]. Terms in parentheses show maximum concentrations and average concentrations in "rich" ashes (gm/ton).

[b] After Deul and Annell [82]. Concentrations in percent of ash.

[c] (1) Harding County, South Dakota; 151 coals; (2) Perkins County, South Dakota; 59 coals; (3) Jefferson County, Colorado; 35 coals; (4) Milam County, Texas; 48 coals.

Goldschmidt [77], chalcophilic elements, which tend to form sulfides and therefore to associate with FeS_x, include arsenic, cadmium, copper, lead, mercury, and selenium, while lithophilic elements, which are concentrated in the silica phase, include potassium, sodium, titanium, yttrium, and zirconium. More recent studies [78–80] indicate that

(i)   B, Be, and Ge are predominantly combined with the *organic* coal substance;

(ii)   As, Cd, Hg, Mn, Mo, Pb, Zn, and Zr occur mostly in association with *inorganic* matter; and

(iii) other elements show varying affinities for association with organic and inorganic matter: Ga, P, Sb, Ti, and V are preferentially allied with elements in (i), while Co, Cr, Ni, and Se occur mostly with the elements in (ii).

Support for these broad assignments has been obtained by fractionating coals by gravity separation between ~1.25 and 2.9 specific gravity and determining the distribution of inorganic elements in the resultant fractions. Figure 2.4.1 illustrates this with reference to germanium (predominantly in organic association) and arsenic (predominantly in inorganic association).

The study of mineral species (or phases) in coal has been greatly aided by the development of electronic *low-temperature* ashing methods [83, 84]. In these, oxygen is activated by passage through a radio-frequency field in which a discharge takes place, and the activated oxygen is used to oxidize the organic coal substance at <150°C, i.e., at temperatures at which no loss of volatile inorganic constituents and little chemical alteration of the mineral matter occur. The inorganic residues can then be identified and quantitatively measured by instrumental techniques, notably x-ray diffraction [85] or infrared spectroscopy in the 650–200 cm⁻¹ region [86]. In some instances, differential thermal analysis [87, 88] and scanning electron microscopy [80, 89] have also been used.

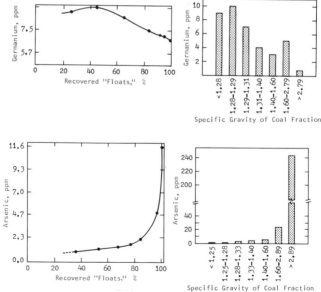

Fig. 2.4.1 Typical distribution of germanium (top) and arsenic (bottom) in coal (Ruch *et al.* [80]. Diagrams on the left show the distributions in the form of washability curves (see Section 8.1), while those on the right display them as histograms that relate the concentrations to particular specific gravity fractions of the coal. (By permission of Illinois State Geological Survey.)

How mineral species (or phases) are disseminated in coal is usually determined by optical microscopy [75, 90], but the general distribution of inorganic matter in coal can also be ascertained by radiographic methods [91]. As a rule, mineral matter contents are found to be quite unevenly partitioned among associated lithotypes and tend to increase in the order: vitrain–clarain–durain–fusain.

For analytical purposes a fundamental distinction must, however, be made between *mineral matter* and the *ash* produced from it under conventional (high-temperature) conditions. As a result of dehydration (e.g., $FeSO_4 \cdot nH_2O \rightarrow FeSO_4$), decomposition (e.g., $CaCO_3 \rightarrow CaO$), and oxidation (e.g., $FeS \rightarrow FeSO_4$) and because of partial loss of volatile constituents (particularly Hg, K, Na, Cl, P, and S), ash is qualitatively and quantitatively quite different from the mineral matter that gave rise to it, and its behavior, which is ultimately determined by its composition, is also different (Note G).

Since introduction of inorganic matter into coal is entirely dependent on vagaries of nature, ash contents can range from <5 to $>35\%$ and show wide variations over relatively small distances. They are also quite unpredictable from other coal parameters. However, experience shows that over 95% of mineral matter is comprised of clay minerals, carbonates, sulfides, and silica, and hence that over 95% of the ash consists of alumina, calcium oxide, iron oxides, magnesia, and silica.[11] Most of the remainder is accounted for by oxides of K, Na, and Ti (which are the principal basic components of the ash) and by chlorides, sulfates, and phosphates (which are acidic components). For most practical purposes, ash analyses are consequently only concerned with a limited range of elements that can be quantitatively determined by conventional "wet" methods [92] or instrumental techniques [93]. But special attention is often also directed to Cl^-, SO_4^{2-}, and phosphates, which, if present in relatively high concentrations, can prove damaging to boiler tubes in industrial steam-raising equipment (see Section 10.7). Interest in trace elements was until recently confined to those deemed to possess particular market value and likely to be present in sufficient concentrations to warrant their recovery. Germanium and uranium are cases in point. (Unusually high concentrations of these elements have been observed near the roofs and/or floors of coal seams and are apparently due to sorption of the elements from percolating waters; [94].) Since the mid-1960s, however, a wider range of trace elements has come under close scrutiny because of their potentially adverse environmental impact [76].

Since detailed ash analyses are in most cases lacking, quantitative relationships between mineral matter and ash (which are required for classification and certain other purposes) are usually calculated from empirical equations that purport to correct weight losses due to chemical changes

---

[11] Most of the inorganic sulfur in FeS, PbS, etc., is lost as SO_2 during ashing (see Section 2.2).

during ashing. The best known of these, due to Parr [95] and used in the ASTM coal classification system (see Section 3.3), are

$$\text{mineral matter} = 1.08[\% \text{ ash} + 0.55(\% \text{ sulfur})]$$

and

$$\text{mineral matter} = 1.1(\% \text{ ash}) + 0.1(\% \text{ sulfur})$$

which are widely employed where great accuracy is not essential. A more comprehensive equation, which takes into account different types of sulfides and carbonates in the mineral matter and also makes allowances for chlorine and sulfur retained in the ash, has been developed by King *et al.* [96]. This gives

$$\text{mineral matter} = 1.09(\% \text{ ash}) + 0.5(\% \text{ sulfur}_{pyr}) + 0.8(\%CO_2)$$
$$- \% SO_{3 \text{ (coal)}} + 0.5(\% \text{ Cl})$$

However, a statistical analysis by Brown *et al.* [97] has shown that some components of the King–Maries–Crossley formula are interrelated, and that it can therefore be simplified to

$$\text{mineral matter} = 1.06(\% \text{ ash}) + 0.55(\% \text{ sulfur}) + 0.74(\% CO_2) - 0.32$$

without significant loss of accuracy.

In industrial practice, information about the *distribution* of mineral matter in coal is helpful when designing coal-cleaning circuits (see Section 8.1), and knowledge of ash *compositions* allows the viscosity characteristics of molten ash (slag) at high temperatures to be calculated.

The temperature range in which an ash first becomes fluid is usually determined by observation of triangular ash cones, which are fabricated by moistening the finely powdered ash with a few drops of an aqueous 10% dextrin solution and then compressing it in a standard mould (see ASTM D 1857; ISO 540). (Alternative British and German standard procedures have been described in BS 1016 and DIN 51730, respectively.) Depending on the particular requirements, the test is conducted in either an oxidizing or a reducing atmosphere, and special note is made of four "critical" temperatures defined as

(a)   the initial deformation temperature (T_i), at which the cone apex becomes rounded,

(b)   the softening temperature (T_s), at which the cone has fused down to a height equal to the width of the base,

(c)   the "hemispherical" temperature (T_h), at which the residual cone height is one-half the width of the base, and

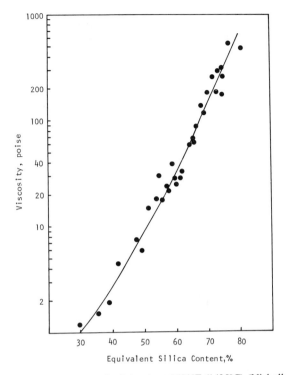

Fig. 2.4.2   Viscosity of molten coal ash in air at 2600°F (1425°C) (Nicholls and Reid [98], Reid and Cohen [99]; by permission of J. Wiley and Sons, Inc.). The different data points refer to different ash compositions.

    (d)   the fluid temperature (T_f), at which the cone has almost completely spread over the ceramic support.

The ash fusion temperature, which is frequently reported as the only datum, corresponds to the softening temperature (T_s).

    The ASTM procedure stipulates that test results must be repeatable to within 50°F (~ 28°C) and reproducible to within 100°F (~ 55°C) (Note H). But since ash fusion temperatures are very sensitive to variations in ash composition,[12] relatively small differences in T_s can be disregarded; and for many purposes, it is in fact sufficient to distinguish broadly between

---

   [12] It should be noted that ash compositions may vary not only from one part of a coal seam to another, but also from one *furnace* region to another. In the latter case, variability arises from the fact that certain ash components, e.g., alkali metals and chlorides, are more volatile than others, and that the ash is progressively depleted of these substances as it moves through the combustion chamber.

(a)   *slagging* coals, which when burned produce ashes that soften below 2200°F ($\sim$1200°C) and can therefore be easily discharged as liquids, and

(b)   *nonslagging* coals, whose ashes soften at >2600°F ($\sim$1425°C) and would generally require removal in solid form.

Coals that produce ashes with softening temperatures between these limits are considered suitable for use in slagging or nonslagging equipment, provided this can be adjusted to operate at the desired heat release rates.

However, where ash is removed as a molten slag, considerable importance also attaches to the viscosity characteristics of the slag, and these can be evaluated from ash compositions [98–100]. As illustrated in Fig. 2.4.2, viscosities at any (constant) temperature in the fluid range are directly related to the so-called equivalent silica content or *silica ratio*, which is defined as

$$\text{silica ratio} = SiO_2/(SiO_2 + \text{total iron as } Fe_2O_3 + CaO + MgO)$$

and the temperature coefficients of slag viscosities at temperatures above the transition from plastic to fluid viscosity, i.e., above the temperature of "critical viscosity" (T_{cv}), can be estimated from Reid–Cohen correlations (see Fig. 2.4.3). The transition temperature T_{cv} itself depends on the $SiO_2 : Al_2O_3$ ratio, the CaO content, and the "ferric percentage"

$$Fe_2O_3/(Fe_2O_3 + 1.11FeO + 1.43Fe)$$

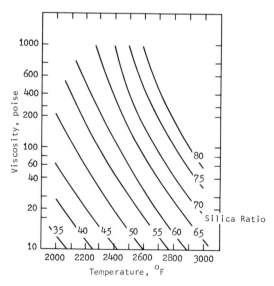

Fig. 2.4.3   Reid–Cohen correlations for calculation of ash viscosities. (After Reid and Cohen [99]; by permission of J. Wiley and Sons, Inc.)

and can similarly be established by reference to detailed correlation diagrams [99] or broadly fixed from its variations with ash fusion temperatures: In a wide-ranging study of US coals, Sage and McIlroy [101] have found T_{cv} to lie some 200°F (~ 110°C) above T_s.

More recently, investigations of relationships between ash compositions and ash fusion temperatures [102–104] have shown that T_s varies inversely with % SiO_2 × % FeO under reducing conditions, and inversely with % SiO_2 × % CaO under oxidizing conditions.

Notes

A. A first international conference on coal petrography, convened in order to reduce the welter of different terms which coal microscopists used at that time, took place in Heerlen (Holland) in 1935; and a second, long delayed by intervening war, was held there in 1951. At the conclusion of the second congress, steps were taken to set up an International Committee for Coal Petrography, and this was formally organized in 1953. The ICCP currently has a membership of some 30 coal-producing and -importing countries, and meets annually for the purpose of developing uniform investigative procedures and petrographic nomenclatures. It has also published or sponsored several authoritative handbooks on coal petrography [11].

B. Because of the greater accuracy of measurement made possible by amplified photo-multipliers, the earlier practice of determining the reflectivity R in air, and thereby obtaining numerically higher values, has been abandoned. National standards for petrographic analysis of coal (such as ASTM D 2798-72) specifically require the use of oil.

C. It should here be noted that *accuracy* is the degree to which an experimental result conforms to a "recognized" or "real" value, while *precision* is a measure of the degree to which replicated measurements conform to each other. Analytical data can thus be precise without necessarily being accurate (or "correct").

D. In most cases, the weight of the bulk sample depends on the total tonnage of the coal being sampled. This is particularly important in commercial transactions.

E. In some instances, hydrogen deficiencies that place a coal below the band shown in Fig. 2.2.3 may also have been caused by oxidation during or after deposition (see Section 5.1).

F. For details of the Münsterland-1 borehole, see M. and R. Teichmüller [62]. This article also offers an excellent discussion of the interplay of time and temperature in coalification processes.

G. This is a particularly important matter in combustion systems and in processes in which the equipment demands certain ash fusion characteristics. Further reference to it is made in Chapter 10 (see especially Section 10.7) and in Sections 12.2 and 12.3.

H. Repeatability refers to measurements by the same analyst, working with the same apparatus, while reproducibility refers to results obtained by different analysts from tests in different equipment.

References

1. F. Muck, "Grundzüge u. Ziele der Steinkohlenchemie," Bonn, 1881.
2. H. Fayol, *Bull. Soc. Ind. Miner.* **15** (2), 546 (1887).
3. M. C. Stopes, *Proc. R. Soc. London* **90B,** 470 (1919).
4. R. Thiessen, *Coal Age* **18,** 1183 (1920); *J. Geol.* **28,** 185 (1920).

5. H. Winter, *Glückauf* **64**, 653 (1928).
6. W. Francis, "Coal," p. 50, 104. Arnold, London, 1961.
7. R. Potonié, *Z. Dtsch. Geol. Ges.* **78**, 357 (1926).
8. A. Duparque, *Rev. Ind. Miner.* **6**, 493 (1926).
9. H. G. A. Hickling, *Proc. S. Wales Inst. Eng.* **46** (6), 911 (1931).
10. M.-Th. Mackowsky, *Fortschr. Geol. Rheinl. Westfalen.* **19**, 173 (1971); *Microsc. Acta* **77** (2), 114 (1975).
11. "International Handbook of Coal Petrology," 2nd ed. Paris, 1963 and Addendum to 2nd ed., Paris, 1971.
12. G. H. Cady, *J. Geol.* **50**, 437 (1942).
13. E. Stach and H. C. Michels, *Geol. Jahrb.* **71**, 113 (1955).
14. H. N. M. Dormans, F. J. Huntjens, and D. W. van Krevelen, *Fuel* **36**, 321 (1957).
15. E. Stach, "Lehrbuch der Kohlenmikroskopie." Berlin, 1935.
16. C. A. Seyler, Survey of National Coal Resources, Paper No. 16. DSIR, London, 1929.
17. M. Teichmüller, *Jahrb. Reichsstelle Bodenforsch.* (*Ger.*) **61**, 20 (1940).
18. E. Hoffman and A. Jenkner, *Glückauf* **68**, 81 (1932).
19. C. A. Seyler, *Proc. Conf. Ultrafine Struct. Coals Cokes* p. 270. BCURA, London, 1944; *Proc. S. Wales Inst. Eng.* **63** (3), 213 (1948).
20. M. Berek, *Z. Kristallogr.* **77**, 1 (1931).
21. J. T. McCartney and L. J. E. Hofer, *Anal. Chem.* **27**, 1320 (1955).
22. K. Kötter, *Brennst. Chem.* **40**, 305 (1959); **41**, 263 (1960).
23. W. Pickhardt and K. Robock, *Brennst. Chem.* **46**, 44 (1965); see also H. Jacob, *Leitz Mitt. Wiss. Tech.* **5**, 65 (1970).
23a. De Vries, *Brennst. Chem.* **49**, 47 (1968).
24. E. Burstlein, *Chal. Ind.* **30**, 351 (1954); **31**, 14 (1955).
25. I. E. Korobchanski, M. D. Kutznetsov, E. Y. E'delman, M. M. Potashnikova, P. I. Korobchanski, and N. P. Sirenko, *Koks Khim.* No. 6, 8 (1956).
26. C. Kröger, *Brennst. Chem.* **37**, 183 (1956).
27. S. Ergun, J. T. McCartney, and M. Mentser, *Econ. Geol.* **54**, 1068 (1959).
28. R. F. Abernethy and J. G. Walters, *Anal. Chem. Ann. Rev.* **41**, 308R (1969).
29. E. A. Hattman and C. Ortuglio, *Anal. Chem. Ann. Rev.* **43**, 345R (1971).
30. R. F. Abernethy, E. G. Tarpley, and R. A. Drogowski, U. S. Bureau Mines Rep. Invest. No. 6440 (1964); M. Roth, *Chem. Eng.* **73**, 83 (1966); H. G. Schäfer and H. Jansen, *Erdöl Kohle* **27**, 436 (1974).
31. T. Mielecki, *Przegl. Gorn.* **5** (1), 28 (1949).
32. P. O. Krumin, Ohio State Univ. Eng. Exp. Station Bulletin 195 (1963).
33. O. W. Rees and W. A. Selvig, *Ind. Eng. Chem. Anal. Ed.* **14**, 209 (1942).
34. G. Ingram, "Methods of Organic Elemental Microanalysis." Chapman and Hall, London, 1962.
35. A. E. Beet and R. Belcher, *Fuel* **17**, 53 (1938).
36. J. Unterzaucher, *Chem. Ing. Tech.* **22**, 128 (1950).
37. W. Radmacher and W. Lange, *Glückauf* **87**, 739 (1951).
38. W. Mantel and W. Schreiber, *Glückauf* **74**, 939 (1938).
39. F. Varrentrapp and H. Will, *Ann. Chem. Pharm.* **39**, 257 (1841).
40. P. J. Jackson, *Fuel* **35**, 212 (1956).
41. A. E. Beet and R. Belcher, *Fuel* **19**, 42 (1940).
42. A. C. Fielding and W. A. Selvig, U. S. Bureau Mines Bull. 492, p. 16 (1951).
43. M. Dolch and H. Will, *Brennst. Chem.* **12**, 141, 166 (1931).
44. W. R. Kirner, *Ind. Eng. Chem. Anal. Ed.* **8**, 57 (1936).
45. H. ter Meulen, *Rec. Trav. Chim. Pays-Bas*, **53**, 118 (1934).

46. R. Wildenstein, *Chal. Ind.* **32,** 293 (1951).
47. H. Guérin and P. Marcel, *Bull. Soc. Chim. Fr.* 310 (1952).
48. M. Z. Schütze, *Anal. Chem.* **118,** 245 (1939).
49. J. Unterzaucher, *Ber. Dtsch. Chem. Ges.* **73B,** 391 (1940).
50. R. F. Abernathy and F. H. Gibson, U. S. Bureau Mines Rep. Invest. No. 6753 (1966).
51. L. C. Bate, *Nucleonics* **21,** 72 (1963).
52. O. U. Anders and D. W. Briden, *Anal. Chem.* **36,** 287 (1964).
53. M. Teichmüller, cited by M.-Th. Mackowsky, *Microsc. Acta* **77** (2), 114 (1975).
54. C. Hilt, *Z. Ver. Dtsch. Ing.* **17,** 194 (1873).
55. A. Strahan and W. Pollard, Mining and Geology Survey, England and Wales (1908).
56. C. S. Fox, *Surv. India Mem.* **57** (1931).
57. A. B. Edwards, *Proc. Aust. Min. Metall. New Ser.* **140,** 206 (1946); **147,** 30 (1947).
58. M. Teichmüller and R. Teichmüller, *in* "Diagenesis of Sediments" (G. Larsen and G. V. Chiligar, eds.). Elsevier, Amsterdam, 1967.
59. R. Kutzner, *Geol. Meldearb., Techn. Univ. Aachen* (1960).
60. H. Damberger, M. Kneuper, M. Teichmüller, and R. Teichmüller, *Glückauf* **100,** 209 (1964).
61. N. V. Lopatin, *Izv. Akad. Nauk USSR Geol. Ser.* **3,** 95 (1971).
62. M. Teichmüller and R. Teichmüller, *in* "Coal and Coal-Bearing Strata" (D. Murchison and T. S. Westoll, eds.), Elsevier, Amsterdam, 1968.
63. K. Patteisky and M. Teichmüller, *Brennst. Chem.* **41,** 79, 97, 133 (1960).
64. E. Berl and A. Schmidt, *Ann. Chem.* **461,** 192 (1928).
65. E. Terres and K. Schultze, *Brennst. Chem.* **33,** 352 (1952).
66. J. P. Schumacher, F. J. Huntjens, and D. W. van Krevelen, *Fuel* **39,** 223 (1960).
67. W. Francis, "Coal." Arnold, London, 1961.
68. M. L. Fuller, *Econ. Geol.* **14,** 536 (1919); **15,** 225 (1920).
69. M. Teichmüller, *Int. Pet. Rev. Ind. Min.* No. Spec. 99 (1958).
70. P. A. Hacquebard, Geological Survey Paper 75-1B, Department of Energy, Mines and Resources, Canada (1975).
71. J. Karweil, *Z. Dtsch. Geol. Ges.* **107,** 132 (1956).
72. N. H. Bostick, *C. R. Congr. Strat. Geol. Carb., 7th, Krefeld* **2,** 183 (1973).
73. K. K. Landes, *Oil Gas J.* May 2 (1966).
74. B. Tissot, Y. Califet-Debyser, H. Deroo, and J. L. Oudin, *Am. Assoc. Pet. Geol. Bull.* **55,** No. 12, 2177 (1971).
75. M.-Th. Mackowsky, *in* "Coal and Coal-Bearing Strata" (D. Murchison and T. S. Westoll, eds.), Elsevier, New York, 1968.
76. F. A. Ayer, *Symp. Proc., Environ. Aspects Fuel Conversion Technol., St. Louis, Missouri* (May 1974); U. S. Environmental Protection Agency, EPA-65012-74-118. *Symp. Proc. Environ. Aspects Fuel Conversion Technol. II, Hollywood, Florida* (December 1975); U. S. Environmental Protection Agency, EPA-600/2-76-149; E. M. Magee, H. J. Hall, and G. M. Varga, U. S. Environmental Protection Agency, EPA-R2-73-249 (1973); R. R. Ruch, H. J. Gluskoter, and E. J. Kennedy, Illinois State Geological Suvey, Environmental Geology Note 43 (1971).
77. V. M. Goldschmidt, *Ind. Eng. Chem.* **27,** 1100 (1935); *J. Chem. Soc.* 655 (1937).
78. P. Zubovic, *Adv. Chem. Ser.* **55,** 221 (1966).
79. H. J. Gluskoter, *Adv. Chem. Ser.* **141,** 1 (1975).
80. R. R. Ruch, H. J. Gluskoter, and N. F. Shimp, Illinois State Geological Survey, Environmental Geology Note 72 (1974).
81. V. M. Goldschmidt, *in* "Geochemistry" (A. Muir, ed.). Oxford Univ. Press (Clarendon), London and New York, 1954.

82. M. Deul and C. S. Annell, U. S. Geological Survey Bull. No. 1036H, p. 155 (1956).

83. H. J. Gluskoter, *Fuel* **44,** 285 (1965); *J. Sediment. Petrol.* **37,** 205 (1967).

84. P. A. Estep, J. J. Kovach, and C. Karr, Jr., *Anal. Chem.* **40,** 358 (1968).

85. C. R. Ward, Illinois State Geological Survey Circ. 498 (1977); L. E. Paulson, W. Berkering, and W. W. Fowkes, *Fuel* **51,** 224 (1972); C. P. Rao and H. J. Gluskoter, Illinois State Geological Survey Circ. 476 (1973); A. F. Rekus and A. R. Haberkorn, *J. Inst. Fuel* **39,** 474 (1966).

86. P. A. Estep, J. J. Kovach, and C. Karr, Jr., *Anal. Chem.* **40,** 358 (1968); J. J. Kovach, *Nat. Meeting, Am. Chem. Soc. Fuel Chem. Div., Minneapolis* (April 1969); C. Karr, Jr., P. A. Estep, and J. J. Kovach, *Chem. Ind. (London)* 356 (1967).

87. S. St. J. Warne, *J. Inst. Fuel* **48,** 142 (1975); **43,** 240 (1970); **38,** 207 (1965).

88. J. V. Gorman and P. L. Walker, Jr., *Fuel* **52,** 71 (1973).

89. H. J. Gluskoter and P. C. Lindahl, *Science* **181,** 264 (1973).

90. M.-Th. Mackowsky, *Proc. Int. Comm. Coal Petrogr.* **2,** 31 (1956); *Arch. Bergbaul. Forsch.* **4,** 5 (1943).

91. A. St. John, *Trans. AIME* **74,** 640 (1926); C. N. Kemp, *Trans. Inst. Min. Eng. (London)* **77,** 175 (1929).

92. See, e.g., ASTM D 2795-72 as well as R. F. Abernethy, M. J. Peterson, and F. H. Gibson, U. S. Bureau of Mines Rep. Invest. 7240 (1969); F. H. Gibson and W. H. Ode, U. S. Bureau Mines Rep. Invest. 6036 (1962); K. Archer, D. Flint, and J. Jordan, *Fuel* **37,** 421 (1958); W. J. S. Pringle, *Fuel* **36,** 257 (1957).

93. R. B. Muter and L. L. Nice, *Adv. Chem. Ser.* **141,** 57 (1975); J. K. Kuhn, W. F. Harfst, and N. F. Shimp, *ibid.* **66** (1975); T. C. Martin, S. C. Mathur, and I. L. Morgan, *Int. J. Appl. Radiat. Isotopes* **15,** 331 (1964).

94. I. A. Breger and J. M. Schopf, *Geochim. Cosmochim. Acta* **7,** 287 (1955).

95. S. W. Parr and W. F. Wheeler, Univ. Illinois Engineering Exp. Station. Bull. 37 (1909).

96. J. G. King, M. B. Maries, and H. E. Crossley, *J. Soc. Chem. Ind.* **55,** 277 (1936).

97. R. L. Brown, R. L. Caldwell, and F. Fereday, *Fuel* **31,** 261 (1952).

98. P. Nicholls and W. T. Reid, *ASME Trans.* **62,** 141 (1940).

99. W. T. Reid and P. Cohen, *ASME Trans. Spec. Sect.* **66,** 83 (1944).

100. H. R. Hoy, A. G. Roberts, and D. M. Wilkins, *J. Inst. Fuel* **31,** 429 (1958).

101. W. L. Sage and J. B. McIlroy, *ASME Trans.* **82,** 145 (1960).

102. O. W. Rees, Illinois State Geological Survey Circ. 365 (1964).

103. P. W. Byers and T. E. Taylor, ASME Paper 75-WA CD-3 (1975).

104. E. A. Sondreal and R. C. Ellman, U. S. Bureau Mines Rep. GFERC-RI-75-1 (1975).

CHAPTER 3

CLASSIFICATION

Efforts to classify coals began over 150 years ago and were at first mainly prompted by a need for simple correlations that might serve to introduce some order into the seemingly chaotic welter of different coals. However, classification on the basis of rank or other properties, such as behavior during combustion, was soon also found necessary for the more systematic coal studies that began to be undertaken in the late nineteenth century and for the expanding coal trade that was largely responsible for triggering the scholarly attention given to coal. Since the early 1900s, classifications have, therefore, not only proliferated through development of national classification systems but have also evolved along divergent lines, with some schemes primarily intended to aid scientific studies and others designed to assist coal producers and users.

3.1 Humic and Sapropelic Coals

Because of the unique properties of the coal-like solids that sometimes formed from putrefying marine biota or massive accumulations of plant

57

spores and pollen grains in lakes or open offshore waters, a fundamental distinction must be made between

(a)   *humic* coals, which developed from terrestrial plant debris that was at least transiently exposed to the atmosphere anu consequently passed through a recognizable peat stage;[1] and

(b)   *sapropelic* coals (or sapropels), which developed from plankton, algae, and/or terrestrial vegetation by putrefaction in entirely anaerobic lacustrine or shallow marine environments (Note A).

The different source materials of sapropels are recognized by dividing these coals into (i) *boghead* coals, characterized by large concentrations of algal or plankton exinites, and (ii) *cannel* coals, which contain a preponderance of fungal spores or pollen.

Sapropels are unstratified and generally resemble durains (or splint coals; see Table 2.1.1) in color and texture. But a characteristic feature is their exceptionally high hydrogen contents, which run to $\sim 8\%$ in cannels and 11–12% in bogheads with $< 90\%$ carbon and only fall to the usual 5–5.5% when their carbon contents exceed $\sim 92\%$. Moreover, mineral matter in sapropels (almost exclusively comprised of clay minerals) is very finely divided and uniformly disseminated.

Since their organic substance is chemically closely similar to materials usually classed as kerogens, sapropels are related to oil shales; and with sufficiently high ash contents, they would, in fact, be regarded as such. However, while their high hydrogen contents would make sapropels especially valuable as sources of synthetic liquid fuels (see, in particular, Sections 13.2 and 14.2), their relative scarcity robs them of industrial significance; and most coal classification schemes are therefore only designed to accommodate humic coals. Even classifications primarily intended to aid studies of coal systematics, i.e., studies of relationships between rank and coal properties, include them only for special, essentially "academic," purposes.

3.2   Classification Parameters

Since carbon contents of coal vary directly or inversely with several other parameters (such as oxygen contents, volatile matter contents, and calorific value), the choice of principal coordinates for classification by rank is a matter of convenience and depends usually on the purposes that the scheme is to serve.

---

[1] Some authors (e.g., Francis [1]) refer to humic coals as "normal" coals. This practice causes confusion since the same term is often used in reference to normal and regional metamorphism.

For schemes mainly intended to facilitate studies of coal systematics, and therefore sometimes referred to as "scientific" classifications, the basic rectangular coordinates, always expressed on a dry, ash-free or dry, mineral-matter-free coal basis, are normally

(a)   percent carbon versus percent hydrogen or oxygen (or, in certain cases, atomic O/C versus H/C ratios), or

(b)   percent volatile matter (VM) or fixed carbon (FC) versus calorific value (Q),

and further refinements are then introduced by combining variants of (a) and (b), for example, by superimposing a VM versus Q plot (at approximately 45° inclination) on a carbon–hydrogen correlation.

In such schemes it is also common practice to introduce correlations of selected physical properties (such as bed moisture contents and/or caking propensities, which are discussed in Section 6.3).

However, because of the inherent complexities of "scientific" classifications, and because the designation of coal classes in an essentially continuous series (such as is delineated in a carbon/hydrogen or VM/Q plot) tends to be arbitrary, "use" schemes generally follow a different course. In most cases, the parameters here are combinations of chemical and physical characteristics, e.g., moisture contents or calorific values and caking properties, and a tabular rather than graphic format is adopted in order to distinguish between and define different classes of coal.

3.3   Classification Systems

Of the numerous "scientific" coal classification systems that start with carbon–oxygen or carbon–hydrogen correlations [2–4], by far the best known—and nowadays the only ones still used in one form or another—are those of Seyler [5] and Mott [6].

The core of Seyler's system is a coal "band" in a carbon–hydrogen plot to which reference has been made in Section 2.2 (see Fig. 2.2.3); and Fig. 3.3.1, which shows the latest version of the complete system, is included here merely as an illustration of how such schemes can be used to define interrelations of coal properties. Strictly speaking, Seyler's classification is applicable only to British Carboniferous coals (for which it was, in fact, specifically designed). Neither the boundaries of the major coal types nor the detailed connections between carbon contents and other parameters reflect *general* coal behavior; and since British coals, except for a small lignite deposit near Bovey Tracey, are all of bituminous and higher rank, the extension of the

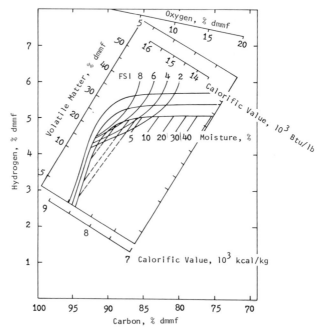

Fig. 3.3.1 Seyler's coal chart (adapted from chart 47B). This version shows relationships between elemental composition, volatile matter contents, moisture contents, and caking properties (expressed in FSI terms; see Section 6.3).

coal band below 85% carbon is of questionable validity. Data for North American lignites indicate very much greater variation of hydrogen contents among immature coals than the band allows and generally point to an average of 4.8 rather than 5.4%. But despite these limitations, the classification— and especially the coal band—affords an important means for visualizing several aspects of coal behavior. Starting points for such application of the system are so-called development lines [7], which show, sometimes in idealized form, how the composition of a coal or a coal constituent will change when altered by specific chemical processes. Figures 3.3.2 and 3.3.3 exemplify such development lines. The former illustrates the effect of progressive metamorphic development on the carbon, hydrogen, and oxygen contents of humic and sapropelic coals [8], while the latter shows how hydrogen contents of the three principal maceral groups (see Table 2.1.2) change as their rank increases [9]. (Other development diagrams are referred to in Section 5.1; see Figs. 5.1.6 and 5.1.7.)

Diagrams of this type have proved valuable in studies of coal diagenesis and are also useful in coal exploration programs, especially where decisions

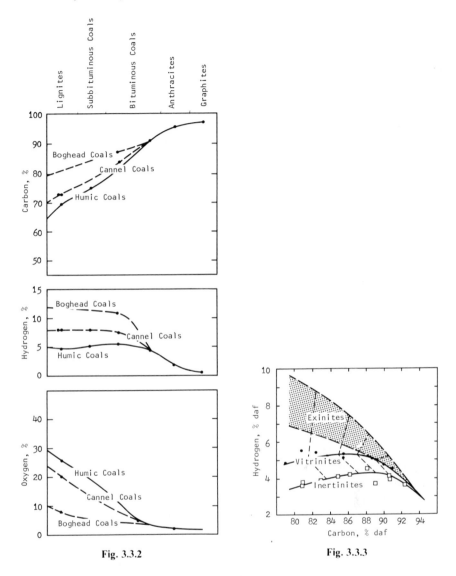

Fig. 3.3.2

Fig. 3.3.3

Fig. 3.3.2 Development lines illustrating the effect of progressive maturation on the composition of humic and sapropelic coals. (After White [8].)

Fig. 3.3.3 Development lines illustrating the effect of progressive maturation on hydrogen contents of individual maceral groups. The dashed lines indicate the compositions of macerals of *equal rank*. (After van Krevelen [9]; by permission of IPC Business Press Ltd.)

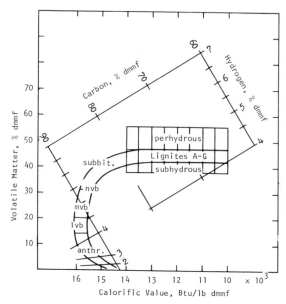

Fig. 3.3.4 Mott's coal classification. (Adapted from Mott [6].)

about the further conduct of the program must be based on the properties of "weathered" outcrop samples (see Section 5.1).

In Mott's classification (Fig. 3.3.4), the basic rectangular coordinates are volatile matter contents and calorific value, and the superimposed carbon–hydrogen correlation is used primarily to differentiate between lignites, among which hydrogen contents are shown to vary from <4.5 to >6.0%. Unlike Seyler's system, Mott's is based on coals from all parts of the globe and is therefore perhaps more generally applicable. However, in view of the considerable scatter of data points for lignites and because there appears to have been no statistical weighting of the samples, it is doubtful whether the course of the coal band below ~80% carbon is, in reality, any more significant than that of Seyler's band in this region.

In contrast to "scientific" classification schemes, systems designed to assist coal users reflect recognition that chemical maturation and physical development of coal do not invariably proceed hand in hand [10], and that parameters particularly suitable for assessing rank changes in immature coals are not necessarily the most appropriate for defining the extent of metamorphic development in mature coals [11]. In many instances, the classification criteria therefore change with rank.

In North America the generally accepted classification is the ASTM

Table 3.3.1

ASTM coal classification by rank

Class and group	Fixed carbon[a] (%)	Volatile matter[a] (%)	Heating value[b] (Btu/lb)
I. Anthracitic			
1. Metaanthracite	>98	<2	
2. Anthracite	92–98	2–8	
3. Semianthracite	86–92	8–14	
II. Bituminous			
1. Low volatile	78–86	14–22	
2. Medium volatile	69–78	22–31	
3. High volatile A	<69	>31	>14,000
4. High volatile B			13,000–14,000
5. High volatile C			10,500–13,000[c]
III. Subbituminous			
1. Subbituminous A			10,500–11,500[c]
2. Subbituminous B			9500–10,500
3. Subbituminous C			8300–9500
IV. Lignitic			
1. Lignite A			6300–8300
2. Lignite B			<6300

[a] Calculated on dry, mineral-matter-free coal; correction from ash to mineral matter made by means of Parr formula.

[b] Calculated on mineral-matter-free coal with bed moisture content.

[c] Coals with heating values between 10,500 and 11,500 Btu/lb are classed as high volatile C if they possess "agglutinating" properties (see Section 6.3) or as subbituminous A if they do not.

scheme, which distinguishes between four coal classes, each subdivided into several groups. High-rank coals (of medium volatile bituminous or higher rank) are classified on the basis of their volatile matter and fixed carbon contents, which are expressed on a dry, mineral-matter-free basis, while low-rank coals are defined in terms of their calorific value, calculated on mineral-matter-free "moist" coal (containing its natural bed moisture). The most recent version of the system (ASTM D 388-66) is shown in Table 3.3.1.

European schemes and the current form of a tentative international coal classification follow similar lines but are rather more elaborate. A typical system, particularly noteworthy because it was the first to use a numerical code for designating the various coal classes and groups, was originally developed in Britain after World War II [12] and later adopted by the National Coal Board. This scheme is shown in Table 3.3.2. (The Gray–King coke types that are used as secondary parameters are further discussed in Section 6.3.)

Table 3.3.2

NCB (Britain) coal classification system

Group	Class	Volatile matter (% dmmf)	Gray–King coke type	Description
100	101	6.1[a]	A	Anthracite
	102	6.1–9.0[a]	A	Anthracite
200	201	9.1–13.5	A–G	
	201a	9.1–11.5	A–B	Dry steam coals
	201b	11.6–13.5	B–C	
	202	13.6–15.0	B–G	
	203	15.1–17.0	B–G4	Coking steam coals
	204	17.1–19.5	G1–G8	
	206	9.1–19.5	A–B[b] A–D[c]	Heat-altered low-volatile bituminous coals
300	301	19.6–32.0		
	301a	19.6–27.5	G4	Prime coking coals
	301b	27.6–32.0		
	305	19.6–32.0	G–G3	Heat-altered medium-volatile bituminous coals
	306	19.6–32.0	A–B	
400	401	32.1–36.0	G9	Very strongly caking coals
	402	>36.0		
500	501	32.1–36.0	G5–G8	Strongly caking coals
	502	>36.0		
600	601	32.1–36.0	G1–G4	Medium caking coals
	602	>36.0		
700	701	32.1–36.0	B–G	Weakly caking coals
	702	>36.0		
800	801	32.1–36.0	C–D	Very weakly caking coals
	802	>36.0		
900	901	32.1–36.0	A–B	Noncaking coals
	902	>36.0		

[a] To distinguish between classes 101 and 102, it is sometimes more convenient to use a hydrogen content of 3.35% instead of 6.1% volatile matter.

[b] For volatile matter contents between 9.1 and 15.0%.

[c] For volatile matter contents between 15.1 and 19.5%.

The international classification of hard coals[2] [13] resulted from studies by the European Economic Commission's coal committee (which included representation from the United States and USSR) and is an elaboration of the NCB scheme and of several similar national systems. The primary classification parameters are volatile matter or calorific value (for coals with, respectively, <35 and >35% volatile matter), and as in the ASTM system,

[2] *Hard* coals are defined as anthracites and bituminous coals but include the subbituminous A coals of the ASTM classification. Coals of lower rank are termed *soft* coals and are sometimes also referred to as brown coals.

Table 3.3.3

International hard coal classification[a]

The first figure of the code number indicates the class of the coal, determined by volatile matter content up to 35% V.M. and by calorific parameter above 35% V.M.
The second figure indicates the group of coal, determined by caking properties.
The third figure indicates the subgroup, determined by coking properties.

GROUPS (determined by caking properties)	ALTERNATIVE GROUP PARAMETERS — Free-Swelling Index	ALTERNATIVE GROUP PARAMETERS — Roga index			CODE NUMBERS									SUBGROUP NUMBER	SUB GROUPS — Dilatometer	SUB GROUPS — Gray-King
			Class 4	Class 5	Class 6								5	>140	>G_9
3	>4	>45	435	535	635										
			334	434	534	634							4	>50–140	G_5–G_8
			333	433	533	633	733						3	>0–50	G_1–G_4
			332 a / 332 b	432	532	632	732	832					2	≤ 0	E–G
2	2½–4	>20–45	323	423	523	623	723	823					3	>0–50	G_1–G_4
			322	422	522	622	722	822					2	≤ 0	E–G
			321	421	521	621	721	821					1	Contraction only	B–D
1	1–2	>5–20	212	312	412	512	612	712	812				2	≤ 0	E–G
			211	311	411	511	611	711	811				1	Contraction only	B–D
0	0–½	0–5	100 (A / B)	200	300	400	500	600	700	800	900		0	Non-softening	A

CLASS PARAMETERS		Class 0	Class 1	Class 2	Class 3	Class 4	Class 5	Class 6	Class 7	Class 8	Class 9
CLASS NUMBER		0	1	2	3	4	5	6	7	8	9
Volatile Matter (dry, ash-free)		0–3	>3–10 (>3–6.5 / >6.5–10)	>10–14	>14–20	>20–28	>28–33	>33	>33	>33	>33
Calorific Parameters[a]		–	–	–	–	–	–	>13950	>12960–13950	>10980–12960	>10260–10980

(Determined by volatile matter up to 35% V.M. and by calorific parameter above 33% V.M.)

[a] United Nations [13].

Table 3.3.4

Position of North American coals in the international hard coal classification[a]

The first figure of the code number indicates the class of the coal, determined by volatile matter content up to 33% V.M. and by calorific parameter above 33% V.M.
The second figure indicates the group of coal, determined by caking properties.
The third figure indicates the subgroup, determined by caking properties

GROUP NUMBER	ALTERNATIVE GROUP PARAMETERS — Free-swelling index	ALTERNATIVE GROUP PARAMETERS — Roga Index	Class 0	Class 1	Class 2	Class 3	Class 4	Class 5	Class 6	Class 7	Class 8	Class 9	SUBGROUP NUMBER	Dilatometer	Gray-King
3	>4	>45					435	535	635				5	>140	>G$_8$
						334	434		634	734			4	>50–140	G$_5$–G$_8$
						333	433		633	733			3	>0–50	G$_1$–G$_4$
													2	≤0	E–G
2	2½–4	>20–45							623	723	823		3	>0–50	G$_1$–G$_4$
									622	722	822		2	≤0	E–G
										721	821		1	Contraction only	B–D
1	1–2	>5–20												2	≤0	E–G
										711	811		1	Contraction only	B–D
0	0–½	0–5	000	100 (A, B)	200						800	900	0	Non-softening	A

CLASS NUMBER	0	1	2	3	4	5	6	7	8	9
Volatile matter (dry, ash-free)	0–3	>3–6.5 (A) / >6.5–10 (B) [>3–10]	>10–14	>14–20	>20–28	>28–33	>33	>33	>33	>33
Calorific Parameters[a]	–	–	–	–	–	–	>13950	>12960–13950	>10980–12960	>10260–10980

(Determined by volatile matter up to 33% V.M. and by calorific parameter above 33% V.M.)

CLASSES (Determined by volatile matter up to 33% V.M.)

[a] Ode and Frederic [14].

these are used to delineate a series of coal classes. But as in the NCB scheme, each class is then divided into groups on the basis of *caking* characteristics (which are defined in terms of free-swelling indices and/or Roga indices); and each group is further split into subgroups in which the coals are characterized by their *coking* properties (as measured by the Gray–King assay or a standard dilatometer; Note B). The current version of the international classification is shown in Table 3.3.3; and Table 3.3.4, based on the work of Ode and Frederic [14], shows how American coals, classified in accordance with the ASTM system, fit into it. In both tabulations the first digit of each code number identifies the class; the second, the group; and the third, the subgroup.

Paralleling its hard coal classification, the coal committee of the European Economic Commission has also developed a tentative international classification of *soft* coals [15]. Such coals are broadly defined as containing less than $10,250 \pm 110$ Btu/lb (23.85 MJ/kg) on a moist, ash-free basis and correspond to the lignites and subbituminous B and C coals of the ASTM system. Because soft coals are still mainly used as a heating and steam-raising fuel, and (to a much lesser extent) in some European countries as sources of chemical raw materials, the classification parameters are bed moisture contents and tar yields, the latter calculated on a dry, ash-free basis. But it seems likely that this format, which is shown in Table 3.3.5, will require substantial modification as soft coals begin to be more extensively used in fuel conversion processes and as petrochemical feedstocks (see Chapters 12–14).

Table 3.3.5

Tentative international classification of soft coals[a]

Group number	Group parameter tar yield (%)	Code number					
40	25	1040	1140	1240	1340	1440	1540
30	20–25	1030	1130	1230	1330	1430	1530
20	15–20	1020	1120	1220	1320	1420	1520
10	10–15	1010	1110	1210	1310	1410	1510
00	<10	1000	1100	1200	1300	1400	1500
Class number		10	11	12	13	14	15
Class parameter Moisture (%)[b]:		20	20–30	30–40	40–50	50–60	>60

[a] United Nations [15].
[b] Moisture determined at 30°C and 96% relative humidity; calculated on ash-free coal.

Notes

A. It might here be observed that sapropels (and oil shales) can, in fact, be best understood as standing halfway between humic coals, which formed from decaying organic matter *on land or in shallow waters*, and petroleum, which formed from such debris in *deep waters*.

B. The distinction between *caking* and *coking* and the methods by which caking properties are assessed are dealt with in Chapter 6 (see, in particular, Section 6.3).

References

1. W. Francis, "Coal." Arnold, London, 1961.
2. F. F. Grout, *Econ. Geol.* **2,** 225 (1907).
3. O. C. Ralston, U. S. Bureau Mines Tech. Paper No. 93 (1915).
4. H. J. Rose, *Trans. Am. Inst. Min. Metall. Eng.* **88,** 541 (1930).
5. C. A. Seyler, *Proc. S. Wales Inst. Eng.* **21,** 483 (1899); **22,** 112 (1900); **53,** 254, 396 (1938).
6. R. A. Mott, *J. Inst. Fuel* **22,** 2 (1948).
7. H. Briggs, *Proc. R. Soc. Edinburgh* **52** (2), 195 (1932).
8. D. White, *Econ. Geol.* **28,** 556 (1933).
9. D. W. van Krevelen, "Coal." Elsevier, New York, 1961.
10. N. Berkowitz, *Fuel* **29,** 138 (1950); **31,** 19 (1952).
11. K. Patteisky and M. Teichmüller, Coll. Int. Pet. Charb., Rev. d l'Ind. Min., No. Spec. 121 (1958).
12. DSIR, Survey Paper No. 58 (1946).
13. United Nations, Publ. No. 1956 II, E.4, E/ECE/247, E/ECE/Coal 110, Geneva (1956).
14. W. H. Ode and W. H. Frederic, U. S. Bureau Mines Rep. Invest. No. 5438 (1958).
15. United Nations, Publ. No. 1957 II, E., Min. 20, Geneva (1957).

CHAPTER 4

PHYSICAL PROPERTIES

During progressive metamorphic development, chemical alteration and increasing compression of the coal substance also combine to change its physical properties; and except in certain anomalous instances,[1] these properties do, indeed, vary so systematically with coal rank that many can be broadly predicted from the carbon or volatile matter content of the coal. Unlike basic chemical parameters, however, they do not always vary monotonically, i.e., in some relatively simple linear or nearly linear manner, with carbon contents. Many pass through a more or less well-defined maximum or minimum (usually between ~85 and 89% carbon) and thereby reflect

---

[1] Such anomalies are mostly found in atypical tectonic environments, e.g., where the coal has been heavily compressed at temperatures insufficient to promote concurrent chemical changes. Examples are afforded by Salt Range (Pakistan) lignites whose bed moisture contents and caking propensities make them resemble bituminous coals [3]. Others present themselves where metamorphic development has been affected by incursions of magma or radioactive minerals (see Section 1.1).

the rank-dependent surface-to-volume ratio of coal or the transition from bituminous to anthracitic coals.

4.1   Porosity and Pore Structure

Few properties influence the behavior of coal more immediately and directly than its pore space. Even in anthracites, this is always so extensive and intricately structured as to make coal something like a solid sponge and endow it with the characteristics of a *solid colloid* [1, 2].

The total void volume (or porosity) of coal can be calculated from its capacity moisture content (see Section 2.2) and, as illustrated in Fig. 4.1.1, falls from a high of ~25 to 30% among lignites and subbituminous C coals to as little as 1 or 2% among coals with 87–89% carbon before increasing again among anthracites.

But more important than the magnitude of this void space is that it is comprised of very small pores that limit the size of (reagent) molecules which can pass through them and consequently make coal a natural molecular sieve. This aspect determines the total surface area[2] ($\bar{S}$) that coal potentially presents to a reagent and also influences the rate at which access to the surface can be gained in any particular case, and has been the subject of numerous studies since the early 1940s. In some, pore size distributions as well as

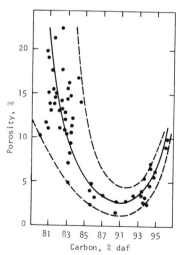

Fig. 4.1.1   The variation of coal porosity with rank. (After King and Wilkins [4].)

[2] It should be noted that this quantity comprises the internal surface area (presented by the pore walls) and the so-called external or "specific" surface area (which depends on the size of the coal particle). Because of the smallness of the pores, the former is always at least one order of magnitude larger than the latter.

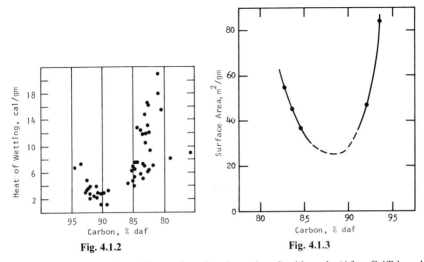

Fig. 4.1.2   Fig. 4.1.3

Fig. 4.1.2   The variation of heats of wetting (in methanol) with rank. (After Griffith and Hirst [9].)

Fig. 4.1.3   The rank dependence of surface areas determined by adsorption of neon at 25°C. (After Bond and Spencer [10].)

the surface areas accessible to different fluids have been measured in efforts to define the *structure* of the pore space.

For measurements of surface areas, use has been made mainly of

(a)   heats of wetting (h.o.w.), i.e., the temperature rises (ΔT) that accompany immersion of thoroughly evacuated, weighed test samples in a known volume of a suitable liquid [5], and

(b)   isothermal sorption of a gas or vapor by the coal [6].

In the former case, provided there is no chemical interaction between the coal and the liquid, the heat of wetting is directly proportional to the extent

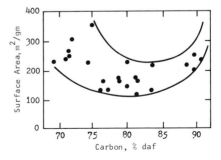

Fig. 4.1.4   The rank dependence of surface areas determined by adsorption of carbon dioxide at 25°C. (After Gan *et al.* [11].)

of the wetted surface, and $\bar{S}$ can be calculated from 1 cal $\simeq$ 10 m^2 (Note A). In the sorption method, the quantity (v) of gas or vapor sorbed at different relative pressures (p/p_0) is plotted against p/p_0, and $\bar{S}$ is evaluated from v at monolayer capacity (i.e., when the entire "available" surface is covered by a monomolecular film) and from the cross-sectional area of the sorbate molecule in its adsorbed state [8].

Typical results obtained by these means (and all at one time or another thought to represent the "real" or *absolute* surface area of coal[3]) are illustrated in Figs. 4.1.2 (heats of wetting in methanol [9]), 4.1.3 (neon sorption at 25°C [10]), and 4.1.4 (carbon dioxide sorption at 25°C [11]).

Notwithstanding the efforts devoted to such measurements, it is however still questionable how much significance can be attached to these (or other substantially similar) data, and numerical values of $\bar{S}$ must be treated with considerable caution. In part, such caution is required because both measuring techniques, when applied to finely pored solids, are subject to experimental errors that may run as high as $\pm 50\%$. But also important is that most published data have been criticized on more specific grounds. Heats of wetting in methanol are now known to be grossly distorted by imbibition and/or polar interaction of methanol with oxygen-bearing functional groups (notably —OH) at the coal surface ([12, 13]; Note B); and estimates of $\bar{S}$ from sorption of neon, carbon dioxide, or other gases depend on how the sorption isotherms are determined and evaluated, and often involve assumptions about how the sorbate molecule reaches the surface and lodges on it.

While it is not too difficult to estimate *relative* surface areas or surface areas *accessible to different sorbates*, the absolute surface areas of coal are consequently still uncertain quantities. All that can be asserted with confidence is that they generally tend to fall from values in the range 100–200 m^2/gm among lignites and subbituminous coals to 50–100 m^2/gm among bituminous coals before again increasing to over 100 m^2/gm among anthracites.

A clear impression of the complexity of the pore space and of its likely impact on coal behavior can, however, be gained from pore size distribution measurements which have revealed the existence of distinctive micro- and macropore systems in coal.

The nature and extent of the micropore system has been explored [7] by measuring heats of wetting (or apparent densities; see Section 4.2) in different liquids and plotting the first derivatives of the resultant curves against the molar volumes of the liquids (Fig. 4.1.5). Such diagrams show

---

[3] These terms imply the assumption that the measuring fluid has completely penetrated all pore spaces.

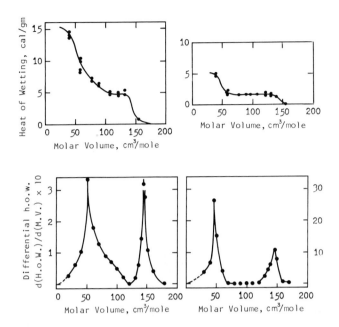

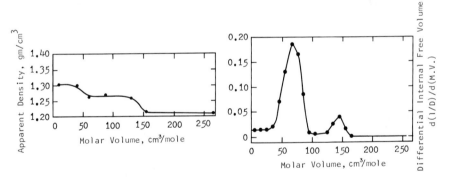

Fig. 4.1.5   Dependence of heats of wetting and densities of coal on the molecular sizes of the measuring fluids. (After Bond and Spencer [7].)

a characteristic bimodal distribution of small pores around ~5 and ~8 Å (0.5 and 0.8 nm), and allow the fraction of $\bar{S}$ presented by pores in any size interval to be determined by integration.

The distribution of pores in the macrosystem has been determined by measuring the penetration of coal by mercury as a function of pressure and calculating pore sizes from

$$r = -(2s/p)\cos\theta$$

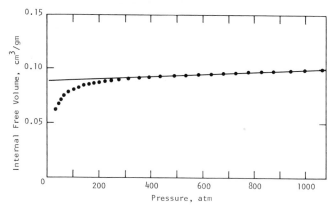

Fig. 4.1.6 Mercury penetration of coal as function of pressure. From the slope of the straight line, a compressibility of 13×10^{-12} cm^2/dyne is found. (After Zwietering and van Krevelen [15]; by permission of IPC Business Press Ltd.)

where r is the pore radius, p the pressure, s the surface tension of Hg, and θ its angle of contact with the coal. By substituting $s = 480$ dyne/cm (4.8 mN/cm) and $\theta = 140°$, this leads to

$$r = 75,000/p$$

in which r is expressed in angstrom units and p in atmospheres. The penetration technique was first used by Ritter and Drake [14] and later applied to coal by, inter alios, Zwietering and van Krevelen [15]. A typical result is shown in Fig. 4.1.6, and a characteristic pore size distribution obtained from such measurements is illustrated in Fig. 4.1.7.

Data such as these have led to the conclusion that

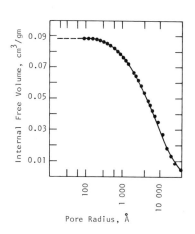

Fig. 4.1.7 A typical pore size distribution in coal, as measured by mercury penetration. (After Zwietering and van Krevelen [15]; by permission of IPC Business Press Ltd.)

(a) the internal surface area of coal is mostly associated with a capillary system in which 40-Å pores, or smaller, are linked by 5–8-Å passages, and

(b) over 90% of the surface area is, in every case, contained within a micropore system that represents between 50 and 80% of the total void space.[4]

4.2 Density

Partly because of the intricately structured void volume of coal, its density varies not only with rank but depends also on how it is measured and therefore carries several different connotations. A distinction must, in particular, be made between

(a) *bulk densities*, which are determined by the average particle (or lump) size, size distribution, and packing density of the coal, and which bear on coal handling, transportation, and storage;

(b) *in-place densities*, which refer to the weight-to-volume ratios of undisturbed coal in a seam and are important parameters for estimating coal reserves;

(c) *apparent densities*, which are conventionally measured by liquid displacement and therefore depend on how completely the pore space of the coal is penetrated; and

(d) *absolute densities*, which represent the *true* densities of the coal substance in its various stages of metamorphic development.

Appropriate methods for determining coal densities depend therefore on what information is sought; and unless they are clearly specified, densities reported in technical literature must be interpreted with great caution.

For practical purposes the most useful value is the apparent density, which can be conveniently measured in pyknometers with water, benzene, or *n*-hexane (Note C). Care must, however, be taken to use thoroughly dried coal and to allow adequate time for equilibration. Even with suitable samples (1–2 gm sized to <3 mm with minimum dust) up to 24 hr may be needed for attainment of constant weight. The mandatory correction for mineral matter, which converts the measured density to the density of the pure coal substance, is then made by means of

$$D_{corr} = \frac{D_a D_{exp}(100 - \% \text{ ash})}{100 D_a - D_{exp}(\% \text{ ash})}$$

[4] Some later work by Walker *et al.* [16] suggests that 75% of the internal free volume in coal with less than ~75% carbon is associated with macropores, but that microporosity comes to predominate as the rank of the coal increases.

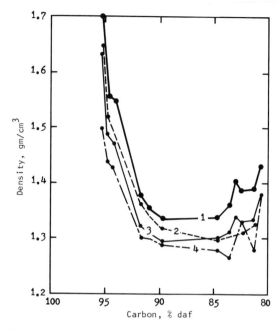

Fig. 4.2.1 Variation of coal densities with rank. Densities were measured with (1) methanol, (2) benzene, (3) water, and (4) *n*-hexane. (After Franklin [17]; by permission of IPC Business Press Ltd.)

where D_{exp} and D_a denote the measured densities of the coal sample and of the ash obtained from it. This correction reflects the fact that the densities of mineral matter and its ash differ very little, but requires a separate determination of the ash density. Where such a measurement cannot be made, D_a may be assumed to have a value of between 2.7 and 3.0 gm/cm³.

Typical apparent densities of coals with over 80% carbon in water, methanol, benzene, and *n*-hexane [17] are shown in Fig. 4.2.1, and the influence of petrographic composition is illustrated in Table 4.2.1, which lists apparent densities of variously metamorphosed macerals in methanol. The consistently higher densities recorded with methanol (Fig. 4.2.1) have been attributed to compression of this liquid in the small pores of the coal [18].[5]

Procedures for determining *absolute* densities resemble those employed for gas or vapor sorption measurements [7, 16, 18, 19] and require that pure helium be used as the dilatometric fluid. The choice of this gas is

[5] Methanol is also strongly sorbed by coal, and Danforth and De Vries [20] have pointed out that the density of the first adsorbed layer of MeOH is itself some 20% greater than the density of liquid MeOH.

Table 4.2.1

Apparent densities of macerals in methanol[a]

Maceral	Carbon, (% daf)	D (MeOH), (gm/cm³)
Vitrinite	83.5	1.345
	85.7	1.334
	88.4	1.317
	88.8	1.368
Exinite	85.5	1.201
	87.4	1.213
	89.1	1.288
	89.3	1.347
Micrinite	86.8	1.463
	88.0	1.415
	89.6	1.414
	89.8	1.413

[a] After Kröger and Badenecker [22].

based on its small effective atomic diameter (1.78 Å) and on the fact that it is not measurably sorbed by coal. Since coal appears to contain very few "blind" pores (Note D), these properties are considered to ensure complete penetration of the šample. Measurements are made on <1-mm particles or small coal pellets after thorough outgassing at 90–120°C in a good vacuum (10^{-4} torr or better) and corrected for mineral matter as in the case of apparent densities.

However, absolute densities determined in this manner do not greatly differ from values recorded with water even when carbon contents exceed ~90% (see Fig. 4.2.2). Density measurements of coal in helium, to obtain D (He), are therefore only of interest where the greatest possible accuracy is required.

Good *estimates* of D (He) can be derived from the heats of wetting (Q) of the coal in methanol. According to Franklin [18], compression of methanol in small pores (see above) is linearly related to heats of wetting in methanol through 1 cal (4.187 J) = 0.0026 cm³, and it is consequently possible to write

$$v = v \text{ (He)} - v \text{ (MeOH)} = \frac{1}{D \text{ (He)}} - \frac{1}{D \text{ (MeOH)}} = 0.0026 \cdot Q \text{ (MeOH)}$$

from which it follows that

$$D \text{ (He)}_{\text{calc}} = \frac{D \text{ (MeOH)}}{0.0026 \cdot Q \text{ (MeOH)} \cdot D \text{ (MeOH)} + 1}$$

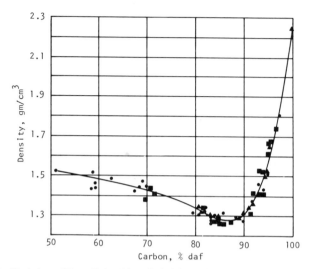

Fig. 4.2.2 Variation of "true" densities of vitrinites with rank. ▲, ■, in helium [18, 27]; ●, in water [26]. (By permission of Elsevier Publ. Co.)

Alternatively, estimates that closely approximate D (He) and D (H$_2$O) can be obtained from the hydrogen contents of coal [17, 23–25] via

$$1/D = 0.54 + 0.043 \cdot \% \, H$$

This correlation holds for all coals with $>80\%$ carbon and also applies to oxidized coals with carbon contents between ~ 73 and 85%.

In-place or so-called *bank* densities are the apparent densities of un-disturbed coal within a seam and are only useful for estimating the tonnages of coal that underlie a particular area. They are always *un*corrected for mineral matter and moisture contents, and must, in fact, be determined on water-saturated samples. In most cases they are calculated from the gross weights and volumes of monolithic specimens; but where such speci-mens are not available, they can also be determined pyknometrically on granular samples after equilibrating them to their respective capacity moisture contents (see Section 2.2). Most convenient for this purpose is a measurement of density by mercury displacement (Note E).

Bulk densities refer to the uncorrected apparent densities of *broken* coal and depend, as already noted, on the size distribution and packing density (or degree of compaction) of the coal, as well as on its mineral matter and moisture contents. They must therefore be determined for each particular case, usually by direct weighing of a known volume of the coal; and being gross measurements, they are always expressed in lb/ft^3 or kg/N m^3 rather than in gm/cm^3.

4.3   Electrical Conductivity

Although often regarded as electrical insulators, high-rank coals are actually intrinsic semiconductors whose conductivities K (or resistivities R^6) can be measured with a Wheatstone bridge [28], an ammeter—voltmeter system [29], or, where R exceeds 10^9 ohms, a fluxmeter [30]. Either monoliths or compressed pellets (or rods) can be used for such measurements. However, since electrical conductivity is very sensitive to ambient conditions and particularly strongly affected by moisture, it is sometimes preferable to determine the *electron excitation energy* Δe from the temperature variation of R, i.e.,

$$1/K = R = R_\infty \exp(\Delta e/2kT)$$

In this expression, R_∞ is the resistivity at $1/T = 0$, k is the Boltzmann constant (1.3805×10^{-16} erg/°K), and T is the temperature of measurement. Δe, the energy barrier that an electron must surmount in order to become conducting, can then be evaluated from a plot of R versus $1/T$.

Some typical values of R and Δe for anthracites are shown in Table 4.3.1 and illustrate that unlike K, which becomes increasingly dependent on the orientation of monolithic specimens as carbon contents increase beyond 93–94%, Δe does not exhibit a corresponding anisotropy. This finding has been taken to imply that conductor properties are associated with charge transfer across aromatic lamellae in the coal (see Section 5.7), and that anisotropy in R is due to variation of the temperature-independent R_∞ term.

Among bituminous coals the very steep inverse variation of R with

Table 4.3.1

Resistivities of anthracites[a]

Carbon (% daf)	R (ohm cm)	Δe (eV)
93.7	4.01×10^7	0.5
94.2	$6.09 \times 10^{4\,b}$	0.34
	$4.97 \times 10^{4\,c}$	
95.0	$1.71 \times 10^{3\,b}$	0.27
	$0.70 \times 10^{3\,c}$	
96.0	6.43^b	0.17
	3.73^c	

[a] After Schuyer and van Krevelen [31].
[b] Measured perpendicular to bedding plane.
[c] Measured parallel to bedding plane.

[6] By definition, resistivity is the reciprocal of conductivity, i.e., $R = 1/K$.

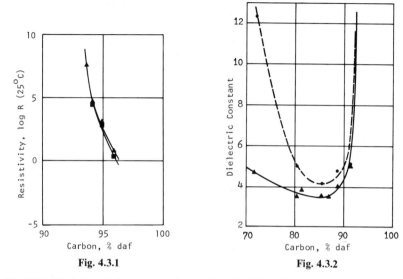

Fig. 4.3.1

Fig. 4.3.2

Fig. 4.3.1   Variation of electrical resistivity with rank. (After Schuyer and van Krevelen [31]; by permission of IPC Business Press Ltd.)

Fig. 4.3.2   Variation of the dielectric constant of coal with rank. ●, air-dried coal; ▲, dry coal. (After Groenewege et al. [32]; by permission of IPC Business Press Ltd.)

carbon contents of anthracites (see Fig. 4.3.1) continues, and resistivities reach values in the range 10^{10}–10^{14} ohm cm.

In passing it might be noted that the semiconductor character of coal is reflected in its dielectric constant. This quantity varies with the square of the refractive index (n), but generally exceeds n^2 if permanent dipoles are present. As a result, the dielectric constant of coal increases rapidly with moisture contents and can be used to monitor moisture contents of coal in preparation plants (see Section 2.2).

The variation of the dielectric constant with rank and the effect of moisture on it are illustrated in Fig. 4.3.2.

4.4   Specific Heats, Thermal Conductivity, and Thermal Expansion

In processes in which coal is exposed to elevated temperatures or required to transmit heat, some importance attaches to its specific heat (C), i.e., its thermal capacity relative to water at 15°C, and to its thermal conductivity. These aspects have received considerable attention since the late

1880s (Landolt and Bornstein [33]) and have been reviewed by McCabe and Boley [34] and by Clendenin *et al.* [35].

For the determination of *specific heats* (normally measured at constant pressure), either the classic "method of mixtures" or a Bunsen ice calorimeter, which utilizes the change in the specific volume of H_2O as it passes from the solid to the liquid state at 0°C, can be employed. In the former, a weighed sample is heated to the desired temperature and quickly transferred into a known volume of water, and the specific heat is calculated from the temperature rise in the liquid [36, 37]. In the Bunsen method, the absorbed heat is calculated from a volumetric measurement [38]. Either method is more accurate than Joule's procedure (in which a suspension of the test sample in water is directly heated by a known quantity of electricity), but care must be taken to avoid, or to appropriately correct for, contributions to the heat change in the liquid from heats of wetting (see Section 4.1 and Note F). Since C varies linearly with moisture contents, it is usually most convenient to determine the specific heat of *dry* coal by extrapolating a plot of C versus % H_2O to $H_2O = 0$.

Specific heats of dry coal decrease uniformly with increasing carbon contents from ~ 0.30 cal/gm at 70% to ~ 0.25 cal/gm at 90%, and then fall more rapidly to ~ 0.20 cal/gm at 95%. (The specific heat of pure graphite is 0.165 cal $\equiv$ 0.691 J per gm.)

Calculation of specific heats from elemental compositions (by Kopp's law) only yields acceptable results for anthracites, but good agreement with experimentally measured values for lower-rank coals can be obtained [39] by evaluating C from an expression of the form

$$C = f \cdot R/2M$$

where f is the number of vibrational degrees of freedom of each atom, R is the gas constant (1.98 cal/°K), and M is the molecular weight. Assuming that each carbon and hydrogen atom contributes, on average, with only 1 degree of freedom,[7] this expression yields values that lie within $\pm 5\%$ of the measured quantities (see Table 4.4.1).

According to Fritz and Moser [40], specific heats can also be computed from volatile matter contents by means of

$$C = 0.242(1 + 0.008\,\mathrm{VM})$$

which is claimed to hold reasonably well for all ranks of coal.

Published values for the *heat conductivity* of coal indicate a range of

[7] This reflects the fact that the specific heat of graphite corresponds to the equipartition value of 1 vibrational degree of freedom.

Table 4.4.1

Calculated and measured specific heats[a]

Carbon (% daf)	C_{exp}	C_{calc}
81.5	0.28	0.27
89.0	0.26	0.26
93.4	0.22	0.23
95.0	0.20	0.21

[a] After van Krevelen [39]; C in cal/gm.

3.3×10^{-4} to 8.6×10^{-4} cal/sec cm °C. Sinnatt and McPherson [41], who measured the conductivities of 8 coals with 30–35% volatile matter (and less than 10% ash) by means of a compound bar method, found values between 3.7×10^{-4} and 5.5×10^{-4}; and Fritz and Diemke [42], who studied 32 coals, including anthracites, reported conductivities of 4.64×10^{-4} to 8.64×10^{-4} cal/sec cm °C. Perhaps because of interfering contributions from variously disseminated mineral matter in different coals, there is no discernible systematic variation with coal rank; but Fritz and Diemke [42] did find a reasonably good correlation between heat conductivities and coal densities.

When using heat conductivity data for process design purposes, it must, however, always be borne in mind that the high porosity of coal makes its conductivity very much dependent on the nature of the gas or vapor that suffuses its internal free space. This is well illustrated by Schumann and Voss [43] who, for dry -10 $+12$-mesh coal with 44% void space, recorded values of 3.27×10^{-4} and 7.03×10^{-4} cal/sec cm °C in air and in hydrogen, respectively. (The thermal conductivities of air and hydrogen per se are 0.63×10^{-4} and 4.6×10^{-4} cal/sec cm °C, respectively.) Other investigators have found heat conductivities in this range to be similarly affected by moisture contents.

More recently, some values have also been reported for the coefficient of *linear thermal expansion* (e_t) of coal. The first detailed measurements of this quantity were made by Bangham and Franklin [44] who used a quartz extensometer to determine e_t for two cold-pressed artifacts with, respectively, 82.4 and 89.7% carbon, and for a monolithic anthracite specimen with 94.2% carbon. Their results are shown in Figure 4.4.1. For the two lower-rank coals, e_t increased progressively from $\sim 32 \times 10^{-6}/°C$ at 30°C to $\sim 58 \times 10^{-6}/°C$ at 330°C; and with the anthracite, for which e_t changed

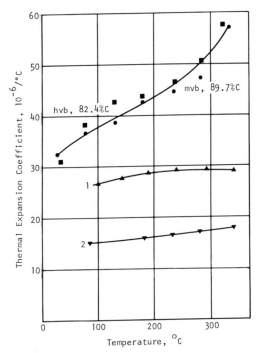

Fig. 4.4.1   Variation of the linear thermal expansion coefficient of coal with temperature. Curves 1 and 2 refer to an anthracite measured perpendicular and parallel to its bedding plane, respectively. (After Bangham and Franklin [44]; by permission of Elsevier Publ. Co.)

little with temperature, it was found to depend on orientation—virtually doubling when it was measured perpendicularly rather than parallel to the bedding plane.

Later measurements on eight vitrinites with 85–96% carbon between 20 and 40°C [45] furnished almost identical results but showed pronounced anisotropy in coals with more than ~87% carbon (Fig. 4.4.2). The mean linear thermal expansion coefficients calculated from Schuyer and van Krevelen's data [45] via

$$\bar{e}_t = e_t/3 \; (\perp) + 2e_t/3 \; (\|)$$

which takes into account the uniaxial symmetry of anisotropic coals, are shown in Fig. 4.4.3. Curve b in this diagram shows corresponding results calculated from the *volume* expansion coefficients between −47 and 0°C [46] by putting $\bar{e}_t = e_{vol}/3$.

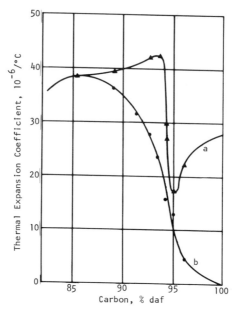

Fig. 4.4.2 Variation of the linear thermal expansion coefficient of coal with rank. The coefficients were measured at between 20 and 40°C (a) perpendicular and (b) parallel to the bedding planes of the test specimens. (After Schuyer and van Krevelen [45]; by permission of IPC Business Press Ltd.)

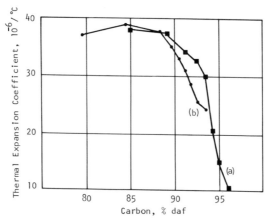

Fig. 4.4.3 Variation of the mean linear thermal expansion coefficient with rank. (a) Data of Schuyer and van Krevelen [45]; (b) data of Joy [46].

4.5 Elasticity, Hardness, and Strength

Very few coals are sufficiently cohesive and uniform to allow fashioning into the substantially homogeneous test pieces needed for conventional measurement of mechanical properties, and *routine* assessments (e.g., determination of strength or grindability) are therefore made by standardized simulation procedures. Nonempirical static and dynamic methods have so far only been successfully used for evaluating the elastic constants of coal. (These constants are identified in Table 4.5.1 where they are defined by the *a* terms of the general strain–energy function *W*; for the sake of simplicity, the definitions refer only to isotropic solids which can be completely described by two independent *a* terms.)

Young's moduli (*E*) from compression and bending of coal rods have been reported by Heywood [47], Inouye [48], and Morgans and Terry [49]; Schuyer *et al.* [50] have derived *E* and other elastic constants from sound velocity measurements via an expression of the form $v = (E/D)^{1/2}$, where *D* is the density of the test piece; and Bangham and Maggs [51] used a vapor sorption technique that permitted *E* to be calculated from the concurrent

Table 4.5.1

Definition of elastic constants[a]

(a) The strain–energy function, which generalizes Hooke's law, is usually written in the form

$$W = a_{11}x_x{}^2 + 2a_{12}x_xy_y + 2a_{13}x_xz_z + 2a_{14}x_xy_z + 2a_{15}x_xz_x + 2a_{16}x_xy_y$$
$$+ a_{22}y_y{}^2 + 2a_{23}y_yz_z + 2a_{24}y_yy_z + 2a_{25}y_yz_x + 2a_{26}y_yx_y$$
$$+ a_{33}z_z{}^2 + 2a_{34}z_zy_z + 2a_{35}z_zz_x + 2a_{36}z_zx_y$$
$$+ a_{44}y_z{}^2 + 2a_{45}y_zz_x + 2a_{46}y_zx_y$$
$$+ a_{55}z_x{}^2 + 2a_{56}z_xx_y$$
$$+ a_{66}x_y{}^2$$

The stress at any point in the system is then obtained by differentiating *W* with respect to the particular deformation, e.g., $X_x = -\partial W/\partial x_x$.

(b) Young's modulus $E = 2[(a_{11}{}^3 + 2a_{12}{}^2 - 3a_{11}a_{12}{}^2)/(a_{11}{}^2 - a_{12}{}^2)]$

(c) Compressibility $k = 3(1 - 2\mu)/E$

(d) Shear modulus $G = a_{11} - a_{12}$

(e) Poisson's ratio $\mu = a_{12}/(a_{11} + a_{12})$

---

[a] (b)–(e) refer to an *isotropic* solid.

linear swelling (x) of the sample. The relevant relationships for this latter method are

$$x = \lambda \cdot \Delta F, \qquad \Delta F = (RT/M\bar{S}) \cdot \int_0^p v \, d \log_e p, \qquad E = 100 D\bar{S}/\lambda$$

where $\bar{S}$ and D are the surface area and density of the test piece; v is the amount of vapor of molecular weight M sorbed by it at a pressure p; and ΔF is the lowering of the free surface energy of the solid that accompanies sorption. The proportionality constant λ can therefore be determined from the slope of a plot of x versus $\int v \, d \log_e p$.

Table 4.5.2 summarizes the results that these methods yielded for coals of different rank.

Since measurements of elastic constants are greatly affected by small cracks and other heterogeneities in test pieces, and the uniformity of coal specimens is often questionable, the "best" data are probably those reported by Schuyer *et al.* These were computed from experimental measurements of the required number of a terms in the general strain–energy function (see Table 4.5.1) and are independently corroborated by compressibility measurements. In contrast, the sorption swelling technique can only provide estimates of the order of magnitude of E. It is, however, noteworthy that all published data show the Young's modulus to be relatively insensitive to coal rank and specimen orientation up to 92–93%. Only among more or less fully developed anthracites does it rise rapidly and come to depend upon whether it is measured parallel or perpendicular to the bedding plane.

The same is broadly true of the shear modulus (G) and of the compressibility (k),[8] which were also calculated from sound velocity measurements [50, 52]. G is generally of the order of 0.3–$0.5E$, and k is almost constant at 2×10^{-11} cm^2/dyne up to 92–93% carbon and only thereafter falls quickly (to about half that value at 96%). The Poisson's ratio derived from these data is likewise roughly constant at ~ 0.35 up to 92–93% carbon and becomes increasingly anisotropic among anthracites (see Fig. 4.5.1).

Compressibilities reported by Kröger and Ruland [53] for three maceral suites stand in excellent agreement with the values reported by Schuyer *et al.* for "whole" coal, but indicate that the compressibility of "whole" coal can be significantly affected by petrographic composition. This influence diminishes with increasing rank (see Table 4.5.3) and disappears at 90% carbon as different macerals progressively lose their distinctive identities (see Section 2.1).

---

[8] The reciprocal of the compressibility, i.e., $1/k$, is sometimes referred to as the bulk modulus.

Table 4.5.2

Young's moduli (E) of coal

	$E\ (\times 10^{10}\ \mathrm{dyne/cm^2})$	
	(a)[a]	(b)[b]
Heywood [47]; coal compositions not specified		
Anthracite no. 1 (UK)	2.6	1.4
Anthracite no. 2 (UK)	2.75	1.45
Bituminous coal ("bright," UK)	2.75	2.15
Bituminous coal ("dull," UK)	3.4	3.5
Bituminous coal (Illinois)	1.4	2.05
Cannel coal (UK)	1.8	1.8
Morgan and Terry [49]		
Anthracite (average value)	4.3[c]	
Bituminous coal (average value)	3.4[c]	
Schuyer *et al.* [50]; carbon, % daf		
81.5	4.50[c]	
85.0	5.27[c]	
89.0	4.53[c]	
91.2	4.65[c]	
92.5	5.19[c]	
93.4	5.69	4.89
95.0	8.69	4.82
96.0	14.33	6.61
Bangham and Maggs [51]; carbon, % daf		
82.2	0.85	
82.8	1.3	0.59
85.3	0.53	
89.6	0.78	0.47
92.4	0.76	0.54
93.6	5.8	1.2

[a] Parallel to bedding plane.
[b] Perpendicular to bedding plane.
[c] Values perpendicular and parallel to bedding plane are identical.

In view of the fair agreement between Young's moduli from static and dynamic tests (see Table 4.5.2), and of reports that nonrecoverable strain observed in strain–time tests accounts for less than 1% of the total deformation [49, 54], criticisms of static measurements on the grounds that such measurements tend to be vitiated by viscoelastic flow of the coal under load can be discounted. However, extensive permanent deformation at ordinary temperatures *does* occur when small coal particles (in the order

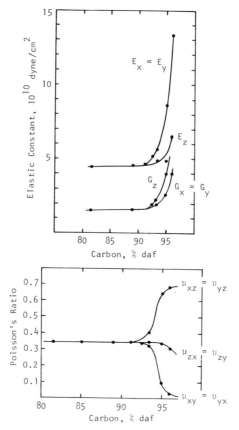

Fig. 4.5.1 Variations of Young's modulus (E), shear modulus (G) and Poisson's ratio with rank. (After Schuyer *et al.* [50, 52]; by permission of IPC Business Press Ltd.)

of 50 μm or less) are subjected to sufficiently high shear forces [55]. Such particles, when dry-mounted between a microscope slide and cover slip, will flow out into rounded, translucent (<0.1-μm thick) films when firm pressure is applied to the cover slip with a mounted needle or microscalpel. Coal rank only governs the color of the deformed films (which changes from yellow to reddish brown as carbon contents increase) and determines the maximum size of particles that can be deformed in this manner without first shattering into smaller fragments. In later work (Berkowitz, unpublished), it was found that this size ranged from approximately 50 μm among lignites to 10 μm among anthracites, and that it was smaller for durains than for vitrains of the same rank.

Measurements of the relative forces (F_c) under which this type of plastic

Table 4.5.3

Compressibilities of coal and coal components[a]

Carbon (% daf)	k (10^{-11} cm^2/dyne)			
	"Whole" coal	Vitrinite	Exinite	Micrinite
81.5	2.07			
83.5		2.62	4.58	1.71
85.0	1.75			
85.7		2.36	3.83	1.94
88.4		2.09	2.92	1.83
88.8		1.94	1.83	1.66
89.0	2.07			
91.2	1.96			
92.5	1.84			
95.0	1.27			
96.0	1.02			

[a] "Whole" coal data: Schuyer *et al.* [50]; maceral data: Kröger and Ruland [53].

or pseudoplastic flow occurs [56] indicate that the behavior of small coal particles resembles the behavior of thermoplastic resins [57]. F_c was found to be extremely sensitive to temperature and adsorbed vapors (including moisture), and with dry coal it increased with rank, rising especially rapidly among anthracites (see Fig. 4.5.2).

There is little doubt that this inherent "plasticity" of small particles,

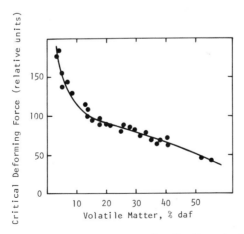

Fig. 4.5.2 Variation of the "critical deforming force" with rank. (After Bangham and Berkowitz [56].)

which contrasts sharply with the essentially elastic brittle behavior of coal lumps, directly affects the efficiency of coal-grinding operations and the significance of coal grindability test data (see below).

For the determination of coal *hardness*, variou̇̇ indentation methods, usually based on impressing a small cone or sphere into the polished surface of the test piece and measuring the size of the resultant permanent indentation, have been employed. Both Shore or Vickers scleroscopes can be used for this purpose [58–60]. In all cases, hardness was found to increase with coal rank toward a broad maximum at $\sim 80\%$ carbon, and then to fall to minimum values between ~ 88 and 90% before very steeply rising again among anthracites (see Fig. 4.5.3). However, since hardness is, by definition, a property of the coal substance, and hardness measurements are therefore only significant when made on uniform test pieces, data such as those shown in Fig. 4.5.3 possess limited value in practice, where size degradation depends primarily on the distribution of actual and incipient flaws in

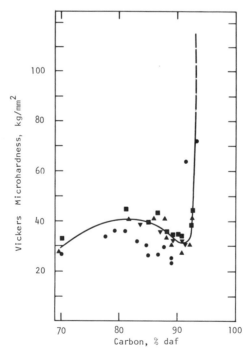

Fig. 4.5.3   Variation of the (Vickers) microhardness with rank. Data are taken from van Krevelen [58], Honda and Sanada [59], Alpern [60]. (By permission of Elsevier Publishing Company.)

the coal. More directly useful information is therefore generally sought from empirical strength tests (which measure a probability of breakage) and from standardized grindability tests (which assess how easily a coal can be pulverized).

Commonly used standard strength tests are exemplified by the ASTM tumbler test (ASTM D 3402-72) and the ASTM drop shatter test (ASTM D 3038-72), which are also used to measure the strength of metallurgical cokes (see Section 11.2). Both allow results to be expressed in terms of a nominal size stability (s) and friability (f), which are defined as $s = 100y/x$ and $f = 100 - s$. In these expressions, x and y are the average screen sizes of the coal before and after the test.

Alternatively, where the grindability is to be measured, recourse is most often made to the Hardgrove test (ASTM D 409-72), which bases itself on the proposition that work done in pulverizing a solid is proportional to the new surface produced (Rittinger's law). In this method, a representative 50-gm ($-16 \ +30$-mesh) sample is subjected to grinding by eight balls, each under a $64 \pm \frac{1}{2}$-lb load, during 60 revolutions in a specified (Hardgrove) mill, and the weight (w) of the -200-mesh fraction is determined. The grindability index (I) is defined as $I = 13 + 6.93w$, and Fig. 4.5.4 shows how this index varies with coal rank. It should, however, be noted that Hardgrove test data are markedly affected by the temperature and moisture content of the sample, and that they may also be influenced by plastic de-

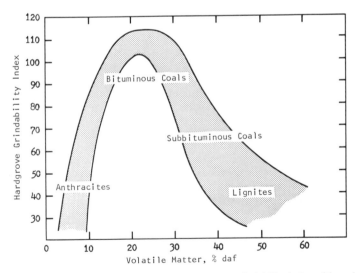

Fig. 4.5.4 Generalized variation of the Hardgrove grindability index with rank.

formation of small particles (see above). These matters—and particularly the fact that deformed small particles will strongly adhere to each other and form aggregates—almost certainly contribute to the progressive decrease of the Hardgrove index with decreasing carbon contents below $\sim 85\%$.

   To gauge the strength of undisturbed in-place coal, which has a direct bearing on mining and mine design,[9] numerous measurements have also been made of the compressive strength of coal cubes. Because of the random distribution of internal flaws, which eventually cause failure, data for any single coal can vary by as much as 30% [61]; but in all cases, the probability of survival of a cube with side x seems to vary inversely with x rather than inversely with the cubic area or volume [62]. Berenbaum [63] has therefore proposed a generalized expression of the form

$$C(x) = e^{x \ln C}$$

where C and $C(x)$ are the probabilities of survival of cubes with unit side and side x, respectively. Theoretical statistical treatments of coal breakage (reviewed by Brown, [64]) have shown that this expression is consistent with the well-known Rosin–Rammler size distribution [65]

$$100 - U(i) = 100 \exp(-bi^n)$$

in which $U(i)$ is the percentage of undersize material, and i, b, and n are constants.

Notes

   A. The approximate equivalence of 1 cal (4.187 J) $\simeq 10$ m^2 was established by measuring the heats of wetting of carbon blacks whose surface areas had been determined by electron microscopy and confirmed by comparing heats of wetting with surface areas measured by sorption techniques. Other measurements indicate that 1 cal $\simeq 10$ m^2 applies to all carbonaceous solids and that it is substantially independent of the nature of the wetting liquid: Bond and Spencer [7], who measured the heats of wetting of a carbon black (with $\bar{S} = 230$ m^2/gm) in 16 different liquids, thus found an average value of 1 cal $\simeq 9$ m^2, with maximum deviations within $\pm 25\%$ and most values within $\pm 10\%$ of this average. It must, however, also be noted that more recent (unpublished) measurements on a Bruceton, Pennsylvania, coal with some 30 liquids indicated much greater variations (J. W. Larsen, private communication).
   B. Imbibition would tend to create new surfaces (through associated swelling of the coal), and interactions of methanol with functional groups are *exothermic* processes. In either event, fictitiously high values of $\bar{S}$ would therefore be recorded.
   C. Other liquids that do not chemically interact with the coal or cause a change in its physical structure (by, e.g., swelling) are, of course, equally suitable, but the results will depend on their respective molar volumes. The liquids here named are those most often used.
   D. This was demonstrated [18] by measuring the absolute densities of a vitrain (with 89.5% carbon) at temperatures between -196 and 100°C and noting that the specific volume, i.e.,

[9] Among other matters, such data allow an estimate of the bearing strengths of coal pillars in mines.

$1/D$, varied linearly with the temperature of measurement. It could therefore be concluded that all pores remained accessible even at the lowest temperatures, where thermal contraction might have been expected to close some. But this view has more recently been challenged on the basis of x-ray measurements on anthracites [21], which indicate that some pores are closed to helium.

E. In some cases, high clay contents can, however, interfere with determination of bank densities by causing abnormally high capacity moisture contents. In such instances, a procedure developed by the U. S. Bureau of Mines and employing dry coal may be helpful.

F. This problem was recognized by Porter and Taylor [36] who attempted to overcome it by using toluene instead of water. But that is, of course, no solution.

References

1. B. A. Onusaitis, *Izv. Akad. Nauk USSR* **5–6,** 17 (1942).
2. D. H. Bangham, *Ann. Rep. Chem. Soc. (London)* **4C,** 29 (1943).
3. N. Berkowitz, *Fuel* **29,** 138 (1950); N. Berkowitz and H. G. Schein, *ibid.* **31,** 19 (1952).
4. J. G. King and E. T. Wilkins, *Proc. Conf. Ultrafine Struct. Coals Cokes, BCURA, London* p. 46 (1944).
5. M. Griffith and W. Hirst, *Proc. Conf. Ultrafine Struct. Coals Cokes, BCURA, London* p. 80 (1944); R. L. Bond and F. A. P. Maggs, *Fuel* **28,** 172 (1949); D. H. Bangham, R. E. Franklin, W. Hirst, and F. A. P. Maggs, *ibid.* **28,** 231 (1949).
6. F. A. P. Maggs, *Proc. Conf. Ultrafine Struct. Coals Cokes, BCURA, London* p. 95 (1944); P. Zwietering and D. W. van Krevelen, *Fuel* **33,** 331 (1954); R. L. Bond, *Nature (London)* **178,** 104 (1956); *Brennst. Chem.* **37,** 233 (1956); R. B. Anderson, W. K. Hall, J. A. Lecky, and K. C. Stein, *J. Phys. Chem.* **60,** 1548 (1956).
7. R. L. Bond and D. H. T. Spencer, *Soc. Chem. Ind. Proc. Conf. Ind. Carbon Graphite, London* p. 231 (1957).
8. S. J. Gregg, *Proc. Conf. Ultrafine Struct. Coals Cokes, BCURA, London* p. 110 (1944).
9. M. Griffith and W. Hirst, *Proc. Conf. Ultrafine Struct. Coals Cokes, BCURA, London,* p. 80 (1944).
10. R. L. Bond and D. H. T. Spencer, *Soc. Chem. Ind. Conf. Ind. Carbon Graphite, London* p. 231 (1958).
11. H. Gan, S. P. Nandi, and P. L. Walker, Jr., *Fuel* **51,** 272 (1972).
12. P. Fugassi, R. Hudson, and G. Ostapchenko, *Fuel* **37,** 25 (1958); L. Robert and S. Pregermain, *ibid.* **42,** 389 (1963).
13. L. Robert and H. Brusset, *Fuel* **44,** 309 (1965); H. Marsh, *ibid.* **44,** 253 (1965).
14. H. L. Ritter and L. C. Drake, *Ind. Eng. Chem. Anal. Ed.* **17,** 782 (1945).
15. P. Zwietering and D. W. van Krevelen, *Fuel* **33,** 331 (1954).
16. H. Gan, S. P. Nandi, and P. L. Walker, Jr., *Fuel* **51,** 272 (1972).
17. R. E. Franklin, *Fuel* **27,** 46 (1948).
18. R. E. Franklin, *Trans. Faraday Soc.* **45,** 274 (1949).
19. P. L. Agrawal, *Proc. Symp. Nature Coal* p. 121. Central Fuel Research Institute, India, 1960.
20. J. D. Danforth and T. De Vries, *J. Am. Chem. Soc.* **61,** 873 (1939).
21. W. V. Kotlensky and P. L. Walker, Jr., *Proc. Carbon Conf., 4th* p. 423. Pergamon, Oxford, 1960.
22. C. Kröger and J. Badenecker, *Brennst. Chem.* **38,** 82 (1957).
23. A. M. Wandless and J. C. Macrae, *Fuel* **13,** 4 (1934).
24. J. W. Whittaker, *Fuel* **29,** 33 (1950).
25. P. G. Sevenster, *J. S. Afr. Chem. Inst.* **7,** 41 (1954).
26. J. A. Dulhunty and R. E. Penrose, *Fuel* **30,** 109 (1951).

27. D. W. van Krevelen, "Coal," p. 314. Elsevier, Amsterdam, 1961.
28. T. Aono and G. Yamauchi, *J. Electrochem. Soc. Jpn.* **19,** 221 (1950). B. Mukherjee, *J. Sci. Res. (India)* **13B,** 53 (1954).
29. H. Akamatu and H. Inokuchi, *J. Chem. Soc. Jpn. Pure Chem. Sec.* **70,** 185 (1949). S. T. Bondarenko and A. I. Yakovleva, *Izv. Akad. Nauk SSSR Otd. Tekh. Nauk* **2,** 132 (1957).
30. A. A. Agroskin, *Khim. Tekhnol. Topl. Masel* **4,** 7 (1959).
31. J. Schuyer and D. W. van Krevelen, *Fuel* **34,** 213 (1955).
32. M. P. Groenewege, J. Schuyer, and D. W. van Krevelen, *Fuel* **34,** 339 (1955).
33. H. Landolt and R. Börnstein, "Physikalisch-Chemische Tabellen," 2nd ed., p. 129. Springer Verlag, Berlin and New York, 1894.
34. L. C. McCabe and C. C. Boley, *in* "Chemistry of Coal Utilization" (H. H. Lowry, ed.), Vol. 1. Wiley, New York, 1945.
35. J. D. Clendenin, K. M. Barclay, J. H. Donald, D. W. Gillmore, and C. C. Wright, Pennsylvania State College, Min. Ind. Exp. Station Tech. Paper 160 (1949).
36. H. C. Porter and G. B. Taylor, *J. Ind. Eng. Chem.* **5,** 289 (1913).
37. E. Terres and A. Schaller, *Gas Wasserfach* **65,** 761, 780, 800, 818, 832 (1922).
38. G. Coles, *J. Soc. Chem. Ind.* **42,** 435T (1923).
39. D. W. van Krevelen, "Coal," p. 419. Elsevier, Amsterdam, 1961.
40. W. Fritz and H. Moser, *Feuerungstechnik* **28,** 97 (1940).
41. F. S. Sinnatt and H. McPherson, *Fuel* **3,** 12 (1924).
42. W. Fritz and H. Diemke, *Chem. Abstr.* **35,** 4939 (1941).
43. T. E. W. Schumann and V. Voss, *Fuel* **13,** 249 (1934).
44. D. H. Bangham and R. E. Franklin, *Trans. Faraday Soc.* **42B,** 289 (1946).
45. J. Schuyer and D. W. van Krevelen, *Fuel* **34,** 345 (1955).
46. A. S. Joy, Annual Rep., Fuel Research Stn., Greenwich, Britain (1955).
47. H. Heywood, *Proc. Conf. Ultrafine Struct. Coals Cokes* p. 172. BCURA, London, 1944.
48. K. Inouye, *J. Coll. Sci.* **6,** 190 (1951); *Bull. Chem. Soc. Jpn.* **26,** 200, 359, 458 (1953).
49. W. T. A. Morgans and N. B. Terry, *Fuel* **37,** 201 (1958).
50. J. Schuyer, H. Dikstra, and D. W. van Krevelen, *Fuel* **33,** 409 (1954).
51. D. H. Bangham and F. A. P. Maggs, *Proc. Conf. Ultrafine Struct. Coals Cokes* p. 118. BCURA, London, 1944.
52. D. W. van Krevelen, A. G. Chermin, and J. Schuyer, *Fuel* **36,** 483 (1959).
53. C. Kröger and H. Ruland, *Brennst. Chem.* **39,** 1 (1958).
54. C. D. Pomeroy, *Nature (London)* **178,** 279 (1956).
55. R. G. H. B. Boddy, *Nature (London)* **151,** 54 (1943); *Proc. Conf. Ultrafine Struct. Coals Cokes* p. 336. BCURA, London, 1944.
56. D. H. Bangham and N. Berkowitz, *Bituminous Coal Res.* **2,** 139 (1945); *Research (London)* **1,** 86 (1947).
57. G. Tammann, "Der Glasszustand," p. 56. Leipzig, 1933.
58. D. W. van Krevelen, *Brennst. Chem.* **34,** 167 (1953).
59. H. Honda and Y. Sanada, *Fuel* **35,** 451 (1956).
60. B. Alpern, *C. R. Acad. Sci. Paris* **242,** 653 (1956).
61. M. M. Protodyakanov, *Izv. Akad. Nauk USSR* No. 2, 283 (1953).
62. I. Evans and C. D. Pomeroy, *in* "Mechanical Properties of Brittle Materials" (W. H. Walton, ed.). Butterworths, London, 1958.
63. R. Berenbaum, *J. Inst. Fuel* **34,** 367 (1961); **35,** 346 (1962).
64. R. L. Brown and F. J. Hiorns, *in* "Chemistry of Coal Utilization," Suppl. Vol., p. 135. Wiley, New York, 1963.
65. P. Rosin, E. Rammler, and K. Sperling, *J. Inst. Fuel* **7,** 29 (1933); E. Rammler, *Braunkohle* **58,** 52 (1951); *Freiberg, Forschungsh.* **A50,** 5 (1956).

CHAPTER 5

CHEMICAL PROPERTIES

Since coal is either used as a heating and steam-raising fuel (see Chapter 10), carbonized to form cokes, chars, and related by-products (see Chapter 11) or converted into "synthetic" gaseous and liquid hydrocarbons (see Chapters 12–14), its most important chemical properties, for practical purposes, are those that determine its response to oxidation and reduction, its behavior during pyrolysis, and its solubility in organic solvents.[1] Detailed studies of these reactions have led to progressive refinement and widening of the scope of coal technology and have also provided important information about the "molecular" structure of coal (see Section 5.7).

5.1 Oxidation with Air or Oxygen

All coals other than anthracites and all coal components other than inertinites (see Table 2.1.2) are so sensitive to oxidation that even minute

---

[1] Pyrolysis and the action of solvents on coal are discussed separately in Chapters 6 and 7.

additions of oxygen, too small to be detected by conventional "wet" methods, will alter their properties. In extreme cases, exposure of freshly mined coal to air at ambient temperatures for as little as a few days will not only cause marked deterioration of caking propensities (see Section 6.3), but also adversely affect its solubility, tar yield, heating value, and other properties.

Although details of the chemical changes responsible for such deterioration are, for the most part, still uncertain, there is agreement that they are ultimately connected with progressive oxidative destruction of non-aromatic configurations in the coal "molecule" (see Section 5.7).

The initial stages of oxidation are characterized by chemisorption of oxygen at readily accessible (aromatic and nonaromatic) surface sites and by formation of acidic functional groups, in particular, —COOH, =CO, and phenolic —OH. If moisture is present or is generated from chemically combined hydrogen in the coal [1], some chemisorbed oxygen will also form peroxide or hydroperoxide complexes, which are capable of oxidizing ferrous thiocyanate [1], titanous chloride [2], and hydroquinone [3]. (These reactions can be used to estimate the concentrations of peroxides in the coal.) However, given enough time, oxidation will gradually reach deeper and begin to degrade the coal substance itself, first converting it into alkali-soluble but otherwise still coal-like solids known as *humic acids* and then breaking those substances down into progressively smaller molecular species.

After causing a slight transient weight gain (due to incorporation of additional oxygen *without* concurrent loss of volatile oxidation products), these processes gradually lower the carbon and hydrogen contents of the coal (toward limiting values of 60 to 65% carbon, 2 to 3% hydrogen[2]) through abstraction of CO, CO_2, and H_2O, and thereafter steadily convert the residual solid into water-soluble benzenoid acids and carbon oxides without further change in its elemental composition. Unless prematurely interrupted, this will continue until all available coal material has been effectively "consumed."

Different ambient conditions leave this course of events qualitatively unchanged. Higher temperatures and/or oxygen partial pressures will, as a rule, merely accelerate the process. But due to thermal instability of some of the primary oxidation products at $> 70°C$, there are important quantitative differences between oxidation at ordinary and elevated temperatures; and these make it helpful to differentiate between three oxidation ranges.

Below $\sim 70°C$, which has been found to be a distinctive transition temperature [1], rates of oxidation are generally low enough to be independent

---

[2] The actual "limiting" compositions tend to vary slightly with coal rank and also depend on the temperature of oxidation. The lower values seem to be associated with high-temperature oxidation (see Figs. 5.1.1–5.1.4).

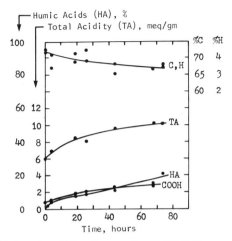

Fig. 5.1.1   Air-oxidation of a subbituminous coal at 100°C, 10% oxygen. (After Jensen *et al.* [5].)

of coal rank; oxidation does not proceed much beyond forming acid functional groups and peroxides; and for all practical purposes the reaction ends when the hydrogen content of the coal has fallen to ~3% [4]. Further oxidation proceeds so slowly that significant concentrations of humic acids develop only over *very* long periods of time.

In the second range, ~70–150°C, this pattern is, for the most part, only modified by thermal instability of peroxides and by the fact that oxidation

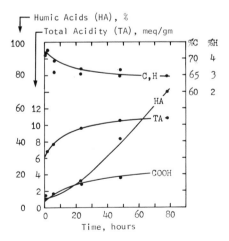

Fig. 5.1.2   Air-oxidation of a subbituminous coal at 225°C, 10% oxygen. (After Jensen *et al.* [5].)

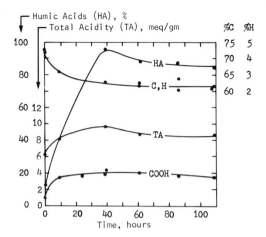

Fig. 5.1.3   Air-oxidation of a subbituminous coal at 275°C, 21% oxygen. (After Jensen *et al.* [5].)

rates are now sufficiently high to be controlled by oxygen transport to the inner coal surfaces (see Section 4.1). Overall oxidation rates become, therefore, progressively more dependent on coal porosity and tend to fall with increasing rank or particle size. Peroxides form only transiently or not at all. And because of the greater instability of functional groups other than peroxides, there is also a reversal of the $CO:CO_2$ ratio in the off-gas, with CO_2 now more abundant than CO.

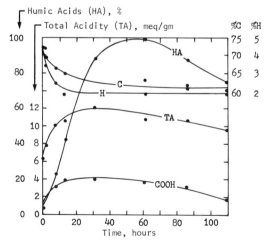

Fig. 5.1.4   Air-oxidation of a subbituminous coal at 300°C, 15% oxygen. (After Jensen *et al.* [5].)

Rapid generation of humic acids occurs only in the third range, i.e., at temperatures above $\sim 150°C$, and rapid (secondary) destruction of these acids is in most cases not observed below $\sim 250°C$. But both reaction rates depend on oxygen partial pressures as well as on the temperature [5]; and degradation of humic acids at $> 250°C$ is also accompanied by some loss of acid functional groups. Figures 5.1.1–5.1.4 illustrate this. Qualitatively, the overall oxidation process at temperatures above $\sim 250°C$ can therefore be schematically represented as

$$
\text{coal} \left\{
\begin{array}{ll}
\longrightarrow \text{CO, CO}_2\text{, H}_2\text{O} & (1) \\[4pt]
\text{humic acids} \left\{
\begin{array}{ll}
\longrightarrow \text{(I)}_a \longrightarrow \text{(I)}_b \ldots \longrightarrow \text{CO, CO}_2\text{, H}_2\text{O} & (2) \\[4pt]
\longrightarrow \text{``regenerated'' coal + CO, CO}_2\text{, H}_2\text{O} & (3)
\end{array}
\right.
\end{array}
\right.
$$

In this scheme, reaction (1) represents the decomposition of acid functional groups on alkali-*in*soluble coal material; (2) illustrates sequential "secondary" degradation of humic acids via various intermediates (I)_a, (I)_b, etc.; and (3) represents thermal decarboxylation and dehydroxylation of humic acids with consequent loss of alkali solubility and reversion to coal-like substances.

Humic acids can be readily extracted from oxidized coal with aqueous alkali and, when "regenerated" by acidification of extract solutions, form brown flocculates that dry to brittle, lustrous black solids (C $\sim$ 55–60%, H $\sim$ 2–3%). The molecular structure of humic acids and their precise relation to coal and alkali-soluble matter in soils (Note A) are still uncertain. However, from their infrared and x-ray diffraction spectra, which closely resemble those of coal, it has been inferred that they are not far removed from their parent molecules in coal; and the wide range of molecular weights reported for them (600–10,000) suggests that they form fairly directly by random oxidative cleavage of the coal "molecule." The observation [7] that humic acids usually still contain some hydroaromatic configurations through which further oxygen attack could occur is consistent with this view.

How such cleavage may be achieved is suggested by the distinctive sequence in which different types of acid functional groups are developed in humic acids. Detailed chemical and spectroscopic studies by Moschopedis *et al.* [8, 9] tend to confirm Tronov's view [10] that oxidation begins with formation of phenols and then proceeds via quinones and acid anhydrides to carboxylic acids; and peripheral molecular changes associated with conversion of coal into humic acids have therefore been postulated [5] as following a scheme such as that shown in Fig. 5.1.5. This sequence resembles the gas-phase oxidation of naphthalene to phthalic and maleic anhydrides

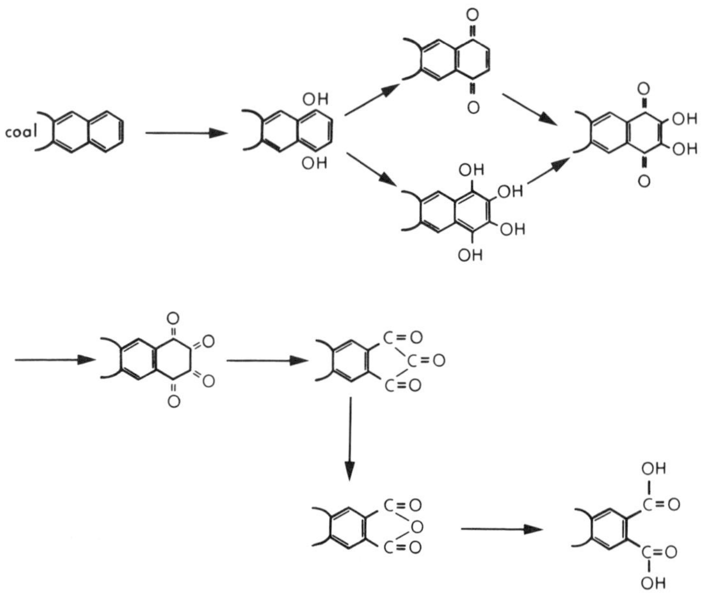

Fig. 5.1.5 Postulated reaction path for conversion of coal into humic acids. Concurrent oxidation reactions are believed to disrupt certain nonaromatic bridge structures, e.g., single —CH$_2$—. (After Jensen *et al.* [5].)

and implies that air-oxidation of coal can degrade aromatic as well as nonaromatic structures.

The structural changes caused by air-oxidation offer several means for determining whether or to what extent a particular coal has been oxidized.

Very slight oxidation, which is sometimes more a matter of scientific than practical interest (Note B), can only be detected by recording the infrared spectrum of the coal and manifests itself in absorption at 3300 cm^{-1} (—OH stretching) and at 1700 cm^{-1} (=CO stretching).[3] The procedure demands considerable care and is critically dependent on thorough drying of the sample (since residual moisture would contribute to the —OH absorption); and when examining low-rank coals, it is normally also necessary to determine the spectrum of the corresponding fresh coal for comparison. However, subject to these precautions, the spectral data are very reliable indicators; in favorable cases, it is even possible to estimate the degree of oxidation from the intensities of the 3300- and 1700-cm^{-1} absorption bands.

More substantial oxidation can be detected and quantitatively estimated

---

[3] Except in lignites, in which =CO configurations are significant components of the *fresh* coal, absorption at 1700 cm^{-1} usually appears as a relatively weak shoulder on the diagnostic 1600-cm^{-1} band. For more detailed references to infrared spectra of coal, see Section 5.7.

by a variety of analytical methods, which include measurement of the
volumes of CO and CO_2 that the coal releases when it is pyrolyzed.

The changes in elemental composition that accompany progressive air
oxidation of coal [11, 12] are illustrated in Fig. 5.1.6, where typical oxidation
tracks have been superimposed on Seyler's coal band (see Sections 2.2 and
3.3). The most sensitive parameter is the hydrogen content; and where
the composition of the fresh coal is known, the amount of oxygen permanently or transiently introduced into the coal during oxidation can be
calculated from the loss of hydrogen ($-\Delta H$) by assuming that abstraction
of one H atom corresponds to addition of one O atom. (This assumption
is based on the fact that "real" oxidation tracks of coal run very nearly
parallel to theoretical hydrocarbon oxidation tracks calculated by suppos-
ing that $+O = -H$.)

Where only the carbon and hydrogen contents of the *oxidized* coal sample
are known, retracing an oxidation track across the coal band also allows
the limits of the probable composition of the corresponding fresh coal to

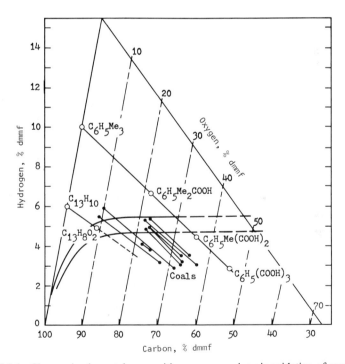

Fig. 5.1.6 Changes in elemental composition accompanying air-oxidation of coal. Empty
circles illustrate analogous composition shifts associated with oxidation of pure hydrocarbons.
(After Francis [4].)

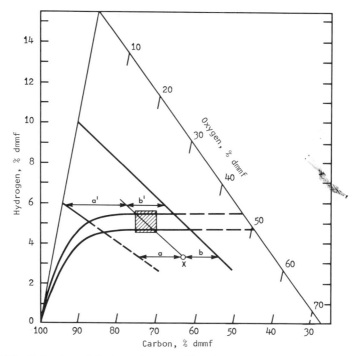

Fig. 5.1.7 Estimation of fresh coal compositions from compositions of "weathered" (oxidized) samples. If x denotes the (known) elemental composition of the oxidized sample, draw a line diagonally upward, such that $a:b = a':b'$. The (shaded) area defined by the intersection of this line with the coal band then indicates the most probable range of compositions of the corresponding fresh coal. The oxidation tracks needed as reference for drawing $a:b = a':b'$ can be taken from Fig. 5.1.6 or computed from (theoretically) oxidizing a pure hydrocarbon, with addition of 1 oxygen atom involving loss of 1 hydrogen atom.

be estimated. This procedure is outlined in Fig. 5.1.7 and is especially valuable where a preliminary exploration program can only recover naturally oxidized ("weathered") outcrop samples (Note C).

Information about the *disposition* of oxygen in fresh and oxidized coal can be obtained by measuring the concentrations of the different types of oxygen-bearing functional groups in the coal. Appropriate methods have been reviewed by Ihnatowicz [13], Blom [14], Kröger *et al.* [15–17], and Ignasiak *et al.* [18].

Concentrations of carboxyl (—COOH) groups are most conveniently estimated by treating the coal with an excess of aqueous calcium acetate and titrating the resultant free acetic acid that forms from

$$2R \cdot COOH + (CH_3 \cdot COO)_2 \cdot Ca \longrightarrow (R \cdot COO)_2 \cdot Ca + 2CH_3 \cdot COOH$$

but this method requires that phenolic —OH, if present or suspected, is first blocked by methylation (see below). Good results have also been obtained by thermal decarboxylation, e.g., by boiling the coal in quinoline with a suitable catalyst (such as copper sulfate or copper carbonate) and gravimetrically determining the carbon dioxide thereby released from the sample.

Total hydroxyl, i.e., phenolic + alcoholic —OH, can be accurately measured by *acetylation*, e.g., by treating a suspension of the coal in pyridine with acetic anhydride at 100°C for 5–6 hr and hydrolyzing the acetylated coal with aqueous barium hydroxide. —OH is then obtained by acidifying the hydrolysate with phosphoric acid, heating the mixture, and determining the resultant acetic acid that distills over.

Phenolic —OH itself has been measured by a *methylation* method that involves esterifying the coal with diazomethane in diethyl ether, saponifying the reaction product (in order to eliminate —COOH), and then estimating the concentration of methoxyl (—OCH_3) by a modified Zeisl procedure [14, 19]. —OH in alcohol configurations can be obtained by subtracting the concentration of phenolic —OH from the total —OH concentration.

Carbonyl groups are more difficult to determine quantitatively because their reactivities vary widely and depend on the particular forms in which =CO exists. Good estimates of total $O_{=CO}$ can, however, be obtained by an oxime formation method that has been described by Blom [14, 19] or by reacting the coal with phenyl hydrazine hydrochloride [13]; and *reducible* quinone groups can be determined by reduction with titanous chloride [14]. Oxygen in *nonreducible* carbonyl structures is usually equated with the difference between total $O_{=CO}$ and oxygen in reducible quinones, but can also be directly estimated by reacting the coal with dinitrophenyl hydrazine [17].

Oxygen not encompassed by these functional group analyses (or contained in peroxides), is termed *unreactive* and generally regarded as an integral component of fresh coal,[4] in which it is considered to exist in various ether configurations, and/or as a heteroelement in ring structures.

Figure 5.1.8 shows how the concentrations of oxygen-bearing functional groups vary with coal rank.

For quality control in industrial operations, there are also several methods by which the overall extent of oxidation of a coal can be estimated. According to Egorova [20] and Khrisanfova [21], the "permanganate number" as defined by Heathcoat [22] can be used for this purpose. (This is the number

---

[4] Depending on its rank, fresh coal does, of course, also contain varying amounts of reactive oxygen in functional groups. Further reference to this is made in Section 5.7.

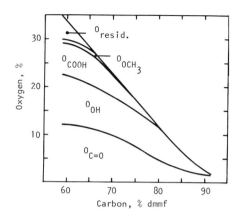

Fig. 5.1.8 Distribution of oxygen in vitrinites of different rank. (After Blom [14].)

of milliliters of aqueous 1 N KMnO$_4$ reduced by 0.5 gm of (dry, ash-free) pyridine-extracted coal in 1 hr at 100°C.) Radspinner and Howard [23] found that a nearly constant fraction of oxygen introduced into coal during oxidation reacts like carboxyl oxygen and can be measured as an incremental volume of CO + CO$_2$ by heating the coal in vacuo at 350°C. And the British National Coal Board has developed an instrument which utilizes the fact that the electric charge acquired by a metal plate when coal particles impinge on it increases with the degree of oxidation of the particles [24].

5.2 Oxidation with Liquid Oxidants

Oxidation of coal with liquid oxidants has been mainly used to investigate coal structure (see Section 5.7), but has also been extensively studied as a relatively simple method for converting coal into potentially useful chemicals. Particular attention has been given to the composition and yields of the mixed aromatic coal acids obtained by such treatment.

Oxidants were commonly nitric or performic acids, alkaline potassium permanganate, dichromates, perchlorates, hypohalites, or aqueous caustic soda—all occasionally used in conjunction with oxygen or hydrogen peroxide; and since the 1920s, a vast technical literature on liquid-phase oxidation of coal suspensions has thus come into being.

Although the diversity of the conditions under which the oxidations were effected makes it difficult to compare the results of different studies (and many reaction products have not yet been identified), it appears that liquid oxidants other than NaOCl (see Section 5.7) attack coal as indiscriminately as oxygen and merely break it down much faster. All first transform the

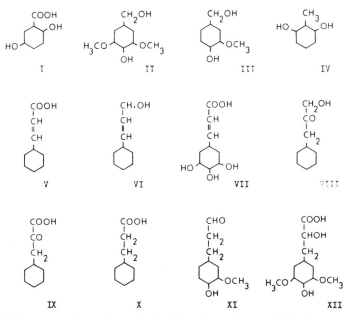

Fig. 5.2.1 Some products of perchloric acid oxidation of coal. I: 2,5-dihydroxy benzoic acid. II: Syringic alcohol. III: Vanillic alcohol. IV: 2-methyl resorcinol. V: Cinnamic acid. VI: Cinnamic alcohol. VII: 3,4,5-trihydroxy cinnamic acid. VIII: Phenyl 3-pyruvic alcohol. IX: Phenyl 3-pyruvic acid. X: Phenyl 3-propionic acid. XI: 3-methoxy,4-hydroxy phenyl 3-propionaldehyde. XII: 3,5-dimethoxy,4-hydroxy phenyl 3-glyceric acid. (After Raj [26].)

coal into alkali-soluble humic acids and then degrade these via partially water-soluble "subhumic acids" to CO, CO_2, and a mixture of acidic low-molecular-weight aromatic and aliphatic compounds. The *aromatic* fraction, which usually accounts for some 32–35% of the organic carbon of the coal, is mostly made up of simple benzenoids (e.g., benzene carboxylic acids) and heterocyclic compounds in which a ring-carbon atom is replaced by N or O. Except in one study,[5] no polycondensed structures (with two or more fused rings) have so far been found; but Raj [26] has shown that the mixed aromatic acids obtained by oxidizing coal with perchloric acid at 50°C contain a broad spectrum of phenols and variously substituted acid phenyl derivatives, many of which carry a 3-carbon side chain (see Fig. 5.2.1; Note D). The *aliphatic* fraction typically represents 10–15% of the

---

[5] By oxidizing a high-rank bituminous coal (C = 90.4%) with oxygen and caustic soda at 270°C under 5–20 MPa pressure, Montgomery *et al.* [25] obtained almost equal proportions of benzene and naphthalene derivatives, as well as small amounts of phenanthrene-based compounds. However, the manner in which the product mixture was separated and subsequently decarboxylated makes it possible that polycondensed compounds formed by ring closure *after* oxidation.

Table 5.2.1

Distribution of carbon among oxidation products (as percentage of carbon in parent coal); Pittsburgh seam[a]

	Oxidation conditions[b]			
	(a)	(b)	(c)	(d)
Carbonate carbon	45.0	50.0 ⎫	65.0	63.0
Oxalate carbon	15.0	13.0 ⎭		
Aromatic carbon	31.0	36.0	35.0	37.0

[a] After Chakrabartty and Berkowitz [30].

[b] (a) Alkaline $KMnO_4$, various conditions; (b) aqueous NaOH + oxygen, 100–250°C, 0.7–2.5 MPa; (c) aqueous $Na_2Cr_2O_7$, autoclave, 250°C; (d) aqueous NaOCl, open flask, 60–65°C.

organic carbon of the coal and has been reported to contain formic, acetic, butyric, oxalic, maleic, succinic, and 1-tartaric acids [25, 27–29].

As is the case when coal is air-oxidized, the severity of the reaction conditions only determines the reaction rates and leaves the course of the oxidation, as well as the general distribution of carbon in the final product mixture, substantially unaffected (see Table 5.2.1). But as illustrated in Table 5.2.2, there is some evidence that the yield of aromatic acids tends to increase with coal rank (Note E).

Although the mixed aromatic acids from coal oxidation have so far attracted little commercial interest, humic acids have been manufactured in small commercial operations by carefully controlled nitric acid oxidation for use as organic nitrogenous fertilizers, soil amendment compounds, and drilling mud additives.

Table 5.2.2

Distribution of carbon among products of oxidation with alkaline $KMnO_4$[a]

% of source carbon appearing as	Lignites	Bituminous coals	Anthracites
Carbon dioxide	45–57	36–42	43
Aliphatic acids	11–31	15–20	9
Aromatic acids	22–34	39–46	48

[a] After Bone *et al.* [31].

5.3 Oxidative Ammoniation ("Ammoxidation")

A special case of oxidation is the interaction of coal with ammonia and oxygen at temperatures above 200°C to form black, infusible, alkali-insoluble solids that may contain as much as 20% chemically combined nitrogen. This reaction, which has been known since the late 1920s [32], has received renewed attention in the 1960s as a possible route to coal-based nitrogenous plant nutrients [33, 34].

Bench-scale studies of oxidative ammoniation [35] show that the reaction proceeds particularly well at 300–350°C and with input of a 2:3 ammonia–oxygen mixture. Under these conditions, fluidized (-48-mesh) subbituminous coal takes up $\sim 10\%$ N within 2 hr and ~ 14–16% within 5 hr. Depending on the actual temperature and size consist of the coal, between 30 and 50% of the input ammonia is incorporated into the coal (as N), and the remainder appears as elemental nitrogen, water, and hydrogen cyanide in the off-gas. Carbon losses to CO and CO_2 vary linearly with oxygen consumption and after 5 hr at 300°C amount to ~ 10–12%.

Unlike the coal from which it forms, the oxidatively ammoniated material yields no tar or condensible oils on pyrolysis and is remarkably unreactive [36]. It cannot, for example, be acetylated, extensively methylated, or reduced, and it also tends to resist further oxidation. While coal, when contacted by hydrogen peroxide or air at 200°C for 72 hr will generally yield

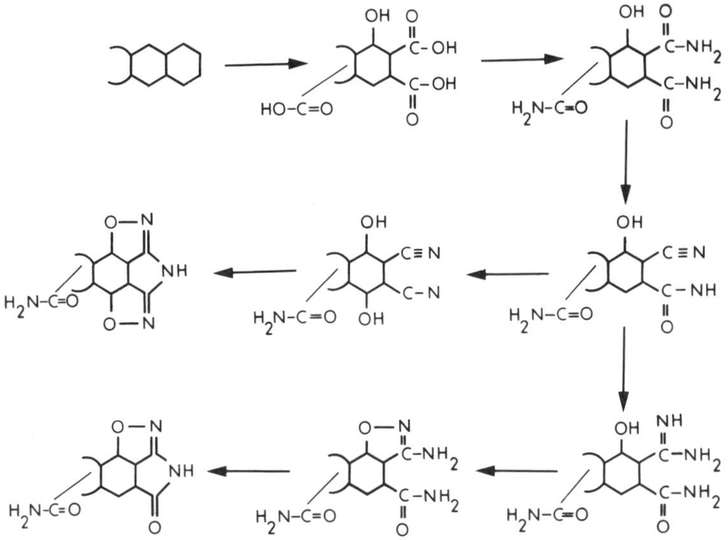

Fig. 5.3.1 Postulated path of the coal–ammonia–oxygen reaction. (After Chakrabartty and Berkowitz [36].)

over 90% of alkali-soluble humic acids, the oxidatively ammoniated product forms only ~20% when so treated. It will, however, lose up to 80% of its nitrogen when hydrolyzed under severe conditions (HCl/270°C under ~24 MPa pressure for 16–18 hr), and it has therefore been hypothesized [36] that nitrogen in oxidatively ammoniated coal exists mainly in stabilized amidine or isoindole configurations. A possible reaction path to such configurations is shown in Fig. 5.3.1.

5.4 Halogenation

The ease with which coal can be halogenated was first demonstrated by Bevan and Cross [37], who found that cannel coal (see Section 3.1) became completely alkali-soluble when shaken with an aqueous chlorine solution, and that bituminous coal could be similarly "solubilized" by treating it with dilute HCl and potassium chlorate.

Subsequent investigations have confirmed these observations and shown that chlorination can be effected by exposing pulverized coal to chlorine [38, 39], passing chlorine gas through a suspension of coal in water or carbon tetrachloride [40], or reacting coal with chlorine dioxide in a pressure vessel [41]. The amounts of chlorine thus introduced into coal vary generally from ~40% in the case of lignites to ~25% for high-rank bituminous coals, and tend to increase with reaction temperatures. Approximately half of the chlorine taken up by the coal can be removed by hydrolysis, e.g., by refluxing the chlorinated material with aqueous NaOH [39] or by extracting it with concentrated ammonia or aqueous sodium carbonate [41], and almost all is released when the solid is heated to 550°C [39].

Halogenation of dry coal with gaseous chlorine is strongly exothermic and accompanied by copious evolution of gaseous hydrogen chloride, and the chlorinated product yields no tar when pyrolyzed [38]. It is therefore thought that chlorine is introduced by concurrent addition and substitution reactions, with the latter abstracting hydrogen from aromatic and hydro-aromatic sites via

There are, however, indications that the properties of the chlorinated products depend on coal type as well as on the extent of chlorination. From bituminous coals, Heathcoat and Wheeler [41] have obtained alcohol-soluble, amorphous yellow solids with 25–31.5% Cl; and from lignites, Russian workers [40] have prepared resinous substances that contained up to 45% Cl and were readily soluble in most organic solvents.

Halogenation of coal with bromine, iodine, or fluorine has been less closely studied than chlorination, but has broadly similar effects. Bromination, which can be accomplished by shaking pulverized coal with bromine water or by passing bromine into a suspension of coal in carbon tetrachloride [42] or chloroform [43], proceeds, like chlorination, by addition and substitution[6] and yields variously soluble solids with up to $\sim 20\%$ Br; and fluorination, if carried sufficiently far, produces a mixture of fluorocarbon oils. Farenden and Pritchard [44] have described a laboratory process that converts all organic carbon of low-rank bituminous coal into fluoro- or fluorochlorocarbon oils, fluorinated hydrocarbon gases, and a small residue of transparent, fusible solids. However, neither this nor any other halogenation has so far been further developed; and in view of the environmental concerns now expressed over fluorocarbons, such further development, except where intended to aid studies of coal chemistry [45], is unlikely.

5.5 Hydrogenation

Unlike oxidation, which, even at low temperatures, occurs spontaneously (see Section 5.1), hydrogenation of coal is relatively difficult, and introduction of sufficient hydrogen to transform coal into *liquid* hydrocarbons (see Chapter 13) is only possible at temperatures above $350°C$, i.e., when the coal substance is actively decomposing (see Chapter 6).

Under milder conditions hydrogenation does not proceed beyond reducing the coal to a more (pyridine-) soluble solid and is therefore of limited interest.

The most that can be achieved by hydrogenation of coal *without* extensive concurrent thermal disruption of its molecular configurations is illustrated by the reaction between coal and lithium in ethylenediamine at $90–100°C$ [46]. (Similar reduction of coal occurs also when it is treated with lithium in ethylamine [47], but the Li–ethylenediamine system appears to be a more powerful hydrogenating agent.) The reaction (Note F) is complicated by the fact that it also incorporates some ethylenediamine and water (which cannot be removed by heating the coal in vacuo at $100°C$); and since lithium tends to interact with ethylenediamine to form N-lithioethylenediamine ($H_2N \cdot CH_2 \cdot NHLi$), which is known to aromatize cyclic dienes, it can sometimes even initiate *de*hydrogenation. But depending on the rank of the coal, up to 55 additional hydrogen atoms per 100 carbon atoms (ex-

---

[6] Like chlorination, bromination tends, therefore, to progressively dehydrogenate the coal substance. Weiler [42] has, in fact, reported that bromination at $65°C$ results in more bromine being eliminated from the coal as HBr than is fixed in the residue.

clusive of hydrogen in chemisorbed ethylenediamine and water) can nevertheless be introduced in this manner. In the case of vitrains [46], the net gain in hydrogen increases from about 20 H atoms per 100 C atoms at 84% carbon to 45 H per 100 C atoms at 91% carbon, and then falls again quite sharply to 20 H per 100 C atoms among anthracites.

Despite these gains, however, the reduced materials are still entirely solid and mainly differ from their precursors in containing higher proportions of carbon in aliphatic =CH— structures (which may be the cause of their enhanced solubility in pyridine). From a utilitarian viewpoint, real significance attaches therefore only to hydrogenation at temperatures between 350° and 550°C ($\sim$660–1020°F).

Practical methods for such "direct" hydrogenation[7] began to be developed in Europe in the early 1920s and typically involved high-pressure processing of coal (or coal–oil slurries) with molecular hydrogen over iron, tungsten, tin, or molybdenum catalysts at 450–550°C (see Section 13.1). However, since the mid-1940s, a number of potentially simpler alternative procedures have been outlined (see Section 13.2), and in some respects the most interesting of these are reactions that utilize transfer of hydrogen from a donor liquid such as tetralin or o-cyclohexylphenol.

First reported by Fieldner and Ambrose [48], transfer processes can be formally represented by

and, like other effective hydrogenation reactions, also proceed only at temperatures above 350°C. Like "classic" processes, they are therefore assumed to involve complex free radical sequences in which pyrolytically generated molecular fragments in the coal stabilize themselves by abstracting hydrogen from the donor [49, 50]. But they do not necessarily require a catalyst or high pressure and allow liquefaction of coal with great hydrogen economy.

Recent kinetic studies by, inter alios, Neavel [51], who employed tetralin as the donor, have shown that sufficient hydrogen for liquefaction is available when a 1:2 coal:tetralin slurry is used (Note G), and that fast, almost complete (sequential) conversion of the coal substance—first to a pyridine-soluble solid and thereafter to benzene-soluble (heavy) liquid hydrocarbons and small amounts of gaseous products—can be effected at 400°C ($\sim$750°F). At this temperature, as illustrated in Figs. 5.5.1 and 5.5.2, vir-

---

[7] "Direct" hydrogenation is further discussed in Chapter 13. So-called indirect hydrogenation processes, which start with coal-derived synthesis gas and consequently involve hydrogenation of *carbon monoxide*, are described in Chapter 12.

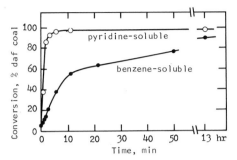

Fig. 5.5.1 Conversion of a high-volatile-C bituminous coal to pyridine- and benzene-soluble matter by interaction with tetralin at 400°C. Yields of soluble matter include small amounts of gaseous by-products. (After Neavel [51]; by permission of IPC Business Press Ltd.)

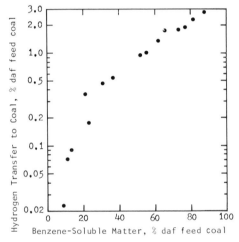

Fig. 5.5.2 Conversion of a high-volatile-C bituminous coal to benzene-soluble matter as function of hydrogen transfer from the H-donor. Coal reacted with tetralin at 400°C; yields of benzene-soluble matter include small amounts of gaseous by-products. (After Neavel [51]; by permission of IPC Business Press Ltd.)

tually 100% solubility in pyridine (and ~60% solubility in benzene) results within 10 min for little more than 1% increase over the original hydrogen content of the coal.[8] Very similar results have been reported by Wiser [50] and Hill *et al.* [52].

Coals up to and including high-volatile-C bituminous rank react at substantially similar rates and to the same ultimate extent; and because

---

[8] Significantly, abstraction of hydrogen from tetralin transforms it almost entirely into naphthalene. As a rule, less than 1% dihydronaphthalene could be detected in the reaction mixture.

such coals always tend to quickly disintegrate in the donor at reaction temperature, prior comminution to particle sizes *below* 2–3 mm has little, if any, beneficial effect on conversion rates. Wide dissimilarities have, however, been found in the behavior of different macerals. While vitrinites and exinites react rapidly and completely, fusinites behave as more or less inert components and appear together with mineral matter in the residue. How much of a given coal can be liquefied by hydrogen transfer methods depends therefore to some extent on its petrographic composition.

5.6 Depolymerization, Alkylation, and Related Reactions

Among the more significant results of chemical studies on coal in the past 10 or 15 years was the discovery of several relatively mild reactions that "depolymerize" the coal substance to more soluble (lower-molecular-weight) material at temperatures far below the onset of active thermal decomposition. Some of these reactions offer potential alternatives to current coal conversion techniques (see Chapters 13 and 14) and are also noteworthy for their contribution to a better understanding of the structure and chemical properties of coal (see Section 5.7).

Depolymerization reactions were first described by Neuworth *et al.* [53] and are based on the ability of phenol, when catalyzed by boron trifluoride, to cleave aryl–alkyl–aryl systems according to

$$Ar_1\!-\!CH_2\!-\!Ar_2 + 2 \underset{\bigcirc}{\bigcirc}\!-\!OH$$

$$\longrightarrow Ar_1H + Ar_2H + HO\!-\!\underset{\bigcirc}{\bigcirc}\!-\!CH_2\!-\!\underset{\bigcirc}{\bigcirc}\!-\!OH$$

The mechanism of such transalkylation or "aromatic interchange" has been discussed by McCaulay [54]. Applied to coal, the reaction proceeds at temperatures as low as 100°C, approaches completion within 4–5 hr, and yields a "depolymerized" product that is up to 75% soluble in phenol. Actual phenol solubilities decrease with increasing carbon content of the coal, falling from 75% in the case of lignites (with 70% carbon) to 9% for low-volatile bituminous coals (with 91% carbon). The extent of "depolymerization" appears to be directly proportional to the number of cleaved $-CH_2-$ bridges, which can be determined from the concentration of di-(hydroxyphenyl)-methane in the product mixture.

Of special interest are (a) the low molecular weights of the phenol-soluble material (which reportedly range from ~390 for depolymerized lignite to

~ 1010 for depolymerized lvb coal) and (b) identification of methylene and isopropyl groups as bridging structures in coal.

A modification of Neuworth's depolymerization technique by Darlage *et al.* [55], who pretreated the coal with 2 *N* nitric acid and then, after reacting it with phenol/BF$_3$, converted phenolic —OH to acetate by means of acetic anhydride, yielded up to 98% chloroform-soluble material.

Substantially increased solubility of coal in organic solvents has also been brought about by *Friedel–Crafts alkylation*, e.g., by reacting coal with an alkyl chloride (such as isopropyl chloride) and AlCl$_3$ in carbon disulfide at 45°C [56]. Up to 10 or 11 alkyl groups per 100 carbon atoms could be introduced in this manner, and up to 70% of the alkylated material was found to be soluble in pyridine. But despite its greater solubility, the reaction product was more highly aromatic than its precursor coal, probably as a result of simultaneous AlCl$_3$-catalyzed condensation, illustrated by

It is, in this connection, relevant to note that AlCl$_3$ has also been found to accelerate formation of graphite from aromatic compounds [57].

Similar aromatization, despite concurrent partial solubilization, occurs when coal is alkylated with olefins, such as ethylene or propylene [58]. However, rapid addition of olefins takes place only when the coal is heated to 300–350°C at ~ 19–20 MPa, and the overall reaction is therefore likely to be materially affected by incipient or active thermal decomposition of the coal.

For practical purposes, greater significance may in these circumstances attach to a *reductive alkylation* method [59] which, as a first step, involves reacting the coal with metallic potassium and tetrahydrofuran (THF) in the presence of naphthalene. Through electron transfer by naphthalene, this produces a coal "anion" that can then be alkylated by treatment with alkyl halides. The overall scheme is represented by

$$\text{coal} + n\text{K} \xrightarrow[\text{naphthalene}]{\text{THF}} \text{coal}^{n-} + n\text{K}^+$$

$$\text{coal}^{n-} + n\text{C}_2\text{H}_5\text{I} \rightarrow \text{coal}\cdot(\text{C}_2\text{H}_5)_n + n\text{I}^-$$

The concentration of coal anions can be determined by reacting the sample with water to form OH$^-$ via

$$\text{coal}^{n-} + n\text{H}_2\text{O} \rightarrow \text{coal}\cdot n\text{H} + n\text{OH}^-$$

Table 5.6.1

Reductive ethylation of coal[a]

	Coal rank (carbon, % daf)			
	71	82	90	96
Before reaction				
Solubility in pyridine, %	3	25	3	N.A.
Solubility in benzene, %	0.5	0.5	0.5	N.A.
After reaction				
"Anion" concentration (n) per 100 C atoms	24	26	12	11
Ethyl groups per 100 C atoms	14	11	9	0.6
Solubility in pyridine, %	43	81	97	N.A.
Solubility in benzene, %	23	74	95	N.A.

[a] After Sternberg *et al.* [59].

and potentiometrically titrating the reaction product against acid. The extent of alkylation can be assessed by employing a radioactive halide and measuring β-emissions from the alkylated end product.

Although such alkylation, as now understood, entails only very limited cleavage of the coal "molecule" (Note H), and the alkylated materials consequently have number-average molecular weights ranging as high as 3300 [59], they are for the most part readily soluble in benzene. This is illustrated in Table 5.6.1, which shows how solubility and the number of alkyl groups that can be introduced into the coal in this manner depend on coal rank.

Subsequent work [60] has, however, raised some questions about the *mechanism* of the reaction. It has been shown that aromatic hydrocarbons can cleave THF in the presence of a metal complex and then generate polymeric substances that embody some of the cleaved residues; and it has also been found that gaseous by-products of Sternberg alkylation contain traces of butane, as well as significant amounts of elemental hydrogen. If, as seems probable, this hydrogen derives from the coal rather than from THF, the reaction is, in effect, a normal carbanion reaction of active hydrogen compounds, exemplified by

$$2R \cdot CH_3 + 2K \rightarrow 2R \cdot CH_2^- K^+ + H_2$$

and is not, as originally thought, a *reductive* alkylation. Some further work is therefore needed in order to determine whether THF and naphthalene play roles other than as solvent and electron transfer agent, respectively.

5.7   The Chemical Structure of Coal

As well as bearing on coal processing and the selection of coal for particular end uses, the chemical properties of coal provide important insights
into its *molecular structure and organization.*

This topic began to attract the interest of coal chemists shortly after the
end of World War I and was for many years pursued by studying the composition of solvent extracts and the nature of coal oxidation products.
Attention was also given to "model compounds," which in some respects
appeared to display coal-like behavior. However, aside from the fact that
they were limited by inadequate analytical techniques (which often precluded meaningful chemical characterization of extract fractions), these
approaches to molecular structure held obvious pitfalls. Since the reaction
paths of coal oxidation were (and for the most part still are) unknown,
relationships between oxidation products and their precursors were usually
speculative;[9] and "model compounds" (which, in any event, only simulated
one or another particular aspect of coal behavior rather than coal itself)
raised, if nothing else, questions about which atomic configurations in their
respective molecules were responsible for "coal-like" behavior and might
therefore be components of the hypothetical coal molecule.

In these circumstances, real progress in elucidating the chemical structure
of coal became possible only with the development of better investigational
tools, notably x-ray diffraction and other spectroscopic techniques, from
the late 1940s on. These new methods of enquiry did not *simplify* exploration of coal chemistry. But they offered means for studying coal without
chemically altering it, provided much previously unobtainable information,
and for the first time allowed verification of experimental data and conclusions by cross-checking against information from other, independent
methods. Although current concepts of molecular structure are still rudimentary and more often than not illustrative rather than definitive, it is
therefore now at least possible to outline the main features of molecular
organization in coal and to identify several significant details.

Much of the work reported in the voluminous literature on coal chemistry
relates specifically to vitrinite which, while the most abundant component
of subbituminous and bituminous coal, is also the most homogeneous
maceral. The structural chemistry of exinites and inertinites has so far received much less attention, and statements about molecular configurations in

---

[9] Some authorities [61] have in fact suggested that oxidation studies have so far contributed
very little to knowledge of coal structure. In this connection see, however, results from selective
NaOCl oxidation of coal, to which reference is made later in this section.

coal must therefore, strictly speaking, be taken as referring only (or mainly) to a particular coal *constitutent*. However, subject to this restriction, a synthesis of evidence from chemical, spectroscopic, and graphical–statistical studies leads to the conclusion that coal can be viewed as a natural mixed polymer akin to a complex synthetic copolymer, and that there exists a *statistical "average coal molecule"* which contains within itself all essential configurations characteristic of the rank of the coal. Like the macromolecules of a synthetic copolymer that combine several monomeric species, individual coal molecules will, of course, differ in size as well as in their internal organization, so that their diverse (aromatic, hydroaromatic, aliphatic, etc.) components will not always necessarily be present in the same relative proportions or arranged in the same pattern. And like an incompletely cross-linked, three-dimensional copolymer, coal can consequently be separated into fractions that differ from each other in physical properties and, depending on peripheral configurations surrounding the molecular "core," may even possess somewhat different chemical properties. But in each case the statistical "average molecule" describes the entire assemblage; and consistent with the concept of progressive metamorphic development of coal (from lignite to anthracite), the *core* structures of the "average molecule," i.e., subunits linked by $—O—$ or single $—CH_2—$ bridges, change relatively little with increasing rank.

Experimental evidence for these conclusions comes, in the first instance, from *x-ray diffraction* spectra, which can be recorded by measuring the reflection (or "scattering") of a monochromatic x-ray beam by powdered coal at different angles of beam incidence (θ). Such spectra are illustrated in Fig. 5.7.1 and display a number of scatter maxima that, on the basis of their angular positions along the $f(\theta)$ coordinate, can be identified with the [002], [10], [004], [11], etc., diffraction bands in the x-ray spectrum of graphite. The strongest is, in all cases, the [002] band, which corresponds to first-order reflection of x-rays from the equidistant planes of graphite or similar aromatic structures.

Since the shape of an x-ray diffraction spectrum is uniquely determined by the spatial arrangement of atoms in the scattering solid, the structure of the solid can, in principle, be quite directly deduced from the relative widths, heights, and positions of the maxima. But in practice, translation of coal spectra into structure information is complicated by the fact that x-ray scattering by amorphous materials is not yet fully understood, and there is still some difficulty in deciding which of the several mathematical procedures available for transforming raw data into structural parameters yields "correct" results. Over the years, some of the inferences drawn from x-ray spectra of coal have therefore been very substantially modified.

The first major x-ray study of coal [62] led to the conclusion that the

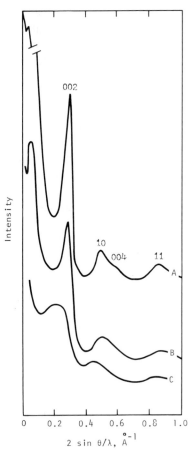

Fig. 5.7.1   Typical x-ray diffraction spectra of coal. Curves A, B, and C relate to coals with 94, 89, and 78% carbon, respectively. For clarity of presentation the three curves have been vertically displaced with respect to each other. Numbers indicate the crystallographic designations of the diffraction maxima. (After Cartz and Hirsch [92]; by permission of J. Wiley and Sons, Inc.)

preponderance of carbon atoms in coal existed in a "turbostratic" system of aromatic (graphite-like) lamellae that, depending on the rank of the coal, had an average diameter of 20 to 23 Å (equivalent to 50–100 polycondensed benzene rings) and were stacked to form "crystallites" with an average height of 14 to 17 Å (equivalent to 4–5 lamellae per crystallite). Later investigations, using more refined experimental and interpretative methods [63–67], have required radical revision of this model. While the diffraction maxima are still associated with partly ordered aromatic nuclei, the average diameter of these entities in coals with 80–90% carbon is now considered to be no larger than 7–7.5 Å (equivalent to 15–18 C atoms or 2–3 condensed benzene rings[10]). On average, only 2–3 nuclei appear to be vertically stacked into crystallites. And as many as 50% of the aromatic nuclei occur singly. By means of a matrix technique developed by Diamond [67], it has also proved possible to calculate histograms that provide an indication of the size distribution of aromatic nuclei in coal and to obtain a rough estimate of the proportion of "amorphous" carbon, i.e., material scattering as single independent carbon atoms (see Fig. 5.7.2). Depending on coal rank, such estimates associate 50–80% of the total carbon with aromatic nuclei, which in view of the high molecular weights of alkylated coal products (see Section 5.6) must be assumed to be extensively interlinked by non-aromatic structures.

This conclusion and the values now assigned to the structural parameters of aromatic nuclei in coals with $>80\%$ carbon (see Fig. 5.7.3) are in good accord with information obtained from radial distribution curves [65]. These curves, which can be derived from x-ray spectra without any assumptions, show little order beyond the region of an average-sized (5–6 Å) aromatic nucleus and thus rule out larger aromatic entities or relatively small aromatic nuclei *uniformly and regularly* linked by hydroaromatic rings.

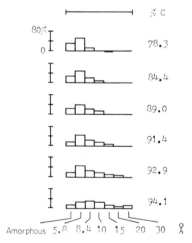

Fig. 5.7.2 Crystallite size histograms for coals of different rank. Sizes are expressed in angstrom units and denote the average diameters of the aromatic platelets. Material scattering as single carbon atoms is classed as "amorphous." (After Diamond [67]; by permission of J. Wiley and Sons, Inc.)

---

[10] Since nonaromatic carbon (or other) atoms directly attached to an aromatic entity contribute to scattering from it, the aromatic nucleus per se is actually somewhat smaller than its calculated diameter or the equivalent number of C atoms indicate.

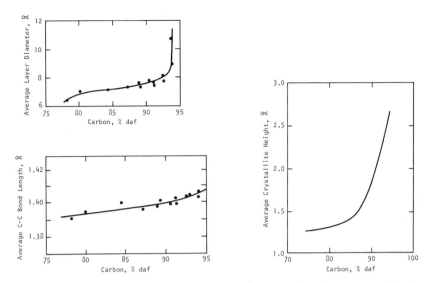

Fig. 5.7.3   The rank-variation of crystallite parameters obtained from x-ray diffraction spectra of coal; after Cartz and Hirsch [92]. (By permission of J. Wiley and Sons, Inc.) These diagrams illustrate changes in the average crystallite diameter, the average crystallite ("stack") height, and the average carbon–carbon bond length brought about by progressive metamorphism.

Since the significance of the "amorphous carbon" term in the histograms obtained by Diamond's method is still unclear, and the procedure may in fact grossly overestimate the fraction of aromatic carbon (f_a) in coal [68], much importance has been attached to the fact that very similar ranges for f_a can be derived from, among other properties, the refractive index, density, and heat of combustion of coal [69, 70]. A basic premise in such graphical–statistical calculations (which have been largely pioneered by van Krevelen and his co-workers and are discussed at length in his book [71]), is that the properties of any substance are additive functions of its component atoms and interatomic linkages; and in some cases, as, for example, when proceeding from refractive indices, it is necessary to make extensive extrapolations of empirical correlations.[11] In order to make

---

[11] The aromatic surface area (*S*), which is another way of expressing the size of the aromatic structural entity, can only be calculated from the refractive increment per aromatic C atom ($l_m C_{ar}$) for coals with less than 85% carbon. For higher-rank coals, the derivation of *S* depends on the empirical finding that the linear relationship between *S* and (100 − % C) can be fairly well extrapolated to an *S* value for graphite that was determined by x-ray spectroscopy. For a rigorous analysis of the concepts and methods used in graphical–statistical studies, reference should be made to Dryden [70].

appropriate allowances for different kinds of atoms, substituents, and interatomic linkages, reliance must also be placed on estimates obtained from
elemental, functional group, and infrared spectroscopic analyses (see below).
However, while the values of f_a thus arrived at (0.70–0.81 for coals with
80% carbon increasing to 0.78–0.90 for coals with 90% carbon) are com-
patible with Diamond's estimates, the corresponding estimates of the number of condensed rings per average aromatic structural unit are 2–3 times
larger than x-ray data allow. As matters stand, all that can be asserted
with respect to aromatic entities in coal is, therefore,

 (a) that aromatic nuclei in all coals other than anthracites are small
and, on average, comprised of no more than three condensed rings (as in
anthracene or phenanthrene); and

 (b) that the fraction of aromatic carbon may range as high as 0.8 but
could quite conceivably be substantially lower. The fact that coal, depending on its rank, incorporates between ~ 20 and 50 H atoms per 100 C
atoms when reacted with lithium in ethylenediamine (see Section 5.5) but
still contains some aromatic structure after hydrogenation indicates that
the *lower* limit for f_a must lie well above 0.2 for coals with 80–85% carbon
and exceed 0.5 in coals with 89–91% carbon. Other evidence, from oxidation studies that are noted later in this section, points to $f_a \sim 0.4$ over the
range to 90% carbon.

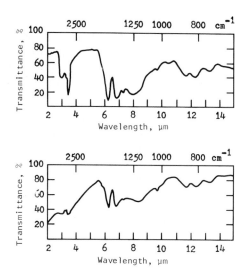

Fig. 5.7.4 Typical infrared spectra of coal. The upper diagram is a spectrum measured with
a thin (~ 0.02 mm) section of coal, while the lower was recorded with a KBr pellet containing
~ 3-mg coal/gm KBr.

Table 5.7.1

Absorption bands in the infrared spectra of coal

Band position (λ)		Assignment
cm^{-1}	μm	
3300	3.0	—OH (stretching), —NH (stretching)
3030	3.3	Aromatic C—H (stretching)
2940	3.4	Aliphatic C—H (stretching)
2925	3.42 ⎫	—CH$_3$ (stretching), —CH$_2$ (stretching)
2860	3.5 ⎬	Aliphatic C—H (stretching)
1700	5.9	C=O (stretching)
1600[a]	6.25	Aromatic C=C (stretching), C=O . . . HO—
1500	6.65	Aromatic C=C (stretching)
1450	6.9	⎰ Aromatic C=C (stretching); —CH$_3$ (asymmetric deformation) ⎱ —CH$_2$ (scissor deformation)
1380	7.25	—CH$_3$ (symmetric deformation); cyclic —CH$_2$
1300–1000	7.7–10.0	⎰ Phenolic and alcoholic C—O (stretching) C$_{ar}$—O—C$_{ar}$ (stretching); C$_{al}$—O—C$_{al}$ (stretching) ⎱ C$_{ar}$—O—C$_{al}$ (stretching)
900–700	11.1–14.3	"Aromatic" bands[b]

[a] See text.

[b] Some tentative specific assignments in this region are

860	11.6 ⎫	
820	12.2 ⎬	Aromatic HCC (rocking) in single and condensed rings
750	13.3 ⎭	
873	11.5 ⎫	
816	12.3 ⎬	Substituted benzene ring with isolated or two neighboring
751	13.3 ⎭	H atoms; *o*-substituted benzene ring
893	11.2 ⎫	Angular condensed ring system; monosubstituted benzene
758	13.2 ⎬	ring; condensed system

The generalized "average molecular structures" that can be broadly deduced from x-ray diffraction data and the elemental compositions of coal (i.e., units composed of small aromatic nuclei linked by hydroaromatic, alicyclic, and/or aliphatic moieties and containing various peripheral functional groups) accord well with evidence from *infrared* spectra of coal [72]. At wavelengths between 750 and 3300 cm^{-1} (~ 3–14 μm), such spectra (see Fig. 5.7.4) show several diagnostic absorptions bands that can be quite unequivocally associated with nonaromatic as well as with aromatic carbon–hydrogen configurations (see Table 5.7.1).

There is still some difference of opinion about the origin of the 1600-cm^{-1} band, which is usually the most prominent feature of the infrared spectrum of coal. This band has been variously assigned to polynuclear aromatic structures connected by predominantly aliphatic =CH$_2$ [73], to hydrogen-bonded or OH-chelated carbonyl groups [74–76], to electron transfer between aromatic carbon "sheets" [77], and to noncrystalline pseudographitic, *but not necessarily aromatic*, C—C configurations [78]. However, more immediately important are the absorption bands at 1380 and 1450 cm^{-1} and bands between 2860 and 3030 cm^{-1}, which establish the existence of aromatic and aliphatic C—H, cyclic —CH$_2$—, and —CH$_3$ in coal. Other spectral features show the presence of ether-linkages (1000–1300 cm^{-1}) and occasional replacement of H by —OH or —NH (3300 cm^{-1}). But there are few, if any, olefinic (C=C) or acetylenic (C≡C) configurations; and in coals with >80% carbon there is also no —COOH or —OCH$_3$. On the assumption that a high proportion of carbon in coal is, in fact, aromatic, the low intensity of the 3030-cm^{-1} band (assigned to aromatic CH) additionally confirms that the aromatic units are extensively substituted or cross-linked [79].

By proceeding from the relative intensities of the 3030- and 2925-cm^{-1} bands, several investigators [74, 80, 81] have sought to derive estimates of the fraction of *aromatic hydrogen* in coal. The methods that must be used for this purpose are fraught with some hazards. They require, first, determination of the optical densities, D_{ar} and D_{al}, which are related to the extinction coefficients e_{ar} and e_{al} for a single C—H bond by

$$D_{ar} = n_{ar} \cdot e_{ar} \qquad \text{and} \qquad D_{al} = n_{al} \cdot e_{al}$$

where n_{ar} and n_{al} are the numbers of the respective bonds; and since the detailed structure of coal is unknown and almost certainly quite variable, e_{ar}/e_{al} must then be experimentally obtained from diverse aromatic/aliphatic "model" compounds. Unfortunately, while the most frequent value of e_{ar}/e_{al} thus found is ~0.5, the actual range extends from ~0.3 to ~1.0 [74]. However, if 0.5 is chosen as the *most probable* value, n_{ar}/n_{al} turns out to increase from ~0.25 in coals with 84% carbon to ~1.0 in coal with 92% carbon; and assuming that (a) all oxygen exists in —OH groups and (b) all aliphatic hydrogen is contained in —CH$_2$— configurations,[12] the fraction of aromatic carbon (f_a) is found to range from ~0.7 at 84–85% C to ~0.9 at 92% C, i.e., in fair agreement with the estimates derived from x-ray diffraction data and from graphical–statistical studies.

Indications of high aromaticity (with values of f_a and H_{ar}/H_{al} similar to those derived from infrared spectra) have also been obtained from *nuclear magnetic resonance* (nmr) measurements on coal and coal extracts (Note I). Applied to liquids or solutions, nmr methods furnish a series of well-defined (0.1–100 mG wide) spectral bands, each of which can be assigned to a

[12] Actually, even if only 60% of all oxygen is assumed to exist in hydroxyl groups and some aliphatic hydrogen is assigned to —CH$_3$, neither n_{ar}/n_{al} nor f_a changes greatly. But both parameters are relatively sensitive to the value taken for e_{ar}/e_{al} [82].

particular H form (e.g., H in C_{ar}—H, —CH_2—, and —CH_3); and even where the variety of aliphatic H forms causes some band overlap, there is generally sufficient separation of H_{ar} and H_{al} bands to make evaluation of H_{ar}/H_{al} a relatively simple matter (Note J). The principal problem arises from an inherent inability to differentiate between H_{ar} and H in —OH groups; and where hydroxyl is present or suspected, a suitable correction— usually based on an independent chemical determination of —OH (see Section 5.1)—is necessary. However, in the case of solids, spin–spin interactions always broaden the nmr bands so much as to effectively "fuse" them into a single (0.1–10 G wide) absorption peak, and interpretation of the spectrum then requires assumptions about H forms in the solid and about their relative contributions to the spectrum.

In practice, such "low-resolution" spectra are analyzed by expressing spin–spin interactions in terms of the *mean square width* or *second moment* (ΔH^2) of the band and assuming that the second moments of particular H forms, which can be calculated from the nmr spectra of appropriate model compounds, are additive. One can then define a ratio a/b by writing

$$(H_{ar} + H_{CH_3})/(H_{CH_2} + H'_{CH_3}) = a/b$$

where H'_{CH_3} denotes H in perimethyl groups, i.e., —CH_3 sited as in 1,8-Me_2-naphthalene. And since measurements on pure compounds [87] have shown that the second moments for H_{CH_3} and H_{CH_2} assume values of, respectively, 27.5, and 9.7, it follows that

$$\Delta H^2 = 9.7a/(a + b) + 27.5b/(a + b)$$

and

$$a/b = (27.5 - \Delta H^2)/(\Delta H^2 - 9.7)$$

For coal that on the evidence of its infrared spectra contains little hydrogen in methyl groups, one can therefore, as a first approximation, write $a/b = H_{ar}/H_{al}$ and then correct this value for H in —OH.

By means of such interpretive techniques, which were later also applied to somewhat improved experimental data from ^{13}C nmr spectroscopy [88], H_{ar}/H_{al} and f_a have been found to increase with coal rank from ~ 0.5 to 1.8 and from 0.7 to 0.9, respectively [89]. Nmr studies of solvent extracts and vacuum distillates of coal [90] have produced similar results and also indicated that the extent of substitution on aromatic carbon atoms falls with increasing coal rank [91].

Synthesis of information derived from elemental and functional group analyses (see Section 5.1), x-ray diffraction, infrared and nmr spectroscopy, molecular weight measurements, and chemical degradation of the coal substance (e.g., by oxidation or depolymerization; see Sections 5.1, 5.2,

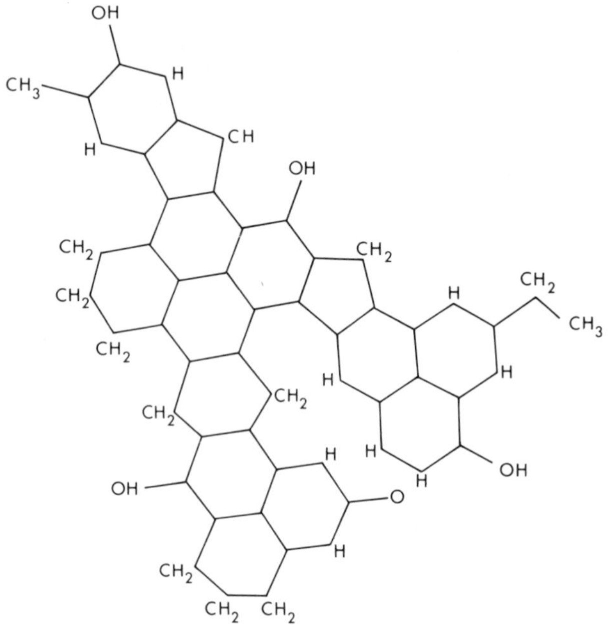

Fig. 5.7.5 "Buckled sheet" molecular structure of coal. (After Cartz and Hirsch [92].)

and 5.6) has enabled several investigators [92–95] to develop *illustrative* skeletal structures for a typical coal—or, more correctly, *vitrinite*—molecule (see Figs. 5.7.5–5.7.7).

The most self-consistent of these hypothesized configurations—in the sense of best reconciling data from different lines of inquiry—appears to be Given's (see Fig. 5.7.6), which embodies the essential features of the Cartz–Hirsch model, but can more easily be brought into better accord with nmr data on hydrogen forms in coal by making the basic aromatic unit a dihydrophenanthrene- rather than a dihydroanthracene-like configuration [93]. It also introduces an important new concept through its incorporation of a tryptycene moiety

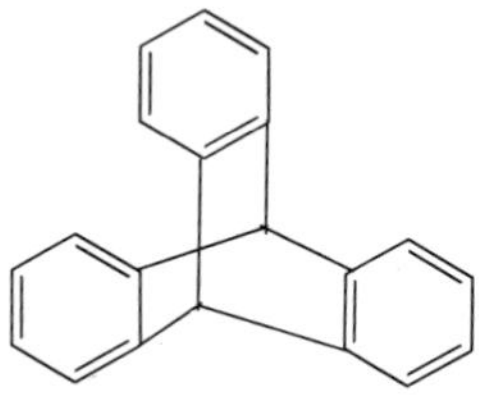

which, in addition to the "buckling" caused by hydroaromatic and alicyclic bridge structures, endows the molecule with an extra spatial dimension.

The skeletal arrangement proposed by Given harmonizes well with the chemical properties of coal at ordinary and elevated temperatures and even

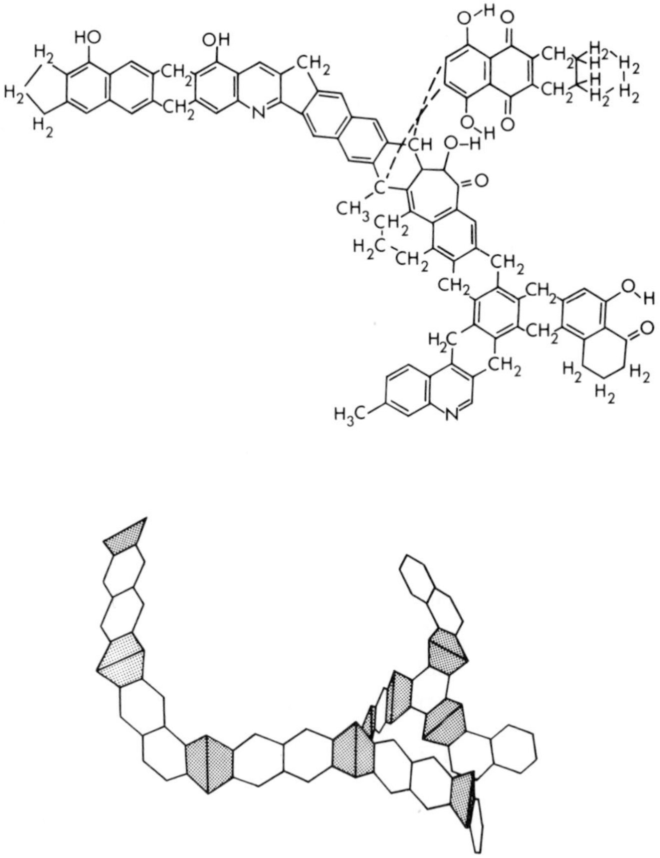

Fig. 5.7.6   Hypothetical coal molecule with ~82% carbon. The lower diagram, in which the shaded areas represent alicyclic structures, shows the spatial configuration of a simplified form of this molecule. (After Given [93].)

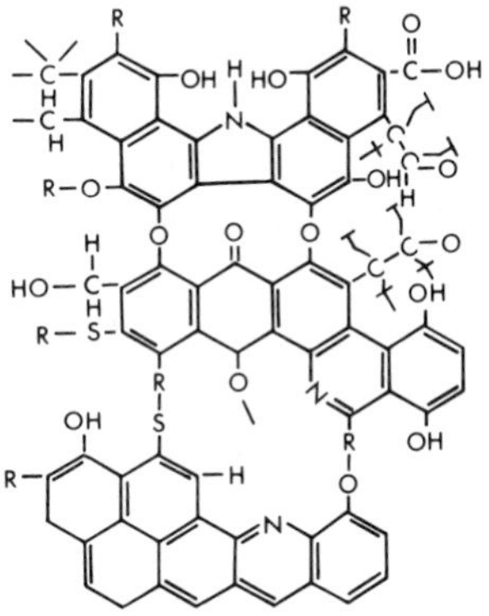

Fig. 5.7.7   Molecular structure of bituminous coal. (After Hill and Lyon [94].)

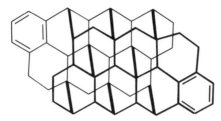

Fig. 5.7.8 Skeletal molecular structure of coal. The structure shown in this diagram is a partly aromatized polyamantane corresponding to $C_{50}H_{50}$. Aromatic rings at the periphery form by bond isomerization. (After Chakrabartty and Berkowitz [99].)

reflects certain optical properties (e.g., the molar refractivity) with some fidelity.[13] But there remains, nevertheless, a distinct possibility that the actual configuration of C atoms is quite different and partly based on non-aromatic *tetrahedral quaternary* C—C bonds (as in diamond) rather than on aromatic graphite-like systems. Ergun and Tiensuu [68] have shown that the x-ray diffraction spectra and other properties of coal would be consistent with a much lower degree of aromaticity than is now generally accepted; and Friedel and Queiser [97] have, in fact, argued that the low intensity and nonspecificity of the *ultraviolet* spectrum of coal implies the existence of an extensive tetrahedral C—C system.

This "minority view" has recently been underscored by studies of the oxidative degradation of a series of coals by aqueous sodium hypochlorite [98]. On the assumption that oxidation of coal by NaOCl follows the path of a normal haloform reaction, it has been concluded [99] that

(a)   in coals with 70–90% C, some 60% of the total carbon exists in nonaromatic $(S_p{}^3)$ valency states, and

(b)   the greater part of the molecular skeleton is made up of a bridged tricycloalkane configuration analogous to a variously extended polyamantane (see Fig. 5.7.8).

Not surprisingly, both conclusions have been challenged [100]. But a predominantly nonaromatic structure based on adamantane

Adamantane, $C_{10}H_{16}$.

Triamantane, $C_{18}H_{24}$.
(Can be considered as a
trimer of adamantane.)

[13] For detailed discussions of the chemical and physical implication of this structure, see Given [93] and Francis [96].

which represents the (monomeric) unit cell of the arrangement shown in Fig. 5.7.8, cannot be ruled on the basis of the experimental evidence now at hand [96]. It is undoubtedly significant that detailed chemical studies on solvent-refined coal (Section 14.2) have indicated the presence of bridged bicyclo- and tricyclo-compounds in such material, but found few condensed structures with more than two rings in it [101].

Notes

A. By an almost 200-year-old convention [6], the term *humic acids* is applied to all dark brown or black alkali-soluble matter, regardless of whether it is produced by air-oxidation of coal or by the action of fungal oxidases on plant material. The degradation products of humic acids are designated as (a) *hymatomelanic acids* (light brown; soluble in alcohols and phenols) and (b) *fulvic acids* (white or yellow; soluble in water as well as in alcohols and phenols).

B. Not infrequently, however, even very slight oxidation can adversely affect the caking properties of a coal and the quality of the coke made from it. Further reference to this aspect is made in Section 6.3.

C. In this connection it must be emphasized that oxidation effects in so-called weathered outcrops can extend as much as $10-15$ m ($30-50$ ft) into the seam. To be certain that a sample is fresh, it must therefore be taken at least ~ 15 m from the outcrop.

D. Compounds with such side chains must, however, be thought of as reaction intermediates. Further oxidation would remove the chains with formation of the corresponding aliphatic acids and leave a benzenoid residue.

E. This evidence is based on relatively early work (when unequivocal identification of reaction products was often difficult) and may require confirmation.

F. Indications are that treatment of coal with lithium in such ammoniacal solvents causes cleavage of some ether linkages and $C-C$ bond scissions of the type $Ar-CH_2 - CH_2-Ar \rightarrow Ar-CH_2- + Ar-CH_2-$.

G. Although most liquefaction studies have used $1:4$ coal:donor slurries, Neavel has emphasized that more than 66% donor in the slurry has no observable effect on the reaction. Even in a $1:2$ coal:tetralin slurry, the donor still contained 40% of its original hydroaromatic hydrogen when 90% of the coal had been converted to benzene-soluble material [51].

H. Available evidence suggests that alkylation splits ether linkages and possibly some $C-C$ bonds in much the same manner as treatment of coal with lithium in, e.g., ethylenediamine (see note F).

I. For a discussion of the theoretical foundations and experimental methods of nmr spectroscopy, reference should be made, for example, to [83–86].

J. If *solutions* are to be investigated by nmr spectroscopy, it is of course essential to eliminate contributions to the spectrum from hydrogen in the solvent. This is achieved by using H-free solvents (such as carbon disulfide or methylene chloride) or fully deuterated solvents (e.g., C_6D_6 or C_5D_5N).

References

1. R. E. Jones and D. A. T. Townend, *Nature (London)* **155,** 424 (1945); *Trans. Faraday Soc.* **42,** 297 (1946); *J. Soc. Chem. Ind. London* **68,** 197 (1949).
2. G. R. Yohe and C. A. Harman, *J. Am. Chem. Soc.* **64,** 1809 (1942).
3. S. R. Rafikov and N. Ya. Sibiryakova, *Izv. Akad. Nauk Kaz. SSR Ser. Khim.* No. 9, 13 (1956); *Chem. Abstr.* **50,** 8992 (1956).

4. W. Francis, "Coal," pp. 493–508. Arnold, London, 1961.
5. E. J. Jensen, N. Melnyk, J. C. Wood, and N. Berkowitz, *Adv. Chem. Ser.* **55**, 621 (1966).
6. C. Vauquelin, *Ann. Chim.* **21**, 39 (1797).
7. S. E. Moschopedis, J. C. Wood, J. F. Fryer, and R. M. Elofson, *Fuel* **43**, 289 (1964).
8. S. E. Moschopedis, *Fuel* **41**, 425 (1962).
9. J. C. Wood, S. E. Moschopedis, and W. den Hertog, *Fuel* **40**, 491 (1961).
10. B. V. Tronov, *J. Appl. Chem. USSR (Engl. transl.)* **13**, 1053 (1940); *Chem. Abstr.* **35** (1966).
11. W. Francis and R. V. Wheeler, *J. Chem. Soc.* **127**, 112 (1925).
12. W. Francis and H. M. Morris, U. S. Bureau Mines Bull. No. 340 (1931).
13. A. Ihnatowicz, *Pr. Gl. Inst. Gorr.* **125** (1952).
14. L. Blom, Ph.D. Thesis, Univ. of Delft, Netherlands (1960).
15. C. Kröger and G. Darsow, *Erdöl Kohle* **17**, 88 (1964).
16. C. Kröger, K. Fuhr, and G. Darsow, *Erdöl Kohle* **18**, 36 (1965).
17. C. Kröger, G. Darsow, and K. Fuhr, *Erdöl Kohle* **18**, 701 (1965).
18. B. S. Ignasiak, T. M. Ignasiak, and N. Berkowitz, *Rev. Anal. Chem.* **2**, 278 (1975).
19. L. Blom, L. Edelhausen, and D. W. van Krevelen, *Fuel* **36**, 135 (1957).
20. O. I. Egorova, *Bull. Acad. Sci. USSR Classe Sci. Tech.* No. 7–8, 107 (1942); *Chem. Abstr.* **38**, 4404 (1944).
21. A. I. Khrisanfova, *Izv. Akad. Nauk SSSR Otd. Tekh. Nauk* **895** (1947); *Chem. Abstr.* **43**, 3593 (1949).
22. F. Heathcoat, *Fuel* **12**, 4 (1933).
23. J. A. Radspinner and H. C. Howard, *Ind. Eng. Chem. Anal. Ed.* **15**, 566 (1943).
24. D. G. A. Thomas, *Br. J. Appl. Phys.* **4**, S 55 (1953).
25. R. S. Montgomery and E. D. Holly, *Fuel* **36**, 63 (1957); **36**, 493 (1957); **37**, 181 (1958).
26. S. Raj, Ph.D. Thesis, Penn State Univ. (1976).
27. G. J. Lawson and S. G. Ward, *Proc. Int. Conf. Coal Sci., 3rd, Valkenberg, Netherlands* (1959).
28. F. Fischer and H. Schräder, *Gesammelte Abh. Kenn. Kohle* **4**, 342 (1919); **5**, 135 (1920); **6**, 1 (1921).
29. W. A. Bone and L. Quarendon, *Proc. R. Soc., London* **110A**, 537 (1926).
30. S. K. Chakrabartty and N. Berkowitz, *Nature (London)* **261** (5555), 76 (1976).
31. W. A. Bone, L. G. B. Parsons, R. H. Shapiro, and G. M. Groocock, *J. Am. Chem. Soc.* **67**, 246 (1945).
32. B. Heimann, British Patent No. 349,001 (1929); French Patent No. 689,041 (1930).
33. A. Lahiri, P. N. Mukherjee, and S. Banerjee, *Planters' J. Agr. (India)* August (1966) P. N. Mukherjee, S. Banerjee, L. V. Ramchandran, and A. Lahiri, *Indian J. Technol.* **4**, 119 (1966).
34. S. K. Chakrabartty, J. C. Wood, and N. Berkowitz, U. S. Bureau Mines Information Circ. 8376 (1967); N. Berkowitz, S. K. Chakrabartty, F. D. Cook, and J. I. Fujikawa, *Soil Sci.* **110** (3), 211 (1970).
35. H. M. Brown and N. Berkowitz, *Chem. Eng. Prog. Symp. Ser.* **64**, 89 (1968).
36. S. K. Chakrabartty and N. Berkowitz, *Fuel* **48**, 151 (1969).
37. E. J. Bevan and C. F. Cross, *Chem. News* **44**, 185 (1881).
38. A. Eccles, H. Kay, and A. McCulloch, *J. Soc. Chem. Ind. London* **51**, 49T (1932).
39. F. J. Pinchin, *Fuel* **37**, 293 (1958).
40. A. E. Kretov, M. I. Savin, M. I. Shenbor, and I. E. Lev, *Ukr. Khim. Zh.* **18**, 305 (1952).
41. F. Heathcoat and R. V. Wheeler, *J. Chem. Soc.* 2839 (1932).
42. J. F. Weiler, *Fuel* **14**, 190 (1935).
43. J. K. Brown, P. H. Given, V. Lupton, and W. F. Wyss, *Proc. Conf. Sci. Use Coal, Sheffield* p. A43 (1958).

44. P. J. Farenden and E. Pritchard, *Proc. Conf. Sci. Use Coal, Sheffield* p. A52 (1958).
45. J. L. Huston, R. G. Scott, and M. H. Studier, *Fuel* **55,** 281 (1976).
46. L. Reggel, R. Raymond, S. Friedman, R. A. Friedel, and I. Wender, *Fuel* **37,** 126 (1958);
    L. Reggel, R. Raymond, W. A. Steiner, R. A. Friedel, and I. Wender, *Fuel* **40,** 339 (1961).
47. P. H. Given, V. Lupton, and M. E. Peover, *Proc. Conf. Sci. Use Coal, Sheffield* p. A38
    (1958); *Nature (London)* **181,** 1059 (1958).
48. A. C. Fieldner and P. M. Ambrose, U. S. Bureau Mines Information Circ. 7417 (1947).
49. G. P. Curran, R. T. Struck, and E. Gorin. Preprints, *Am. Chem. Soc. Petr. Chem. Div.,*
    C-130-148 (1966); *Ind. Eng. Chem. Proc. Design Dev.* **6,** 166 (1967).
50. W. H. Wiser, *Fuel* **47,** 475 (1968).
51. R. C. Neavel, *Fuel* **55,** 237 (1976).
52. G. R. Hill, H. Hairiri, R. I. Reed, and L. L. Anderson, *Am. Chem. Soc. Adv. Chem. Ser.*
    55, 427 (1966).
53. L. A. Heredy and M. B. Neuworth, *Fuel* **41,** 221 (1962). L. A. Heredy, A. E. Kostyo, and
    M. G. Neuworth, *Fuel* **42,** 182 (1963); **43,** 414 (1964); **44,** 125 (1965).
54. D. A. McCaulay, *in* "Friedel-Crafts and Related Reactions" (G. A. Olah, ed.), Vol. 2,
    Part 2, p. 1062. Wiley (Interscience), New York, 1964.
55. L. J. Darlage, J. P. Weidner, and S. S. Block, *Fuel* **53,** 54 (1974).
56. C. Kröger and H. de Vries, *Liebig's Ann.* **652,** 35 (1962).
57. E. Ota and S. Otani, *Chem. Lett.* **241** (1975).
58. C. Kröger, H. B. Rab, and B. Rabe, *Erdöl Kohle Erdgas Petrochem.* **16,** 21 (1963).
    C. Kröger, *Forsch. Ber. Land. Nordrhein-Westfalen* No. 1488, Westdeutsch. Verlag,
    Cologne, 1965.
59. H. W. Sternberg, C. L. Delle Donne, P. Pantages, E. C. Moroni, and R. E. Markby,
    Fuel **50,** 432 (1971).
    H. W. Sternberg and C. L. Delle Donne, *Fuel* **53,** 172 (1974).
60. S. K. Chakrabartty, Private communication (1977).
61. I. G. C. Dryden, *in* "Chemistry of Coal Utilization" (H. H. Lowry, ed.), Suppl. Vol.
    Wiley, New York, 1963.
62. H. E. Blayden, J. Gibson, and H. L. Riley, *Proc. Conf. Ultrafine Struct. Coals Cokes*
    p. 176. BCURA, London, 1944.
63. D. P. Riley, *Proc. Conf. Ultrafine Struct. Coals Cokes* p. 256. BCURA, London, 1944.
64. R. E. Franklin, *Acta Crystallogr.* **3,** 107 (1950).
65. J. B. Nelson, *Fuel* **32,** 153, 381 (1954).
66. P. B. Hirsch, *Proc. R. Soc. London Ser. A* **226,** 143 (1954); *Phil. Trans.* **252A,** 68 (1960);
    Proc. Conf. Sci. Use Coal, Sheffield p. A29 (1959).
67. R. Diamond, *Acta Crystallogr.* **10,** 359 (1957); **11,** 129 (1958); see also W. Ruland and
    H. Tschamler, *Brennst. Chem.* **39,** 363 (1958).
68. S. Ergun and V. H. Tiensuu, *Nature (London)* **183,** 1668 (1959); *Fuel* **38,** 64 (1959); *Acta
    Crystallogr.* **12,** 1050 (1959).
69. H. R. G. Chermin and D. W. van Krevelen, *Fuel* **33,** 338 (1954); D. W. van Krevelen,
    H. R. G. Chermin, and J. Schuyer, *Fuel* **38,** 483 (1959).
70. I. G. C. Dryden, *Fuel* **34,** 536 (1955).
71. D. W. van Krevelen, "Coal." Elsevier, Amsterdam, 1961.
72. See, e.g., H. Tschamler and E. de Ruiter, *in* "Chemistry of Coal Utilization" (H. H. Lowry,
    ed.), Suppl. Vol. Wiley, New York, 1963.
73. H. S. Rao, P. L. Gupta, F. Kaiser, and A. Lahiri, *Fuel* **41,** 417 (1962).
74. J. K. Brown, *J. Chem. Soc.* 744 (1955).
75. S. Fujii and F. Yokayama, *Nenryo Kyokaishi* **37,** 643 (1958).
76. M. M. Roy, *Fuel* **36,** 249 (1957).

77. R. M. Elofson and K. F. Schulz, Preprints, *Am. Chem. Soc. Fuel Chem. Div.* **11**, 513 (1967).
78. R. A. Friedel, J. A. Queiser, and G. L. Carlson, Preprints, *Am. Chem. Soc. Fuel Chem. Div.* **15** (1), 123 (1971).
79. J. K. Brown and P. B. Hirsch, *Nature* (*London*) **175**, 229 (1955).
80. K. Kojima, K. Sakashita, and T. Yoshino, *Nippon Kagaku Zasshi* **77**, 1432 (1956).
81. Y. Osawa, H. Sugimara, and S. Fujii, *Nenryo Kyokaishi* **48**, 694 (1969).
82. I. G. C. Dryden, *Fuel* **37**, 444 (1958).
83. F. Bloch, *Phys. Rev.* **70**, 460 (1946).
84. J. E. Wertz, *Chem. Rev.* **55**, 829 (1955).
85. E. R. Andrew, "Nuclear Magnetic Resonance." Cambridge Univ. Press, London and New York, 1955.
86. E. E. Schneider, "Nuclear Moments." Academic Press, New York, 1958.
87. E. R. Andrew and R. G. Eades, *Proc. R. Soc. London Ser. A* **216**, 398 (1953); **218**, 537 (1953); C. L. M. Bell, R. E. Richards, and R. W. Yorke, *Brennst. Chem.* **39**, S30 (1958).
88. H. L. Retcofsky and R. P. Friedel, *Adv. Chem. Ser.* **55**, 503 (1966); *Inter. Conf. Coal Sci., 7th, Prague* (1968); *Fuel* **47**, 487 (1968).
89. P. C. Newman, L. Pratt, and R. E. Richards, *Nature* (*London*) **175**, 645 (1955); R. E. Richards and R. W. Yorke, *J. Chem. Soc.* 2489 (1960).
90. J. K. Brown, W. R. Ladner, and N. Sheppard, *Fuel* **39**, 79 (1960); W. R. Ladner and A. E. Stacey, *Fuel* **40**, 295 (1961); J. F. M. Oth and H. Tschamler, *Fuel* **42**, 467 (1963); W. R. Ladner and A. E. Stacey, *Fuel* **44**, 71 (1965); K. D. Bartle and J. A. S. Smith, **44**, 109 (1965); **46**, 29 (1967).
91. J. K. Brown, W. R. Ladner, and N. Sheppard, *Int. Conf. Coal Sci., 3rd, Valkenberg, Netherlands* (1959).
92. L. Cartz and P. B. Hirsh, *Phil. Trans. R. Soc. A* **252**, 557 (1960).
93. P. H. Given, *Fuel* **39**, 147 (1960); **40**, 427 (1961).
94. G. R. Hill and L. B. Lyon, *Ind. Eng. Chem.* **54** (6), 36 (1962).
95. B. K. Mazumdar and A. Lahiri, *J. Sci. Ind. Res.* (*India*) **21B**, 277 (1962).
96. W. Francis, "Coal," pp. 736–750. Arnold, London, 1961.
97. R. A. Friedel and J. A. Queiser, *Fuel* **38**, 369 (1959).
98. S. K. Chakrabartty and H. O. Kretschmer, *Fuel* **51**, 160 (1972); **53**, 132 (1974); *J. Chem. Soc. Perkin* **1**, 222 (1974).
99. S. K. Chakrabartty and N. Berkowitz, *Fuel* **53**, 240 (1974).
100. R. Hayatsu, R. G. Scott, L. P. Moore, and M. H. Studier, *Nature* (*London*) **257**, 378 (1975); F. R. Mayo, *Fuel* **54**, 273 (1975); R. G. Landolt, *ibid.* **54**, 299 (1975); G. Gosh, A. Banerjee, and B. K. Mazumdar, *ibid.* **54**, 294 (1975); T. Aczel, M. L. Gorbaty, P. S. Maa, and R. H. Schlosberg, *ibid.* **54**, 295 (1975); S. K. Chakrabartty and N. Berkowitz, *Nature* (*London*) **261** (5555), 76 (1976); *Fuel* **55**, 362 (1976).
101. Electric Power Research Institute, California, Rep. AF-480, Project 410-1 (July 1977).

CHAPTER 6

BEHAVIOR AT ELEVATED TEMPERATURES: DECOMPOSITION AND CARBONIZATION

When heated to progressively higher temperatures in an inert (substantially oxygen-free) atmosphere, coal decomposes with evolution of water, tar, and gas and leaves a solid residue whose composition and properties vary in direct dependence on the heat treatment temperature. In the end, if taken above $\sim 2200°C$, the residue becomes, for all practical purposes, a pure carbon with all the essential characteristics of microcrystalline graphite.

Although usually associated with a temperature range ($\sim 350–500°C$) in which devolatilization proceeds particularly rapidly, thermal decomposition actually begins at much lower temperatures and can, as illustrated by a typical (cumulative) weight loss curve (see Fig. 6.0.1), be divided into three stages.

In the first stage, which commences well below 200°C (Note A), decomposition is fairly slow and manifests itself primarily in release of small quantities of "chemically combined" water, oxides of carbon, and hydrogen

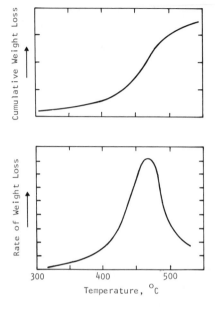

Fig. 6.0.1   Typical variation of cumulative weight loss (top) and weight loss rates (bottom) with temperature.

sulfide. These products are generally considered to result from detachment of substituent functional groups, e.g.,

$$\text{C}_6\text{H}_5\text{-COOH} \longrightarrow \text{C}_6\text{H}_5\text{-H} + CO_2$$

and from limited condensation reactions, e.g.,

$$\text{C}_6\text{H}_5\text{-OH} + \text{H-C}_6\text{H}_5 \longrightarrow \text{C}_6\text{H}_5\text{-C}_6\text{H}_5 + H_2O$$

or

$$\text{C}_6\text{H}_5\text{-COOH} \longrightarrow \text{C}_6\text{H}_5\text{-}\overset{\displaystyle O}{\underset{}{\text{C}}}\text{-} + OH^- + \text{C}_6\text{H}_5 \longrightarrow$$

$$\text{C}_6\text{H}_5\text{-}\overset{\displaystyle O}{\underset{}{\text{C}}}\text{-C}_6\text{H}_5 + H_2O$$

However, above $\sim 200°C$ some benzylic carbon begins to isomerize to form methyl phenyl derivatives [1], and traces of alkyl benzenes are evolved. Few details of these low-temperature processes are known, but that they are significant is evidenced by the fact that they alter the original structure of the coal sufficiently to influence its subsequent thermal behavior. Pre-heating at ∼200°C thus tends to destroy any caking propensities that the coal may possess and also increases its solubility in organic solvents.

The second stage, which is sometimes termed *active* thermal decomposition, begins between ∼350° and 400°C and ends near 550°C, with a formal decomposition temperature (T_d) conventionally set at the point at which the weight loss curve suddenly turns upward. Figure 6.0.2 illustrates how T_d varies with coal rank. As a rule, about 75% of all "volatile matter" ultimately released by the coal, including *all* tar and lighter condensible hydrocarbons, are evolved in this temperature interval. But product compositions, including those of the solid residues, are markedly dependent on ambient conditions (e.g., pressure and rate of heating). This sensitivity is attributable to the fact that "active" decomposition involves extensive fragmentation of the coal molecule and concurrent random recombination of the free radical species thus formed.

If the coal possesses caking properties, it will also pass through a characteristic transient "plastic state" between ∼350 and 450°C and thereby form a more or less porous *coke* (see Section 6.3).

The final stage of decomposition, represented by the relatively flat upper portion of the weight loss curve and termed *secondary degasification*, is characterized by gradual elimination of heteroatoms (principally H and O) and ends, strictly speaking, only when the char is transformed into a *graphitic* solid. In practice, however, the composition and quantity of volatile matter released above 800–850°C are of such little interest that studies of coal pyrolysis, other than those specifically concerned with crystallographic changes in the solid residue, are not carried beyond 900–1000°C. The principal by-products of third-stage decomposition are water, oxides of carbon, hydrogen, methane, and traces of C_2 hydrocarbons, and its most important feature is the progressive "aromatization" of the char, i.e., the

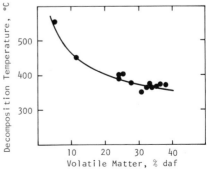

Fig. 6.0.2 Variation of the decomposition temperature T_d with coal rank.

development of increasingly large hexagonal carbon platelets. Where the residue is a coke, heat treatment up to 900° or 1000°C also leads to marked increases in its mechanical strength.

Although "pyrolysis," "thermal decomposition," and "carbonization" are often used interchangeably, it should be observed that "pyrolysis" refers to a process which *involves* extensive thermal decomposition (and subsequent carbonization of the residual solid), and that "carbonization" (i.e., progressive enrichment of the solid in carbon) is more properly used in reference to changes that occur when that solid is pyrolyzed at temperatures above ∼500 or 550°C.

6.1 The Kinetics of Thermal Decomposition

Since the nature and amount of volatile matter depend on the type and rank of the coal (see Section 2.2), there is a direct relationship between the carbon content (or atomic H/C ratio) of a coal and the extent to which it decomposes at any particular temperature. As carbon contents increase, active thermal decomposition occurs in progressively higher temperature intervals, and maximum weight loss rates become smaller; and because different macerals in any one coal also generate different amounts of volatile matter, similar displacements are observed when exinites, vitrinites, and inertinites of the *same* rank are pyrolyzed (see Figs. 6.1.1 and 6.1.2).

However, like other destructive distillation processes, even *active* thermal decomposition can be arrested at any point by interrupting the heating program and holding the temperature constant. The coal will then asymp-

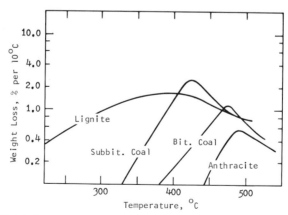

Fig. 6.1.1 Weight loss rate curves for coals of different rank. (After van Krevelen *et al.* [2]; by permission of IPC Business Press Ltd.)

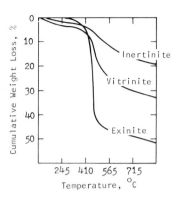

Fig. 6.1.2 Cumulative weight loss curves for different maceral entities of the same rank. The maceral entities used for these measurements were isolated from a high-volatile bituminous coal. (After Kröger and Pohl [3]; by permission of J. Wiley and Sons Inc.)

totically approach an equilibrium weight that depends on the particular temperature as well as on the nature of the coal, and further release of volatile matter thereafter occurs only if the temperature is raised. Figure 6.1.3 shows some typical $-\Delta w/t$ curves and illustrates the fact that equilibria are generally not reached in less than 20–25 hr.

This behavior pattern, which is characteristic of decomposition processes comprised of a series of sequential, partly overlapping reactions, has an important bearing on carbonization practices (see Chapter 11) and other high-temperature coal processing and has been explored in several kinetic studies (e.g., van Krevelen *et al.* [2], Fitzgerald [4], Hanbaba *et al.* [5], Chermin and van Krevelen [6], Berkowitz and Mullin [7], Shepatina *et al.* [8]). Rate data were usually obtained from thermogravimetric measurements; but in some cases, instead of recording isothermal weight loss curves (such as those shown in Fig. 6.1.3), weight losses were measured while the temperature was being raised at a constant rate. On the assumption that instantaneous rates of decomposition, as reflected in weight losses, are functions of the remaining fraction (f) of undecomposed coal, one can then write

$$-df/dt = kf^n$$

where $-df/dt$ is the rate of decomposition, k is the velocity constant, and n is the order of the reaction. Substituting $k = k_0 e^{-A/T}$ and introducing a term x ($=dT/dt$) for the rate of temperature rise, this yields

$$-df/f^n = (k_0/x)e^{-A/T}\,dt$$

where A is related to the gas constant (R) and the activation energy (E) of the reaction by $A = E/R$. And by a process of approximation, it can then be shown that the reaction order n is 0, 1 or 2 when the integral of $-df/f^n$ is $1 - f$, $\ln f$, or $(1/f) - 1$, respectively.

Using this approach, van Krevelen *et al.* [2], Jüntgen and co-workers [5], and others have found thermal decomposition of coal to be a first-order

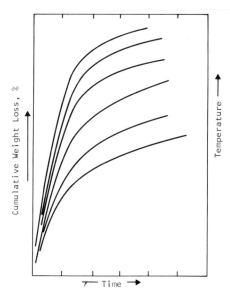

Fig. 6.1.3   Isothermal weight loss curves recorded at different temperatures.

process with apparent activation energies of 105–335 kJ/mole, and therefore concluded that decomposition rates are controlled by rupture of C—C and other covalent bonds. While this conclusion is undoubtedly correct in principle, its derivation from thermogravimetric measurements has been challenged on the grounds that experimentally measured weight loss rates reflect *diffusion* rather than *formation* of volatile matter [9]. Analyses of $-\Delta w/t$ curves showed that initial rates of volatile matter release from coal at constant temperature are very much greater than allowed by first-order equations that fit the extended upper portions of these curves and furnished apparent activation energies of 12.5–15 kJ/mole. It was therefore postulated that fast decomposition (by bond scission) creates a transient high-pressure gradient in the micropore system of the coal that sustains diffusion over prolonged periods of time. Compliance with first-order kinetics was ascribed to formal identity of a first-order reaction equation with Fick's diffusion law [9].

This view, which explicitly distinguishes between *decomposition* and *devolatilization* and sees the latter as a diffusion-controlled process, is supported by the low activation energies (20–25 kJ/mole) reported from a number of other pyrolytic studies [7, 8] and by the dependence of weight loss rates on rates of heating (see Fig. 6.1.4). As noted later (see Section 6.3), devolatilization is also influenced by the particle size of the coal.

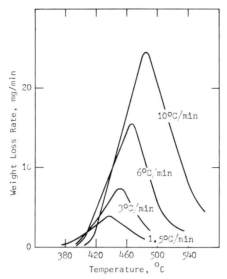

Fig. 6.1.4   Typical dependence of weight loss rates on heating rates.

6.2   Liquid and Gaseous Products of Thermal Decomposition

The yields of liquid and gaseous decomposition products obtainable from any particular coal can be determined in the laboratory by quantitative "destructive distillation" of the coal into a suitable condenser train and gas buret; and several such assay methods have been standardized to furnish results that can be directly correlated with industrial practices. The most important of these are

(a)   the Fischer assay, which was originally developed in Germany by Fischer and Schräder [10] and later, with minor modifications [11, 12], adopted in the United States as well as in a number of European countries, and

(b)   the Gray–King assay, which originated in Britain [13] and is now extensively used there and elsewhere as a means for classifying coals (see Table 3.3.2).

Both are usually conducted at 500 or 600°C (935 or 1110°F), even though they tend at these temperatures to underestimate gas volumes and to show *maximum* tar plus "light oil" yields; and in the hands of experienced operators both are sufficiently precise to yield mass balances of $100 \pm 0.3\%$. The Gray–King assay can also be carried out at 900°C ($\sim$1650°F), and then provides information about the quality of the solid residue.

For evaluation of coal on a semitechnical scale, considerable use has been made of the BM–AGA retort [14], which was developed by the US Bureau of Mines and the American Gas Association. And for more fundamental studies, several vacuum retorts and short-path "molecular stills" have been employed [15, 16].

Typical assay data, which illustrate how yields of gas, water, tars, and light oils tend to vary with coal type and rank, are shown in Table 6.2.1 and exemplify the competition for hydrogen between oxygen and carbon during thermal decomposition of coal. With increasing rank (and diminishing oxygen contents), yields of tar and light oil thus rise progressively at the expense of water, and only fall off again as carbon contents exceed $\sim 87\%$ and hydrogen contents begin to decline. Anthracites, with $H < 4\%$, therefore produce little or no tar, while sapropels (i.e., cannels and bogheads), which contain unusually high proportions of hydrogen (see Section 3.1), furnish between two and four times as much tar as their humic counterparts.

This connection between elemental coal compositions and tar yields is reflected in empirical formulas that have been developed from assay data in order to "predict" product yields from the hydrogen and oxygen contents of coal. Parry [18] has reported good agreement with experimental data from

$$\% \ H_2O = 0.35(\% \ O)^{1.3}$$

where the water yield is expressed as a percentage of the dry, ash-free coal; and Francis [19] gives

$$\% \ (\text{tar} + \text{light oils}) = 5.48(\% \ H_e)^{1.5}$$

where ($\% \ H_e$) is an "excess" hydrogen quantity approximately equal to 3.5.

However, since tars (including light oils) form mostly by random recombination of free radicals generated during active thermal decomposition of the coal substance, tar yields and, more importantly, tar *compositions* depend also on *how* the coal is pyrolyzed. Particularly important in this regard is the maximum temperature to which the tar vapors are exposed before they are condensed.

Low-temperature (LT) tars from subbituminous and bituminous coals, which by definition do not experience temperatures above $\sim 700°C$, are relatively fluid, dark brown oils that contain high proportions of mono- and dihydric phenols, methyl-substituted pyridines, *n*-paraffins, and olefins, but which are characteristically so heterogeneous that they rarely hold any one component in amounts greater than 0.5%. The same is true of lignite LT tars, except that these often also contain up to 10% of paraffin waxes,

Table 6.2.1

500°C Fischer assays of different coals[a]

Rank	Number of samples	Tar (gal/ton)[b]		Light oil[c] (gal/ton)		Water (gal/ton)		Gas (scf/ton)[d]	
		Range	Average	Range	Average	Range	Average	Range	Average
lvb	17	6.3–12.7	8.6	0.7–1.6	1.0	1.1–6.6	3.2	1600–1960	1760
mvb	30	9.7–25.6	18.9	1.0–2.3	1.7	2.8–7.0	4.1	1390–2240	1940
hvAb	134	22.9–40.7	30.9	1.5–3.3	2.3	3.0–9.2	6.0	1690–2360	1970
hvBb	11	24.3–43.1	30.3	1.6–3.4	2.2	10.2–13.1	11.1	1660–2420	2010
hvCb	7	18.5–38.8	27.0	1.3–2.7	1.9	12.0–19.1	15.9	1560–2070	1800
Subbituminous A	5	18.4–24.4	20.5	1.4–1.9	1.7				
Subbituminous B	7	13.2–16.7	15.4	1.1–1.6	1.3	23.3–30.4	27.8	1830–2760	2260
Cannel	7	53.7–108.3	73.5	3.7–7.4	5.1	2.0–4.8	3.7	1500–2120	1810

[a] After Selvig and Ode [17].
[b] 1 (US) gal/ton $\simeq$ 4.2 liters/tonne.
[c] Light oils are usually reported separately but are, in industrial practice, the first distillate cuts of condensible by-products.
[d] 1000 scf (= 1 Mcf) $\simeq$ 31.2 m^3.

which endow them with a butterlike consistency and cause them to solidify at temperatures as low as 6–8°C.

At temperatures above 700°C, on the other hand, the "primary" tar vapors formed from coal material are very quickly aromatized by dealkylation, dehydroxylation, and related condensation reactions; and high-temperature (HT) tars, which are usually recovered as by-products of coke manufacture (see Section 11.2), are therefore very different from, and much more homogeneous than, LT tars. The light oil fractions consist predominantly of benzene, toluene, and xylenes; and the tars themselves are bitumen-like viscous mixtures that contain high proportions of 2–5 ring polycondensed aromatics as well as a variety of phenolic tar acids and tar bases comprised of pyridines, picolines, quinolines, and some anilines. Many of these compounds occur in sufficiently high concentrations to make their isolation from the light and middle oil fractions of the tar distillates (see Section 11.3) commercially attractive.

Table 6.2.2 illustrates the differences between LT and HT tars and also shows how extensively retort *design*, which determines how long the "primary" tar vapors are exposed to high temperatures, affects tar composition. The extremes in this tabulation are provided by A (a low-temperature retort) and D (a coke oven).

The yields and compositions of tars obtained from individual petrographic components of coal under similar conditions are generally as would be expected from their proximate and elemental analyses (see Table 6.2.3),

Table 6.2.2

Compositions of tars plus light oils produced from high-volatile bituminous coal in different retorts[a]

	A[b]	B[b]	C[b]	D[b]	E[b]
Aromatics	15.56	79.62	74.31	85.26	63.04
Naphthenes	8.00	1.85	1.72	0.21	3.62
Mono-olefins	16.26	1.61	1.89	1.64	2.33
Dienes	1.36	3.03	4.39	2.48	2.58
Cyclo-olefins	9.55	1.64	4.69	5.37	1.16
Paraffins	46.53	5.84	7.17	0.34	22.37
Indene	0.15	0.79	0.33	1.13	0.72
Carbon disulfide	0.06	0.39	0.38	0.40	0.06
Thiophenes	0.66	0.52	0.62	0.67	0.33
Others	1.87	4.71	4.49	2.50	3.19

[a] After Roberts *et al.* [20] and Claxton [21].
[b] A, low-temperature retort; B, horizontal retort; C, intermittent vertical chamber oven; D, coke oven; E, continuous vertical retort.

Table 6.2.3

**Condensible products from vacuum pyrolysis
of an exinite and vitrinite**[a]

	Exinite	Vitrinite
% Water	3.36	2.06
% Light oils	2.79	0.99
% Heavy oils	29.78	2.28
Heavy oil composition:		
% Acids	0.3	2.0
% Phenols	3.8	21.6
% Bases	1.6	6.4
% Neutral oils	90.5	70.0

[a] After Macrae [22].

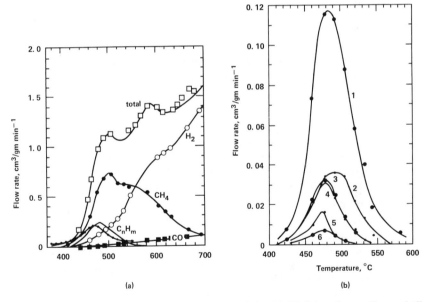

(a)   (b)

Fig. 6.2.1   Rates of gas generation during pyrolysis. (a) Vitrinite of a medium-volatile bituminous coal, heated at 1.8°C/min. (b) Hydrocarbon gases generated during pyrolysis of a medium-volatile bituminous coal at 1.8°C/min: curve 1, C_2H_2; curve 2, C_2H_4; curve 3, C_3H_8; curve 4, C_3H_6; curve 5, C_4H_8; curve 6, C_4H_{10}. (After Fitzgerald [4]; by permission of IPC Business Press Ltd.)

and differences between different macerals are most marked in lower-rank coals in which forms of oxygen play an important role in decomposition processes.

Gas that is coproduced with tar vapors or formed at higher temperatures is mostly a mixture of carbon monoxide, hydrogen, C_1 and C_2 hydrocarbons, hydrogen sulfide, and ammonia, the last being particularly significant in industrial operations in which it is usually recovered as ammonium sulfate from the aqueous liquor fraction of the condensate. But the relative amounts of these constituents vary markedly with the temperature at which the coal is pyrolyzed and are also, though to a lesser degree, influenced by the rank of the coal and by the extent to which secondary vapor-phase dealkylation, etc., of the tar contributes noncondensible products to the mixture. How gas compositions usually tend to vary with the pyrolysis temperature is shown in Fig. 6.2.1.

6.3   The Plastic Properties of Coal

Over a temperature interval that coincides approximately with active decomposition, certain coals pass through a *transient plastic state* in which they successively soften, swell, and finally resolidify into a more or less distended cellular *coke* mass. Such coals are termed *caking* coals, while those that do not become plastic on heating and therefore leave a weakly coherent or noncoherent *char* are said to be noncaking.[1]

As a general but by no means invariable rule, pronounced caking properties, which are uniquely important in carbonization practice (see Section 11.2), are only encountered among high-volatile-A, medium-volatile, and low-volatile bituminous coals (see Table 3.3.1), and tend to be particularly strongly expressed among mvb coals (with C = 86–89%). There is, however, no sharp dividing line between caking and noncaking coals; and among caking coals, behavior in the plastic range can vary widely. An appraisal of caking and related rheological properties is consequently far from simple and usually requires measurement of several fundamentally different parameters.

A broad qualitative assessment of caking propensities is nowadays most often made by means of an empirical *free-swelling test* that evolved from

---

[1] There is considerable confusion, especially in the older technical literature, with respect to the use of *caking, coking,* and *agglutinating.* In what follows, the term *caking* is reserved for coals that are found to possess plastic (or "agglutinating") properties in laboratory tests. *Coking* will be used in reference to caking coals in which plastic properties are so strongly expressed as to make them suitable for conversion into metallurgical and other industrial cokes (see Section 11.2). Coking coals are therefore, in effect, strongly caking coals.

the British crucible swelling test and was later embodied in a number of national standards [23]. As outlined in the ASTM procedure (D 720-67), this entails heating ∼1 gm of −60-mesh coal in a silica crucible to 820 ± 2°C in $2\frac{1}{2}$ min, and determining the free-swelling index (FSI) by comparing the resultant "button" with a series of standard profiles (see Fig. 6.3.1). A non-coherent, pulverulent residue is assigned an index of 0, and indices of 2 or 3 are usually taken as implying that the coal is only marginally (or weakly) caking.

An alternative method, which is used in some European countries and assesses the mechanical strength rather than the distension of a coke button, is the *Roga test* [24]. In this, ∼1 gm of −0.2-mm coal is mixed with 5 gm of −0.4-mm anthracite, compacted for 30 sec under 6 kg, and heated to 850°C in 15 min. The resultant button is then weighed (w), screened to remove −1-mm material, reweighed (a), and rotated in a 20-cm-diam drum for three 5-min periods, with −1-mm particles being removed after each. The Roga index (R) is calculated from

$$R = \frac{\frac{1}{2}(a + d) + b + c}{3} \frac{100}{w}$$

where b, c, and d are the weights of the residual button after the first, second, and third period of rotation and can assume any value between 0 and 80.

Where so desired, caking propensities can also be assessed by comparing the pyrolyzed residues from 600°C Gray–King assays with a series of "standard cokes" designated as A, B, . . . , G (see Fig. 6.3.2). Coals that

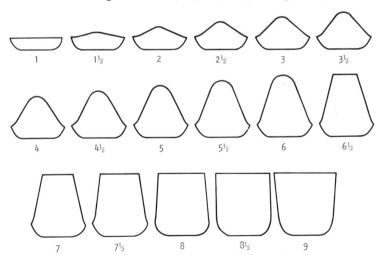

Fig. 6.3.1   Standard coke button profiles and corresponding free-swelling indices, approximately 2/3 full scale.

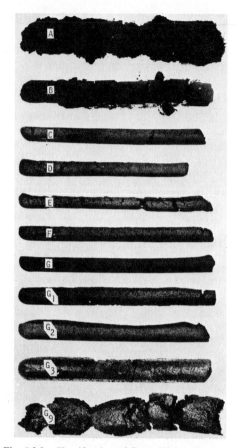

Fig. 6.3.2 Classification of Gray–King coke types.

yield residues more distended than a G coke are retested in admixture with electrode carbon, and subscripts (i.e., 1, 2, etc.) are used to indicate the number of parts of electrode carbon in 20 parts of the mixture from which a G-type residue is obtained. This procedure has now almost entirely replaced an earlier test that measured the "agglutinating value" of coal [25] by determining the maximum sand-to-coal ratio in a mixture from which a coke button capable of supporting a 100-gm weight could be made. This ratio ranged from 0 to 35.

Developed and modified in the light of experience to provide practical guides to the behavior of coal in industrial operations, the FSI, Roga index, and Gray–King coke type are now widely used as alternative classification parameters (see Section 3.3) and for laboratory screening of "candi-

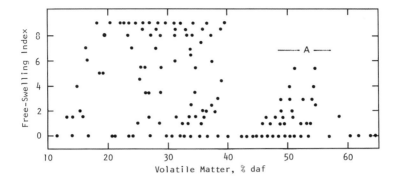

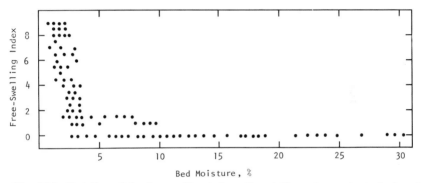

Fig. 6.3.3 Variation of the free-swelling index with volatile matter contents (top) and capacity moisture contents (bottom) of coal. The region identified by A in the upper diagram contains atypical (low-moisture) lignites from Pakistan. (After Berkowitz [26]; by permission of IPC Business Press Ltd.)

date" coals for particular end uses. Moreover, all three indices afford convenient means for correlating caking propensities with coal rank indicators. This is illustrated by Fig. 6.3.3, which shows the variation of the FSI with volatile matter and capacity moisture contents [26],[2] and by Fig. 6.3.4, which shows correlations between the FSI, Roga index, Gray–King coke type, and dilatation (see below). However, neither the FSI nor any other test of caking power offers much information about the complex rheological behavior of coal in its plastic state, and such information must be separately

---

[2] Aside from their practical value for predicting caking properties from other rank indicators, these correlations have considerable significance for theories of coal plasticity.

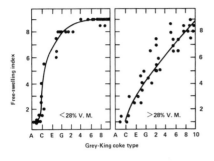

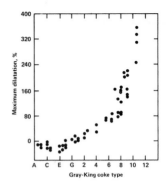

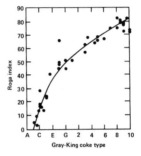

Fig. 6.3.4   Relationships between different coal plasticity (or caking) indices. (After van Krevelen and Huntjens [27]; by permission of Elsevier Publishing Company.)

sought by recourse to suitably standardized plastometric or dilatometric methods.

For evaluating the apparent *fluidity* of plastic coal, a modified constant-torque Gieseler plastometer [28] is now preferred over earlier (variable-torque) instruments [29–31], and the standard US variant has been detailed in ASTM D 1812-69. This instrument (see Fig. 6.3.5) measures the rotation of a specially designed rabble-armed stirrer in a compacted 5-gm coal charge while the coal is being heated from 300°C at 3 ± 0.1°C/min and records the fluidity in dial divisions per minute (dd/min). "Critical" points are the *softening* temperature T_s, the temperature of *maximum fluidity* T_m, and the *resolidification* temperature T_r. T_s and T_r are identified with the temperatures at which the fluidity equals 20 dd/min and the width of the plastic range is defined by $T_r - T_s$. A typical Gieseler curve is shown in Fig. 6.3.6.

Volume changes that accompany the heating of a caking coal through its plastic range are conveniently measured with a dilatometer, and several such instruments have, in fact, been specifically developed for this purpose.

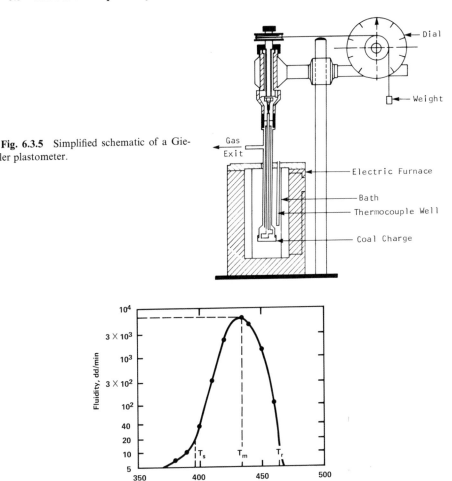

Fig. 6.3.5 Simplified schematic of a Gieseler plastometer.

Fig. 6.3.6 Typical (Gieseler) fluidity curve. T_s, T_m, and T_r denote the softening, maximum fluidity, and resolidification temperatures. T_s and T_r are conventionally set at points at which the fluidity equals 20 dd/min.

The best known and most widely used are the Audibert–Arnu[3] [32], Sheffield [33], and Ruhr [34] dilatometers. All operate on the same basic principle, i.e., they record the vertical displacement of a piston that rests on a compacted coal charge, but differ in such details as piston weight and diameter, and the degree of compaction of the charge. For measurements in the Audibert–

[3] Some steps are now being taken by the International Organization for Standardization to standardize the Audibert–Arnu dilatometer with a view to having it eventually employed as the only (or at least strongly preferred) instrument for dilatometric measurements on coal.

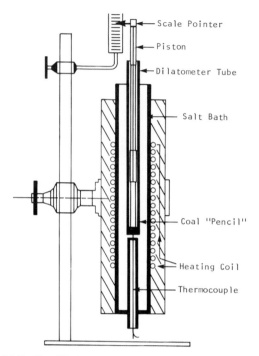

Fig. 6.3.7 Simplified schematic of an Audibert–Arnu dilatometer.

Arnu dilatometer (see Fig. 6.3.7), the coal is compressed into a slightly tapered 6.5-mm-diam, 60-mm-long "pencil," inserted into an 8-mm-i.d. metal tube, and surmounted by a 7.8-mm-diam piston that, together with its extension rod, places a 150-gm weight on the coal charge. On the other hand, in the Sheffield dilatometer, the coal is charged into a $\frac{5}{8}$-in. ($\sim$15.9-mm-) i.d. tube and compacted by dropping a 130-gm weight onto it, and the counterbalanced piston, which has a substantially smaller diameter than the tube, can penetrate the coal mass when it attains sufficiently high fluidity. Because of these variations, dilatometer curves (see Fig. 6.3.8) are only directly comparable when they are recorded with the same instrument and under identical conditions.

In principle, dimensional changes in a coal charge can be measured as functions of time by conducting tests at a series of constant temperatures, and the dilatometer curves then obtained take forms such as those shown in Fig. 6.3.9. However, in most cases interest centers more on the behavior of the coal while its temperature is being raised at a constant rate (see Fig. 6.3.8), and curves recorded under such conditions characterize the coal by

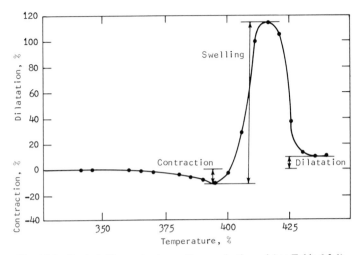

Fig. 6.3.8   Typical dilatometer trace of a per-plastic coal (see Table 6.3.1).

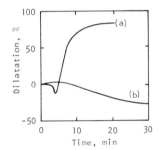

Fig. 6.3.9   Typical isothermal dilatometer traces of a low-volatile bituminous coal. These traces were recorded at (a) 426°C and (b) 463°C. (After van Krevelen *et al.* [35]; by permission of Elsevier Publishing Company.)

defining the extent of contraction (*c*), dilatation (*d*), and swelling (*s*), as well as the temperatures at which these changes begin or end.

Since *c*, *d*, and *s*, as usually measured, depend on the dimensions of the coal charge and on the diameter of the tube that contains it, their numerical values are, of course, only relative. Van Krevelen *et al.* [35] have therefore suggested that it is preferable to calculate a "true" expansion (*e*), which is of particular significance in coke oven practice where it determines the pressures that the coal will exert on oven walls, from

$$e = (v_0/v_i)(1 + d) - 1$$

In this expression, v_i is the initial volume of the coal charge and v_0 the volume under the piston at any particular time or temperature. On the assumption that e is an additive function, it is then possible to calculate dilatometer curves for binary or ternary coal blends from the traces of their individual components. In this connection, however, it should be borne in mind that coals with substantially similar caking propensities (as reflected in the FSls, Roga indices, or Gray–King coke types) can show very different dilatometric behavior. Hoffmann and Hoehne's classification of dilatometer curves (see Table 6.3.1) exemplifies this variety. For practical purposes, it is

Table 6.3.1

Classification of dilatometric behavior[a]

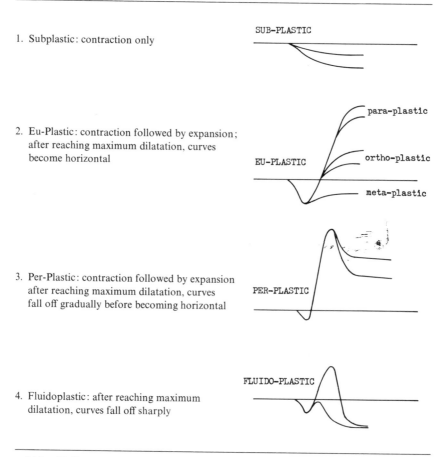

1. Subplastic: contraction only

2. Eu-Plastic: contraction followed by expansion; after reaching maximum dilatation, curves become horizontal

3. Per-Plastic: contraction followed by expansion after reaching maximum dilatation, curves fall off gradually before becoming horizontal

4. Fluidoplastic: after reaching maximum dilatation, curves fall off sharply

[a] After Hoffmann and Hoehne [36].

therefore mandatory, even when the plastic properties of a caking coal have been carefully measured in the laboratory, to test the coal in experimental ovens before actually using it for coke making. (Some experimental ovens are briefly noted in Section 11.2.)

The need for such further testing is, in any event, underscored by the extra-ordinary sensitivity of plastic properties to changes in ambient conditions and by their susceptibility to modification:

(a)   Increasing heating rates increase the maximum Gieseler fluidity and extent of swelling, and simultaneously raise the temperatures at which maximum fluidity and the onset of swelling are observed (see Figs. 6.3.6 and 6.3.8).

(b)   Preheating the coal in a protective (inert) atmosphere at temperatures as low as 200°C for extended periods of time progressively diminishes fluidity, swelling, and related caking indices.

(c)   A similar, if less drastic, diminution of all aspects of plasticity is observed when the coal is increasingly finely comminuted; and even a strongly caking coal (with FSI >6–7) will only yield a barely coherent coke button if sufficiently finely ground and very slowly heated.

(d)   Plastic properties of weakly and moderately caking coals with high ash contents (e.g., coals with FSI 3–5 and $>10\%$ ash) can be substantially enhanced by reducing the mineral matter contents of the coals (see Section 8.1).

(e)   Oxidation quickly and progressively narrows the plastic range $(T_r - T_s)$, reduces the maximum fluidity, and eventually completely destroys any caking propensities. (The fact that *slight* oxidation will occasionally seem to increase the FSI of a coal is attributed to reduction of fluidity and to the consequent ability of the coke button to expand without, at the same time, tending to collapse upon itself.)

(f)   Mild hydrogenation, which does not significantly alter the molecular structure of the coal (see Section 5.5), causes converse effects, i.e., it broadens the plastic range and increases fluidity and swelling.

(g)   Manifestations of plasticity can be suppressed by pyrolyzing the coal in vacuo or enhanced by heating it under high pressure.

This dependence of coal plasticity on ambient conditions has been utilized in industrial operations to destroy caking properties where they would be deleterious or to improve them where, as in coke making, they are critical for commercial success. But it has also created difficulties in formulating a comprehensive, self-consistent *theory of plasticity*. Although earlier views [37, 38], which envisaged homogeneous melting and concurrent decomposition of the melt, have now been discarded in favor of hypotheses that regard softening and swelling as consequences of thermal decomposition, there is still some disagreement about mechanisms (which must account for softening and swelling in individual small particles as well as in closely packed assemblies of such particles).

Contemporary views are summarized in the concept that decomposition of parts of the coal substance creates fluid material (variously referred to as primary tar or "thermobitumen") that acts as a plasticizer and softens the mass by facilitating relative motion of solid matter (Note B). As the temperature rises, this fluid material gradually escapes by evaporation and/or further thermal decomposition into volatile products and, by "bubbling" through the plastic mass, then causes it to swell; and the residue resolidifies when most of the thermobitumen has been volatilized. Schematically, this sequence can be represented by

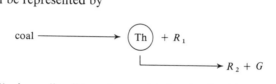

where R_1 is "primary" solid coal material that is plasticized by thermobitumen, and R_2 and G denote solid matter and gas formed by subsequent decomposition of thermobitumen. Noncaking coals would, on this basis, either form an insufficiency of thermobitumen or not retain it long enough to become fluid.

On the assumption that all transformations of solid to fluid coal and thence to a coke are first-order changes, Chermin and van Krevelen [40] have put this sequence in the form

$$dC/dt = k_1C, \qquad dM/dt = k_1C - k_2M, \qquad dG/dt = k_2M$$

where M, equivalent to Th, is termed *metaplast*, and C is the concentration of solid coal material. And if it is further assumed that fluidity is directly proportional to the metaplast concentration, this formulation does indeed lead to expressions that reflect most of the experimental observations with considerable fidelity. But it is also important to note its deficiencies. It says nothing about the physical relationship between metaplast and solid matter and consequently fails to account for the fact that the ultrafine pore structure of the coal is substantially retained throughout the plastic range (see below). It does not, in itself, make it clear why the temperature of maximum fluidity should increase with the heating rate. And it fails to explain why otherwise very similar plastic masses often show widely divergent swelling behavior.[4] It is, therefore, in many respects more helpful to suppose [42] that swelling is a predominantly intra- rather than interparticle phenomenon and governed by internal pressures that arise from disparities between rates of formation

---

[4] Significantly, there is no relationship between the FSI and the maximum Gieseler fluidity [41].

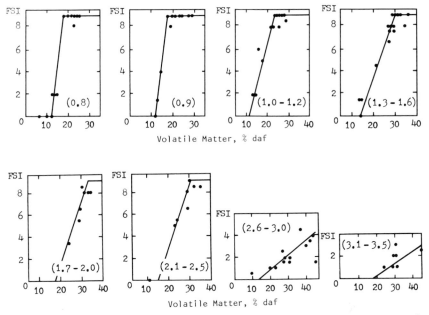

Fig. 6.3.10 Variation of the free-swelling index with volatile matter contents among coals of similar porosity. Numbers in parentheses refer to the capacity moisture contents, which are functions of porosity (see Section 4.1). (After Berkowitz [26]; by permission of IPC Business Press Ltd.)

and of disengagement of volatile material (see Section 6.1). At sufficiently low temperatures, when formation and diffusion of gas would be accompanied by formation and exudation of liquid matter, fluidity can then be associated with plasticizer-assisted mobility of molecules, molecular aggregates, and small particles.

An important feature of this hypothesis is that it identifies caking coals with highly compacted, low-porosity coals, and this is borne out by the fact that swelling indices increase rapidly (from zero) as capacity moisture contents fall below ~5% (see Fig. 6.3.3; Note C). Also significant in this regard is the observation [26] that the free-swelling indices of coals with approximately equal moisture contents (and, hence, similar porosities) vary directly with their volatile matter contents (see Fig. 6.3.10).

6.4 Solid Products of Thermal Decomposition

As active thermal decomposition proceeds and is followed by secondary degasification, the solid residues are progressively aromatized and homogenized by growth of pseudographitic "rafts" via reactions of the form

and heating to increasingly higher temperatures therefore gradually obliterates differences between initially quite dissimilar coals. After exposure to temperatures in the order of 800 or 900°C, chars and cokes, whatever their source, differ from each other in the main only in such matters as appearance and cohesive strength.[5]

This convergence to common properties is exemplified by the manner in which the elemental compositions of chars and cokes from different coals vary with the heat treatment temperature. So long as that temperature lies below the onset of active decomposition (as defined by T_d), changes in elemental composition are similar to those caused by metamorphic development (see Section 1.2) and shift the solid toward slightly higher carbon contents without significantly reducing its hydrogen content. Above T_d, however, pyrolysis is accompanied by extensive loss of hydrogen as well as of oxygen; and when plotted into a %C versus %H or a H/C versus O/C diagram, elemental compositions are then found to move *out* of the coal band and to shift directly toward the origin of the coordinate system. Figure 6.4.1 illustrates these shifts and the consequent increasing similarity of residues from different coals.

Since active decomposition creates molecular structures that can order themselves better than their precursors, reflectances (R_0) and densities of pyrolyzed residues also begin to increase quite regularly as T_d is passed—although with respect to densities it is important to note the spurious decreases in high-temperature chars and cokes that arise from diminishing accessibility of the pore systems of these solids. In view of these effects, it is sometimes helpful to *calculate* the helium density of the char or coke, and according to Agrawal [43], this can be done by using

$$\frac{1}{D} = \frac{1}{D_0} + \frac{0.443 - (1/D_0)}{100 - C_0}(C - C_0)$$

[5] As already noted (see Section 6.3), these aspects depend on the caking propensities of the precursor coals and on the conditions under which they are pyrolyzed.

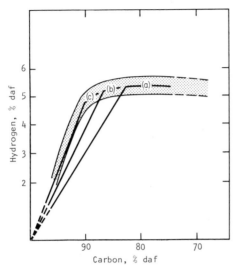

Fig. 6.4.1 Effects of progressive pyrolysis on the carbon and hydrogen contents of coal chars. In this schematic (a), (b), and (c) refer to a lignite, a high-volatile bituminous coal, and a low-volatile bituminous coal, respectively. The shaded band represents Seyler's coal band, and the breaks in the curves are, typically, roughly coincident with the decomposition temperatures T_d of the coals.

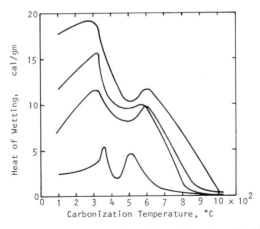

Fig. 6.4.2 Variations of heats of wetting of coal chars (in methanol) with carbonization temperatures. (After Cannon *et al.* [44].)

where D_0 and C_0 are the helium density and carbon content of the coal, and D and C the corresponding values for the pyrolyzed residue.

Properties that directly reflect growth and ordering of carbon "rafts" do not change significantly until temperatures of 500 to 550°C are exceeded (Note D). Thermal and electrical conductivities thus begin to increase, and the electron excitation energy to fall, only after heat treatment of the residues at ~ 600°C. But particularly noteworthy—and of great practical significance —is the fact that the ultrafine pore structure of the parent coal is substantially retained until pyrolysis has been carried to ~ 650–700°C, and that accessible surface areas only decrease sharply in *high*-temperature chars and cokes. This preservation of porosity, which is illustrated in Fig. 6.4.2, is as characteristic of strongly caking coals (which become "fluid" in their plastic ranges) as of noncaking coals [44] and makes it evident that coal neither "melts" nor becomes dispersed in a continuous fluid phase when it attains plasticity. It also accounts for the fact that the reactivities of low-temperature chars and cokes are not greatly different from those of their respective precursor coals.

Notes

A. Between 100 and 180°C the only detectable processes are loss of capillary-condensed moisture and discharge of occluded gases (such as methane and carbon dioxide).

B. An analogy might be seen in hand specimens of oil sand [39], which become relatively fluid and tend to flow when the temperature is raised sufficiently to liquefy the bitumen that surrounds each sand grain.

C. As noted in Section 4.1, the capacity moisture contents or, for that matter, any relatively constant fractions thereof (such as air-dried moisture contents) are directly proportional to coal porosities. It should also be observed that the relationship shown in Fig. 6.3.3 holds even for otherwise "abnormal" coals, such as lignites with highly developed caking propensities [26].

D. This is consistent with the postulated growth mechanism, since evolution of elemental hydrogen does not become prominent until 500 to 550°C is passed (see Fig. 6.2.1).

References

1. S. K. Chakrabartty and N. Berkowitz, American Chemical Society Fuel Chemistry Division, 174th Meeting, Chicago, Illinois (August 1977).
2. D. W. van Krevelen, C. van Heerden, and F. J. Huntjens, *Fuel* **30**, 253 (1951).
3. C. Kröger and A. Pohl, *Brennst. Chem.* **38**, 102 (1957).
4. D. Fitzgerald, Arthur Duckham Fellowship Rep., 8th, Publ. No. 516. Institute Gas Engineers, London (1957); D. Fitzgerald and D. W. van Krevelen, *Fuel* **38**, 17 (1959).
5. P. Hanbaba, H. Jüntgen, and W. Peters, *Brennst. Chem.* **49**, 369 (1968).
6. H. A. G. Chermin and D. W. van Krevelen, *Fuel* **36**, 85 (1957).
7. N. Berkowitz and W. J. Mullin, *Fuel* **47**, 63 (1968); N. Berkowitz and W. den Hertog, *ibid.* **41**, 507 (1962); L. F. Neufeld and N. Berkowitz, *ibid.* **43**, 189 (1964).
8. E. A. Shepatina, V. V. Kalyuzhnii, and Z. F. Chukhanov, *Dokl. Akad. Nauk USSR* **72**, 869 (1950); *Chem. Abstr.* **44**, 10294 (1950).
9. N. Berkowitz, *Proc. Symp. Nature Coal* p. 284. Central Fuel Research Institute, India, 1959; *Fuel* **39**, 47 (1960).

10. F. Fischer and H. Schräder, *Z. Angew. Chem.* **33**, 172 (1920).
11. J. B. Goodman, M. Gomez, V. F. Parry, and W. S. Landers, U. S. Bureau Mines Bulletin No. 530 (1953); J. B. Goodman, M. Gomez, and V. F. Parry, U. S. Bureau Mines Rep. Invest. No. 5383 (1958).
12. P. O. Krumin, Ohio State Univ. Studies, Engineering Experimental Station Bulletin 165, p. 51 (1957).
13. T. Gray and J. G. King, DSIR Fuel Research Tech. Paper No. 1 (1921); J. G. King and L. J. Edgecombe, *ibid.* No. 24 (1930).
14. A. C. Fieldner and J. D. Davis, U. S. Bureau Mines Monograph 5 (1934).
15. H. Guérin and R. Margel, *C. R. Acad. Sci. Paris* **231**, 1502 (1950); H. Guérin and P. Marcel, *Bull. Soc. Chim. Fr.* 1212 (1956).
16. B. Sun, C. H. Ruof, and H. C. Howard, *Fuel* **37**, 299 (1958); J. K. Brown, I. G. C. Dryden, D. H. Dunevein, W. K. Joy, and K. S. Pankhurst, *J. Inst. Fuel* **31**, 259 (1958).
17. W. A. Selvig and W. H. Ode, U. S. Bureau Mines Bulletin No. 571 (1957).
18. V. F. Parry, U. S. Bureau Mines Rep. Invest. No. 3482 (1939).
19. W. Francis, "Coal," 2nd ed., p. 424. Arnold, London, (1961).
20. A. L. Roberts, J. H. Towler, and B. H. Holland, Research Comm. GC 31, Gas Council, Council, Britain, 1956.
21. G. Claxton, "Benzoles: Production and Uses." National Benzole and Allied Products Assoc., London, 1961.
22. J. C. Macrae, *Fuel* **22**, 117 (1943).
23. ASTM D 720-67, American Society for Testing and Materials, Philadelphia, Pennsylvania; BS 1016, 1942, Britain; N. F. M 11-001, 1944, France (foreign standards available from American National Standards Institute, New York.
24. B. Roga, *Z. Oberschles. Berg. u. Hüttenmänn. Ver. Kattowice* **74**, 565 (1931).
25. L. Campredon, *C. R. Acad. Sci. Paris* **121**, 820 (1895); A. A. Meurice, *Chal. Ind.* **4**, 45 (1923).
26. N. Berkowitz, *Fuel* **29**, 138 (1950).
27. D. W. van Krevelen and F. J. Huntjens, *C. R. Congr. Int. Chim. Ind., 31st, Liège* T. 1.426 (1958).
28. K. Gieseler, *Glückauf* **70**, 178 (1934).
29. J. D. Davis, *Ind. Eng. Chem. Anal. Ed.* **3**, 43 (1931).
30. G. Echterhoff, *Erdöl Kohle* **8**, 294 (1955).
31. A. F. Boyer and J. Lahouste, *Rev. Ind. Minér.* **35**, 1107 (1954).
32. E. Audibert, *Rev. Ind. Miner.* **6**, 115 (1926); cf. also U. N. Publ. II. E.4, E/ECE/247 (E/ECE/Coal/110) (1956); ISO/TC/27, Doc. 221 (1958).
33. R. G. Davies and R. A. Mott, *Fuel* **12**, 294 (1933); R. A. Mott and C. E. Spooner, *ibid.* **16**, 4 (1937).
34. H. Hoffmann, *Öl u. Kohle* **40**, 531 (1944).
35. D. W. van Krevelen, H. A. G. Chermin, H. N. M. Dormans, and F. J. Huntjens, *Brennst. Chem.* **39**, S 77 (1958).
36. H. Hoffmann and K. Hoehne, *Brennst. Chem.* **35**, 202, 236, 269, 298 (1954).
37. E. Audibert, *Fuel* **5**, 229 (1926).
38. A. Gillet, *Nature (London)* **167**, 406 (1951).
39. N. Berkowitz and J. G. Speight, *Fuel* **54**, 138 (1975).
40. A. G. Chermin and D. W. van Krevelen, *Fuel* **36**, 85 (1957).
41. S. K. Chakrabartty and N. Berkowitz, *Fuel* **51**, 44 (1972).
42. N. Berkowitz, *Fuel* **28**, 97 (1949).
43. P. L. Agrawal, *Proc. Symp. Nature Coal* p. 121. Central Fuel Research Inst., India, 1959.
44. C. G. Cannon, M. Griffith, and W. Hirst, *Proc. Conf. Ultrafine Struct. Coals Cokes* p. 131. BCURA, London, 1944.

CHAPTER 7

THE ACTION OF SOLVENTS ON COAL

Development of solvent extraction as a technique for investigating coal compositions and producing waxes, resins, or other coal derivatives of potential commercial value dates from Bedson's discovery in 1902 that bituminous coals are substantially soluble in hot pyridine [1] and has since then spawned a voluminous technical literature that has been reviewed by Bakes [2], Kiebler [3], Dryden [4, 5], and Kröger [6].

Much of the early work centered on efforts to isolate a presumed *coking principle*, i.e., substances then believed to be responsible for caking properties, and is now only of interest because it demonstrated the occasional presence in coal of small amounts of loosely associated material that differed from the bulk of the coal substance. An example is the classic series of studies by Wheeler *et al.* [7, 8] who exhaustively extracted coal with boiling pyridine and fractionated the regenerated soluble solids by sequential selective extraction in schemes such as that shown in Fig. 7.0.1. Subsequent analyses showed that the γ_1-fraction was mostly comprised of hydrocarbons (e.g., paraffins,

158

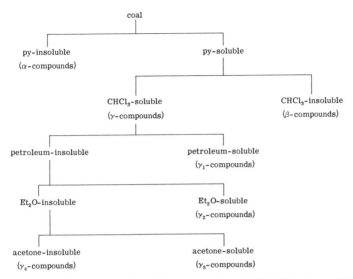

Fig. 7.0.1 Classic fractionation of pyridine extracts of coal. (After Wheeler *et al.* [7, 8].)

naphthenes, and terpenes), while the γ_2-, γ_3-, and γ_4-compounds were resin-like substances with $C = 80-89\%$, $H = 8-10\%$, and also revealed an unexpected close chemical similarity between the insoluble residue and the β-compounds (see Section 7.3).

But from the 1920s on, increasing attention was also directed to solvolysis, i.e., solvent extraction under conditions in which the coal suffers incipient or active thermal decomposition [9, 10], and this laid the foundations for much of the modern work on solvent refining of coal (see Section 14.2) and hydrogenation in H donor systems (see Section 5.5).

7.1 Specific and Nonspecific Solvents

Except under solvolytic conditions (see Section 7.4), common organic solvents, such as benzene and alkyl benzenes or methanol, acetone, chloroform, and diethyl ether, dissolve little of the coal material per se and as a rule only extract waxes and/or resins that are sometimes occluded in the coal "matrix" in the form of resin bodies or present as macroscopic flakes. *Substantial* extract yields, often as high as 35–40%, can only be obtained by using pyridine, certain heterocyclic bases, or primary aliphatic amines (which may, but need not, contain aromatic or hydroxyl substituents). Secondary and tertiary aliphatic amines tend to be much less effective, apparently because more than one alkyl group on the amine presents serious steric hindrance to interaction between the solvent and coal.

Dryden [11] has pointed out that the potency of effective coal solvents can be associated with the existence of an unshared electron pair on a nitrogen or oxygen atom in the molecule (which makes the solvent behave as a polar fluid) and designated them as *specific* (coal) solvents (Note A). But correlations derived from solution theory possess only partial validity when applied to coal: Some ineffective (i.e., "nonspecific") solvents, of which methanol is an example, can swell coal almost as much as the most potent specific ones. The approximately linear relationship between extract yields and the internal pressures of the solvents ([3, 12] breaks down when the solvents are used at temperatures below their normal boiling points [11]. And there are no discernible connections between solvent power and the dipole moment, dielectric constant, and surface tension of the solvent, such as have been observed with respect to solubility of high polymers [13].

Also, certain solvent *pairs* have been found to possess much greater potency than either solvent used alone. Examples are a 70:30 benzene–ethanol mixture and equimolecular mixtures of acetophenone–methyl-formamide or methylcyclohexanone–dimethylformamide [14].

At temperatures at which the coal thermally decomposes, the most effective solvents are α- or β-naphthols, tetralin, anthracene, and phenanthrene. At atmospheric pressure these produce from bituminous coals extract yields that vary directly with the boiling point of the solvent; but this relationship does not hold for subbituminous coals which are generally much less soluble in α-naphthol, phenanthrene, etc., than in β-naphthol. According to Orchin *et al.* [15], aromatics with angular ring systems (e.g., phenanthrene) are particularly good solvents for high-volatile bituminous coals.

7.2   Extract Yields

When contacted by a solvent or its hot vapor, all but the most mature coals will imbibe fluid, swell, and, depending on circumstances, disintegrate and disperse as well as dissolve. Extract yields depend, therefore, as much on coal composition and extraction procedures as on the nature of the solvent.

Other matters being equal, yields of soluble material obtained by extraction with specific solvents decrease with increasing rank and become negligible when carbon contents exceed 90%. Figure 7.2.1 illustrates this with data for ethylenediamine [16]. However, actual variations of extract yields with rank also depend, to some extent, on the particular solvent. With pyridine and piperidine, yields change more slowly than Fig. 7.2.1 suggests; with ethanolamine they begin to fall steeply sooner (at 84–85% rather than

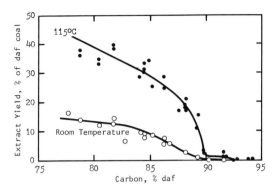

Fig. 7.2.1 Ethylenediamine extraction of coal at room temperature and at the boiling point of the solvent. Variation of extract yields with coal rank. (After Dryden [16]; by permission of IPC Business Press Ltd.)

at $\sim 87\%$ carbon); and pyridine extract yields have been reported to pass through a broad maximum between 80 and 85% carbon [16].

Solubilities in nonspecific solvents such as benzene or benzene–ethanol mixtures appear to follow qualitatively similar courses, except that extract yields are very much smaller ($<5\%$ even for low-rank coals) and therefore grossly affected by the presence or absence of adventitious fossil waxes and resins (see Section 7.3). Yields approaching those obtained with specific solvents accrue only when extraction is carried out under relatively severe solvolytic conditions.

But, particularly when assessed with specific solvents, which magnify the effect by virtue of producing fairly high extract yields, coal solubilities are also influenced by petrographic compositions. Exinites and vitrinites are, as a rule, much more readily soluble than inertinites, and fusinites are virtually insoluble. Predominantly "bright" coals, rich in vitrinites, will therefore deliver substantially greater extract yields than "dull" coals of similar rank. These differences tend, however, to diminish with increasing rank and to disappear among coals with $>91\%$ carbon as macerals lose their identities (see Section 2.1).

The other factors bearing on extract yields have to do with extraction conditions.

Below the onset of active thermal decomposition of the coal, extract yields vary directly with extraction temperature. This effect is usually most pronounced with nonspecific solvents [17], but even as potent a solvent as ethylenediamine will deliver almost three times as much extract from bituminous coal at its normal boiling point (115°C) as at room temperature [18]. Extract yields obtained with a series of primary aliphatic amines have, in fact, been found, as a first approximation, to vary with the extraction temperature rather than with any other solvent property [19].

However, even near the boiling point of a solvent, the process of coal dissolution is protracted unless there is direct contact between coal and solvent. Soxhlet extraction thus proceeds much more slowly than extraction by shaking the coal with the simmering solvent, and essentially complete extraction depends on using suitably comminuted (−72-mesh or finer) coal. As would be expected from the relationship between coal porosity and rank (see Section 4.1), the effect of particle size on extract yields and extraction rates is greatest among bituminous coals.

7.3   The Composition of Coal Extracts

Like extract yields, extract compositions depend on the particular coal/ solvent system and on extraction conditions.

Near their normal boiling points, *nonspecific* solvents tend, as already noted, to extract coal selectively and to dissolve primarily waxy and resinous substances that, although derived from the primordial plant debris, are not integral components of the coal substance. For the most part, the resultant extracts are therefore amenable to further resolution and will then yield chemically identifiable products. Extraction of peats and lignites with benzene or benzene–methanol mixtures, followed by fractionation of the crude extract with diethyl ether or hot ethyl alcohol, thus yields valuable montan waxes and resins and has, in fact, been commercially used for their production. The waxes consist mostly of aliphatic long-chain (C_{24}–C_{32}) acids and their esters, and of C_{24}–C_{32} alcohols, while the resins, which resemble colophy or kauri gums, are derivatives of mono- and dibasic C_{12}–C_{20} acids with aromatic/hydroaromatic nuclei (e.g., abietic acid). Typical elemental compositions and properties of the two fractions are shown in Table 7.3.1.

As coal rank increases, the amounts of extractable waxes and/or resins fall quickly; and since nonspecific solvents can also take small quantities of undifferentiated coal-like material into solution, bituminous coal extracts are generally less "specialized" (as well as smaller). However, even these extracts will reflect solvent selectivity in possessing higher H/C ratios than the parent coal and absorbing much more selectively in the infrared.

In contrast, *specific* solvents, which dissolve a large fraction of the coal, do so nonselectively; and while different solubilities of petrographic constituents can impact on the compositions of "whole" coal extracts, the extracted material almost always resembles the insoluble residue so closely as to be virtually indistinguishable from it. Unless present in exceptionally large amounts, "specialized" matter (such as waxes and resins) is entirely

Table 7.3.1

Montan waxes and resins

	Waxes[a]	Resins
% carbon	80.5	80.0
% hydrogen	13.0	11.0
% oxygen	6.5	9.0
Melting point, °C	80	65–70
Acid number	23	33
Saponification number	95	76
Ester number	72	43

[a] Wax:resin ratios have been reported to range from a high of 5.2 to as little as 0.25 [20].

masked by the bulk of undifferentiated coal "substance"; and the elemental compositions of the extracts, as well as those of its properties as are associated with the chemical aspects of the dissolved material, e.g., magnetic and spectroscopic properties, are all almost identical with those of the original coal.

In practice, the extracts are therefore as intractable as coal and do not to any significant degree facilitate studies of coal structure.

It seems reasonable to attribute this near-identity of extracted matter and residue to the fact that specific solvents *disperse* coal material as well as take it into true (molecular) solution. Although the average molecular weights of pyridine and similar coal extracts have been reported [21] as no greater than 1200, and γ-fractions of pyridine extracts were found to have molecular weights of less than 600, there is persuasive evidence that such extracts are, in fact, colloidal suspensions. There are obvious difficulties in differentiating between such suspensions and "true" solutions of high polymers, especially since any test of optical clarity depends on the resolution that the optical equipment can achieve. It is, however, significant that ethylenediamine extraction of coal at room temperature produces solutions in which "dissolved" matter exists mostly as discrete particles with diameters in the hundreds of angstroms [22] and thus appears to have molecular weights in the order of 10^6. Also noteworthy is that pyridine (and similar) extract solutions, when allowed to stand under nitrogen as protection against atmospheric oxidation, commonly form light precipitates within hours or days.

These observations and the close direct relationship between extract yields and the quantity of solvent that the coal will imbibe [16, 23] lend support to the view [24] that coal is built up of micelles that vary in size and strength of mutual bonding rather than in chemical composition. Provided that the coal is first sufficiently swollen by the solvent, the smaller and less tightly held micelles would then be removed by relatively mild extraction, and more potent solvents and/or higher extraction temperatures would release progressively larger micelles. Dryden [25] has discussed this concept in some detail.

7.4 Solvolysis

Solvolysis refers to the action of solvents on coal at temperatures at which the coal substance decomposes and, in practice, relates in particular to extraction at temperatures between 200 and 400°C (Note B).

Enhanced solubility at these temperatures (and the corresponding solvent equilibrium pressures) was observed as early as 1916 by Fischer and Gluud [27], who extracted subbituminous and bituminous coals with benzene at 250°C and then recorded extract yields some 10–20 times greater than those obtainable by Soxhlet extraction at the normal boiling point of benzene. But more significant was the subsequent demonstration [28, 29] that anthracene oil and tetralin at 400°C could each take almost *all* of the coal substance into (colloidal) solution. This led soon afterward to development of a commercial extraction process (see Section 14.1) and also forms the basis of recent work on solvent refining of coal (see Section 14.2).

Since the nature and extent of pyrolytic processes in coal depend on temperature, the yield and, to a lesser degree, the composition of soluble matter extracted under solvolytic conditions also vary with extraction temperatures. If a suitable solvent (such as anthracene oil) is used, extract yields from bituminous coal can be tripled by raising the temperature from 200 to 250°C and nearly tripled again by raising it to 350°C, and as extraction temperatures rise, the extracts come to increasingly resemble the asphaltene-like products obtained from pyrolysis of coal in short-path "molecular stills" [30].

As in milder forms of solvent extraction, however, yields are also always critically dependent on the choice of solvent. At temperatures below ∼350°C, solvent potency appears to be solely determined by the ability of the solvent to swell, peptize, and promote depolymerization of the coal; and the most effective solvents are aromatics and hydroxylated aromatics (such as phenol, naphthalene, α- and β-naphthols, phenyl phenols, anthracene, or phenanthrene). But above ∼350°C, where the coal actively decomposes and thereby generates free radicals that will tend to recombine into relatively high-molecular-weight species unless otherwise stabilized, a large extract yield— approaching 100% of the coal substance—accrues only from the use of a solvent that can donate hydrogen to the coal.[1] Such solvents include, in

[1] Formation of large amounts of soluble matter under severe solvolytic conditions can therefore be identified with the initial stages of coal hydrogenation in H donor systems (see Section 5.5).

addition to tetralin and anthracene oil, 5-hydroxy-tetralin and *o*-cyclohexyl-phenol. Compounds that do *not* donate hydrogen, e.g., naphthalene, cresol, diphenyl, or *o*-phenylphenol, will generally only dissolve 20–30% of the coal at 400°C [31].

Notes

A. It should be emphasized that the designation of a solvent as a *specific solvent* refers only to its potency for coal and not to its ability to extract (as, e.g., benzene does) a particular coal constituent. As noted in Section 7.3, specific solvents are, in fact, singularly nonselective, in the sense that they extract substances that are chemically almost or wholly identical with the insoluble residue.

B. Van Krevelen *et al.* [26] have used the term "extractive disintegration" to describe solvent extraction of coal at temperatures between 200 and 400°C. At temperatures above 400°C, pyrolytic condensation reactions in the coal make it progressively less soluble, and extract yields will again fall quickly.

References

1. P. P. Bedson, *J. Soc. Chem. Ind.* **21,** 241 (1902).
2. W. E. Bakes, Department of Scientific and Industrial Research, Britain, Fuel Research Tech. Paper No. 37 (1933).
3. M. W. Kiebler, *in* "Chemistry of Coal Utilization," Vol. 1, p. 715. Wiley, New York, 1945.
4. I. G. C. Dryden, *Fuel* **29,** 197, 221 (1950).
5. I. G. C. Dryden, *in* "Chemistry of Coal Utilization," Suppl. Vol. p. 237. Wiley, New York, 1963.
6. C. Kröger, *Erdöl Kohle* **9,** 441, 516, 620, 839 (1956).
7. R. V. Wheeler and M. J. Burgess, *J. Chem. Soc.* 649 (1911); D. T. Jones and R. V. Wheeler, *ibid.* 313 (1915); 707 (1916).
8. C. Cochram and R. V. Wheeler, *J. Chem. Soc.* 700 (1927); 854 (1931).
9. W. A. Bone and R. J. Sarjant, *Proc. R. Soc. London Ser. A* **96,** 119 (1920); W. A. Bone, A. R. Pearson, and R. Quarendon, *ibid.* **105,** 608 (1924); W. A. Bone, L. Horton, and L. J. Tei, *ibid.* **120,** 523 (1928).
10. S. W. Parr and H. F. Hadley, *Fuel* **4,** 31, 49 (1925).
11. I. G. C. Dryden, *Fuel* **30,** 39 (1951); *Chem. Ind.* (*London*) **30,** 502 (1952).
12. J. M. Pertierra, *Anal. Fis. Quim.* (*Madrid*) **37,** 58 (1941).
13. W. Ostwald and H. Ortlov, *Kolloid Z.* **59,** 25 (1932).
14. S. M. Rybicka, *Fuel* **38,** 45 (1959).
15. M. Orchin, C. Golumbic, J. B. Anderson, and H. H. Storch, U. S. Bureau Mines Bull. No. 505 (1951); C. Golumbic, J. B. Anderson, M. Orchin, and H. H. Storch, U. S. Bureau Mines Rep. Invest. No. 4662 (1950).
16. I. G. C. Dryden, *Fuel* **30,** 217 (1951).
17. M. K. D'yakova and V. V. Surovtseva, *Zh. Prik. Khim.* (*Leningrad*) **28,** 65 (1955).
18. I. G. C. Dryden, *Nature* (*London*) **162,** 959 (1948); **163,** 141 (1949).
19. I. G. C. Dryden, *Fuel* **30,** 145 (1951).
20. H. Steinbrecher, *Braunkohle.* **12,** 40 (1926).
21. W. F. K. Wynne-Jones, H. E. Blayden, and F. Shaw, *Brennst. Chem.* **33,** 201 (1952).
22. L. Kann, *Fuel* **30,** 47 (1951).

23. I. G. C. Dryden, *Nature (London)* **166,** 606 (1950).
24. J. K. Brown, *Fuel* **38,** 55 (1959).
25. I. G. C. Dryden, *Fuel* **31,** 176 (1952).
26. A. P. Oele, H. I. Waterman, M. L. Goedkoop, and D. W. van Krevelen, *Fuel* **30,** 169 (1951).
27. F. Fischer and H. Gluud, *Ber. Dtsch. Chem. Ges.* **49,** 1460 (1916).
28. A. Gillet and A. Pirlot, *Bull. Soc. Chim. Belge* **41,** 511 (1932); **44,** 504 (1935); **47,** 518 (1938); **51,** 23 (1942).
29. A. Pott and H. Broche, *Glückauf* **69,** 903 (1933).
30. H. C. Howard, *in* "Chemistry of Coal Utilization," Suppl. Vol., p. 357. Wiley, New York, 1963.
31. A. C. Fieldner and P. M. Ambrose, U. S. Bureau Mines Information Circ. 7446 (1948).

UPGRADING, HANDLING, AND PROCESSING OF COAL

CHAPTER 8

BENEFICIATION: CLEANING, DRYING, AND BRIQUETTING

Strictly speaking, *beneficiation* means *upgrading* and can therefore be applied to any form of processing that improves the quality of a coal or makes it easier to handle, transport, and store. In most instances, however, the term is used in a more restricted sense to mean cleaning and/or drying (Note A), and where demand for *formed fuel* exists, it may also refer to briquetting.

8.1 Cleaning

Whether or how extensively a coal is cleaned before use depends entirely on the purposes for which it is mined and on the specifications it must meet in the marketplace. But some reduction of the mineral matter (or "ash") content of the raw ("as mined") coal is generally mandatory if the coal is intended for coke making (see Section 11.2) or if it contains relatively large amounts of inorganic sulfur. In the former case, cleaning may serve to improve the caking propensities of the coal (see Section 6.3),[1] and in the latter, it can make the coal an environmentally acceptable boiler fuel where it would otherwise be barred. If the coal must be transported to distant markets, prior removal of inorganic matter may also bring economic benefits through a lowering of the effective transportation costs per ton of usable carbon or per million Btu (1 Btu = 1.055 kJ).

Because of large density differences between the organic coal substance and mineral matter associated with it (see Section 4.2), cleaning is normally (and in industrial operations, always) effected by gravity separation methods. "Chemical" cleaning, to which brief reference is made later, is still only experimental. But how readily or completely a coal can be cleaned in this manner depends on how the mineral matter is distributed in it. In autochthonous coals, which developed at or near the growth site of the parent vegetation (see Section 1.1), inorganic material is for the most part concentrated along bedding planes and in major fractures, and such material can normally be quite easily separated. On the other hand, in allochthonous coals, which formed from transported plant debris, much of the mineral matter consists of variously altered silts that are colloidally dispersed throughout the coal substance; and to the extent that it is so disseminated, it is virtually impossible to remove by physical means (Note B). The amenability of a coal to cleaning is consequently a characteristic that must, like any other intrinsic coal property, be specifically determined.

Regardless of the scale on which they are conducted, determinations of washability are made by float-and-sink tests, i.e., by measuring the weight percentages of the coal sample that will float on liquids of different specific gravities in the range 1.2–2.2. Table 8.1.1 lists some of the more common fluids used for this purpose. Where only preliminary information about washability is required, it suffices to determine the ash contents of the "floats" fraction and to plot a yield-versus-ash-contents curve (see Fig. 8.1.1). However, where it is necessary to predict the gravity at which the coal would have to be cleaned in order to contain a specified percentage of ash, and to obtain information about "middlings" fractions (i.e., near-gravity material that will remain in suspension at a particular specific gravity), it is generally better to measure and analyze floats *and sinks*. This allows calculation of middlings fractions from $100 - (f + s)$, and the construction of a washability curve of the type illustrated in Fig. 8.1.2. Here, curve 1 shows how the ash contents of the cleaned coal vary with float yields; curve 2 relates these yields to the specific gravities at which the separations were

---

[1] To some extent, this improvement is also due to simultaneous enrichment of the clean coal in reactive macerals.

Table 8.1.1

Common media for float-and-sink tests

		Specific gravity
Aqueous solutions of inorganic salts		
Calcium chloride	40%	1.40
Zinc chloride	35%	1.35
	38%	1.40
	46%	1.50
	52%	1.60
	61%	1.75
Calcium nitrate	45%	1.42
Potassium carbonate	50%	1.54
Zinc oxide	75%	2.40
Organic liquids		
Carbon tetrachloride		1.60
CCl_4/benzene mixtures		1.20–1.60
CCl_4/bromoform mixtures		1.60–2.40
Suspensions		
Pulverized barytes in water, stabilized with clay		Up to 1.80

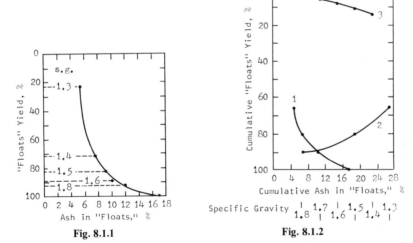

Fig. 8.1.1   **Fig. 8.1.2**

Fig. 8.1.1 Typical (washed coal) yield-versus-ash-contents curve. (By permission of J. Wiley and Sons Inc.)

Fig. 8.1.2 Typical coal washability curve. At, e.g., a specific gravity of 1.5, a floats yield of $\sim 80\%$ with an ash content of $\sim 6.5\%$ would be obtained, and the middlings fraction would amount to $\sim 11\%$. At a specific gravity of 1.3, the floats yield would be $\sim 65\%$, the ash contents of the floats would be $\sim 4.5\%$, and the middlings fraction would approach 20%. (By permission of J. Wiley and Sons Inc.)

effected; and curve 3 gives the percentages of middlings at these specific gravities. Thus, by proceeding horizontally from a cleaning specific gravity x on curve 2 to curve 1, the corresponding yield y and ash content A of the floats can be read off; and by moving vertically from x to curve 3, the quantity z of middlings at the particular specific gravity can be found.

An alternative form of presenting washability data is the so-called M or mean value curve [1]. Although superficially more complex to construct than Fig. 8.1.2, this is particularly useful for describing the washability characteristics of binary and ternary coal blends.

In interpreting results of float-and-sink tests, it must, however, always be borne in mind that separation of "clean" coal from high-ash "discard" material is critically affected by the size consist of the raw coal sample. Where washability characteristics are to be assessed with a view to establishing design and operating parameters for a cleaning plant, it is therefore necessary to either

(a)   conduct float-and-sink analyses on a series of closely sized samples of the coal, e.g., 4×1 in., $1 \times \frac{1}{4}$ in., and $\frac{1}{4} \times 0$, and to compute a prorated composite from the projected size consist of the feed coal to the plant; or

(b)   make use of distribution curves [1, 2] for other variously sized coals.

The latter are plots of distribution coefficients $\bar{c}$ versus cleaning specific gravities, with $\bar{c}$ defined as the weight percentage of the total raw coal sample that separates as "clean" coal in a limited specific gravity range. Thus, if f and s, respectively, are the weight percentages of raw coal that appear as floats and sinks between specific gravities 1.40 and 1.50.

$$\bar{c}_{1.45} = 100f/(f + s)$$

The specific gravity at which $\bar{c} = 50\%$, i.e., at which equal weights of raw coal appear in floats and sinks, is the so-called *Tromp cut point*.

According to Symington *et al.* [3], distribution curves are characteristic of the size distribution of the coal and the cleaning equipment, but otherwise virtually independent of the nature of the coal. While this contention appears to be borne out by practical experience with western European and North American Carboniferous coals—and distribution curves have sometimes served as bases for performance guarantees on plant equipment—it is still doubtful whether such curves can, in fact, be directly transposed to coals in which mineral matter may be quite differently disseminated.

For commercial coal cleaning, jigs, heavy media cells (see Note C), or cleaning tables are most commonly used, and flotation methods are, as in mineral dressing practices, generally reserved for processing coal *fines*, i.e., < 10-mesh particles.

In *jigs*, the coal is subjected to upward and downward water currents as it moves across an inclined perforated screen. This causes the coal, regardless of its size consist, to stratify and the lower specific gravity ("clean") material to pass to the surface. Different jig designs, some capable of cleaning coal down to 48 mesh, have been described in several authoritative articles (e.g., Bird and Mitchell [4], Yancey and Greer [5], Citron [6]); and Massmann [7] has discussed the more important factors that govern jig performance.

In *heavy media processes*, separation of "clean" coal from a "discard" is effected by circulating an appropriately concentrated suspension (usually magnetite in water) through a vessel to which raw coal is continuously admitted. Clean coal and discard are extracted near the top and bottom of the vessel, respectively, and magnetite in each of the two fractions is removed on rinsing screens. Recovered magnetite is then freed from residual fine coal in magnetic separators before being recycled. Suitable designs of heavy media cleaning cells for coal have been discussed by Geer *et al.* [8] and Whitmore [9].

Cleaning tables make use of the fact that heavy particles carried by flowing water will settle faster than light ones, and are therefore designed to separate clean coal from discard while the feed travels horizontally from one end of the table to the other. Discard material is retained by riffles which divert it to the side of the table. Conventional tables do, however, have relatively low throughput capacities, and modern versions are therefore built into stacked, multideck units [10, 11].

In *flotation*, separation of clean coal is achieved by passing air upward through a suspension of coal to which a frothing agent (commonly methyl isobutyl carbinol at 0.1 lb/ton of feed coal) has been added. The lighter "clean" coal particles then attach themselves to the air bubbles and rise with them to the top where they are skimmed off with the froth.

Beyond these traditional cleaning methods, however, two more recent developments, both particularly well suited for processing coal fines, must be noted.

One of these is the *compound water (CW) cyclone* [12], which evolved from the earlier DSM hydrocyclone and separates clean coal from a heavier discard by centrifugal forces in a vortex (see Fig. 8.1.3). The cut point is adjustable through a vortex finder. Several versions of this device have been successfully tested in small coal-cleaning installations and found capable of treating coal down to −200 mesh. A single 24-in. CW cyclone operated with an inlet driving pressure of 22 psi can process in excess of 130 tons of coal per hour [13].

The other development is a *spherical agglomeration* technique [14] which employs flotation principles but goes beyond flotation by yielding a *consolidated* cleaned product rather than clean fines. In this procedure, small particles down to −200 mesh are separated from discard by rapid agitation

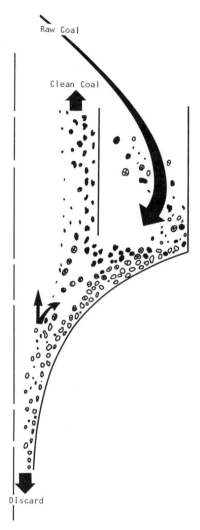

Fig. 8.1.3   Schematic illustrating the operating principles of a hydrocyclone.

of an aqueous suspension of coal to which a water-immiscible liquid hydro-carbon that preferentially wets the coal has been added. (A suitable liquid is diesel oil which, depending on the nature of the coal, can be used alone or with surface conditioning agents.) The flocs of clean coal that thereby form are then taken off near the top of the cleaning cell and more firmly agglomerated by a few minutes' slower stirring in a holding tank.

Since the coal substance is always more or less extensively intergrown

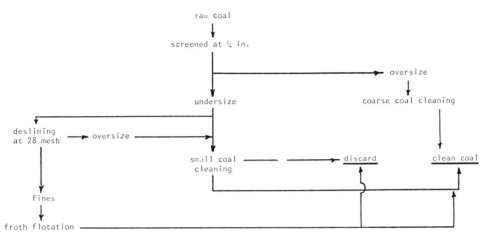

Fig. 8.1.4 Simplified flow diagram of a modern coal-cleaning plant.

with inorganic material—and even very fine grinding can, as a rule, only liberate *part* of the mineral matter—physical cleaning methods will, of course, always tend to misplace product by sending some low specific gravity material into the discard fraction and some high-ash discard to the clean coal. But the extent to which this happens depends on the particular equipment and operating conditions, as well as on the nature of the coal; and cleaning efficiencies can therefore only be determined from performance data, i.e., by comparing actual yields and ash contents of the cleaned products with those predicted from washability tests or distribution curves. A convenient and widely used formula for this purpose, proposed by Fraser and Yancey [15], is

$$E = (R/R_{\text{theor}})(A_1 - A_2)/(A_1 - A_3)$$

where R and R_{theor} are the actual and predicted yields of "clean" coal, and A_1, A_2, and A_3 are the ash contents of the feed coal, of the actually produced clean coal, and of the predicted clean coal, respectively.

In practice, cleaning efficiencies tend to fall as the range of sizes in the feed is broadened to include small coal and coal fines; and most plants therefore use *separate* circuits for cleaning $+\frac{1}{4}$-in., $-\frac{1}{4}$-in., and -28-mesh (or equivalent) material. Figure 8.1.4 shows a schematic flow sheet for this mode of operation. In addition, some plants also reprocess the primary discards at a higher specific gravity into middlings[2] and rejects fractions, and thereby

[2] These products should not be confused with the other kinds of middlings, i.e., the near-gravity material that may be found between floats and sinks in washability tests (see above).

recover in the former a fuel that, notwithstanding ash contents as high as 40%, can be used under boilers in electricity-generating facilities.

Efforts to develop *chemical cleaning* techniques have been mostly directed toward selective removal of sulfur from coal and have, for this purpose, tested oxidation, reduction, microbial oxidation, leaching with weak organic acids, and sequestering of sulfur by H donors and/or S acceptors (e.g., MnO) at ambient and elevated temperatures. But while some of these methods have yielded promising results in bench-scale work, none has so far proved sufficiently effective or economically promising to attract commercial interest.

8.2   Drying

Other factors being equal, the "per ton" value of a coal is ultimately determined by its useful carbon content or the net heat obtainable from it; and partial drying before use, like partial removal of mineral matter from it, can therefore carry significant technical and economic benefits with it. However, since inherent moisture contents high enough to make a substantial impact on coal performance are confined to low-rank coals which are, at present, primarily used for generation of electric energy, artificial drying[3] is, in practice, restricted to

(a)   washed bituminous coals, which must usually meet fairly rigorous user specifications, and

(b)   lignites and subbituminous coals used in the manufacture of briquettes (see Section 8.3) or of other "specialty" products (e.g., absorbent carbons).

In addition, drying may also be considered where low-rank coals with "as mined" moisture contents in excess of 10% are produced for use under boilers but must be shipped over long distances. In such cases, some pre-drying may be desirable in order to lower transportation costs and improve the performance of the coal in the combustion chamber—though whether this is feasible depends also on other factors, notably hazards from auto-genous heating (see below and Section 9.2).

Operational experiences from industrial installations and the diversity of equipment now available allow these needs to be met in several different ways.

---

[3] Some loss of moisture does, of course, occur naturally when freshly mined coal is stored in the open. But such natural drying is relatively slow, except at the surface of a storage pile.

For removal of surface moisture from washed coarse bituminous coals and anthracites, it normally suffices to provide a brief period of drainage on screens. Dewatering of washed small coal or coal fines from flotation circuits can be accomplished in cyclones or centrifuges, which, where required, are preceded by conventional thickeners. And where moisture contents must be reduced to lower levels than can be conveniently reached by these means, recourse is made to thermal drying in rotary kilns or fluidized bed dryers.

Low-rank coals (which are rarely, if ever, washed before use) are dried in rotary kilns, cascade dryers, or, where the size consist of the coal permits, entrainment dryers.

But unless drier coal is specifically required for immediate further processing (as, e.g., in briquetting), care must be taken to avoid drying to levels much below the *air-dried* moisture content of the coal (see Section 2.2). Aside from being unproductive, since the coal will then quickly reabsorb moisture on exposure to the atmosphere, such excessive drying can adversely affect the quality of the coal and may, in certain instances, also create serious fire and dust explosion hazards.

In bituminous coals, the deleterious effects of overdrying arise mainly from increased dustiness of the coal. This tends to compound handling difficulties and to increase "windage" losses during haulage (see Section 9.1), and may also pose technical problems in some end uses. For example, during carbonization (see Sections 11.1 and 11.2) excessive dust will enter the gas and tar by-product streams and thereby complicate their further processing. However, in the case of lignites and subbituminous coals, overdrying can have more serious consequences. Here, it will not only accentuate the pronounced natural tendency of the coal to decrepitate as moisture levels fall below the capacity moisture content (see Section 2.2) but also greatly increase the risk of subsequent autogenous heating. This matter is further discussed in Section 9.2.

8.3 Briquetting

Suitably prepared pulverized coal can be compacted into strong, homogeneous (formed fuel) briquettes; and with various refinements, this technique has been used since the mid-1800s to upgrade a variety of low-rank coals and coal fines that would otherwise have little commercial value [16]. But details of briquetting processes are closely tied to the nature of the feed coal, and a distinction must, in particular, be made between briquetting with and without binders.

Briquetting of coal *without* prior addition of a binder that promotes agglomeration of the coal particles is, in practice,[4] restricted to relatively soft unconsolidated lignites or so-called brown coals and demands careful attention to the size consist and the moisture content of the feed sent to the press. To facilitate close packing of particles during compression, it is generally essential to crush the coal to <4 mm, with 60–65% <1 mm [17]; and for maximum briquette strengths, moisture contents must be adjusted to values that vary inversely with the briquetting pressure. In theories of binderless briquetting [18, 19], this variation is seen as showing that the area of direct contact between contiguous particles is extended by capillary forces, with excess moisture physically militating against establishment of the maximum contact areas otherwise attainable at the particular pressure (Note D).

Aside from drying the coal to the appropriate moisture content, it is, of course, also essential to ensure that moisture is uniformly distributed; and in commercial operations, coal leaving the dryer is therefore very slowly cooled before dispatch to the briquetting press.

Briquetting is usually conducted at 38–65°C (∼100–150°F), i.e., at temperatures at which advantage can be taken of enhanced plasticity of small particles; and since the bulk of binderless lignite briquettes is used for residential heating, the preferred press is the Exter extrusion press [16, 22]. In modern versions of this machine, the coal is compacted under ∼1400 kg/cm² by a reciprocating ram that pushes it into and through an increasingly constricted channel by repeated forward strokes and admits additional coal during each return stroke. A simplified schematic of this mode of operation is shown in Fig. 8.3.1. Kaiser [23] and Mayer [24] have discussed the performance of the press in relation to various alternative designs of its working components.

Output from the Exter press is normally in the form of a string of 20 × 6 × 4-cm briquettes, each weighing ∼550 gm; but on occasion, 4 cm³ *briquolets*, weighing ∼43 gm, are manufactured.

Although the high pressures necessary for satisfactory compaction of more mature coals make the production of binderless *fuel* briquettes from such coals uneconomic, smaller quantities are manufactured for other purposes, e.g., preparation of "shaped" activated carbons. These operations extend the principles of lignite briquetting by using much more finely comminuted coal (commonly 100% −200 mesh) and pressures in the order of

---

[4] Since the parameters (e.g., hardness, elasticity, and plasticity) that govern compressibility change gradually with coal rank and could, in any event, be modified (e.g., by presoftening the coal through sorption of solvents), the distinction between coals that can and cannot be briquetted without binders is somewhat arbitrary and determined by economic rather than technical considerations.

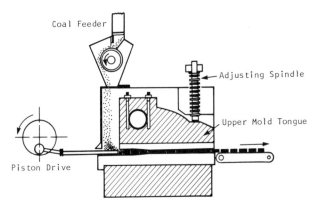

Coal Feeder

Adjusting Spindle

Upper Mold Tongue

Piston Drive

Fig. 8.3.1 Schematic illustrating the working principles of the Exter briquetting press.

15–20 tons/in.^{2} (2100–2800 kg/cm^2) but require special mold designs and fairly complex mold charging–discharging cycles [25]. Haigh [26] has also commented on the need for special steels that can withstand these severe operating conditions.

In briquetting coal *with* a binder, the manufacturing sequence is

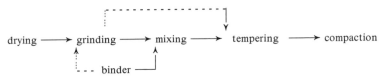

drying ⟶ grinding ⟶ mixing ⟶ tempering ⟶ compaction

binder

(where the broken lines indicate an alternative); and while close attention must be paid to the quality of the binder and to how it is incorporated into the coal, the total process is distinctly more flexible than briquetting without binders.

In most cases the coal is dried to <4% moisture (which from experience has been found to minimize binder requirements) and crushed to <1 mm (with 75–80% <0.5 mm), but somewhat coarser coal is acceptable when briquettes larger than domestic egg-shaped "ovoids" (see below) are made. After addition of the binder, the mixture is homogenized by "tempering" in a so-called pug where it is heated to 95–100°C by injection of superheated steam and agitated by rotating paddles; and before dispatch to the presses, it is cooled to a few degrees above the softening point of the binder in order to endow it with appropriate plasticity for compaction.

The binder is generally either a pitch or a bitumen and, if the former, crushed to 65–90% <0.5 mm before being mixed with the coal. In some instances it is then introduced into the *coarse* coal, i.e., between coal drying and grinding, while in others it is added to the coal *after* grinding (as shown above). On occasion, it is melted and sprayed into the coal. Bitumen is always applied in liquid form.

Binder proportions depend on the nature of the coal and range from 5 to 8 wt % in the case of pitch, or from 8 to 10 wt % when a bitumen is used. But there is little unanimity as to what constitutes a suitable binder, and there is also still much disagreement about the criteria that should be used in binder selection. In the case of pitch, it is customary to specify the softening point (which different users place between 68 and 79°C) as well as the "free" carbon, volatile matter, and ash contents; and there appears to be some consensus that pitches from high-temperature coal tars are generally preferable to residues from low-temperature tar distillation (see Sections 6.2 and 11.3). However, attempts to correlate briquette quality with pitch compositions (as reflected in, e.g., the proportions of toluene-, pyridine-, and/or carbon disulfide-insoluble matter) or with pitch properties (e.g., softening point, ductility, and viscosity characteristics) have yielded conflicting results and occasioned quite diverse recommendations respecting binder selection. For example, while Wood [27] found the "best" briquettes, tested 24 hr after compaction, to be those made with pitches of *low* aromaticity and containing little pyridine-insoluble matter, other investigators [28] have concluded that the most effective binders are characterized by *high* aromaticity.

Bitumens used as briquetting binders are variously air-blown residues from distillation or thermal cracking of crude oil and are specified in terms of their softening points (52–93°C), penetration (10–15 at 250°C), and ductility. But as in the case of pitch, these and other conventional methods of characterization are thought to be inadequate [29]; and since briquettes tend to break far more frequently along particle/binder interfaces than across the coal particles themselves, it has been suggested [30] that a more meaningful criterion of binder suitability may be afforded by the energy of adhesion ($W_{1/s}$) of the binder on the coal surface. Preliminary measurements of this quantity via Young's equation

$$W_{1/s} = Y_{1/a}(1 + \cos \theta)$$

where $Y_{1/a}$ is the surface tension of the binder against air and θ is the contact angle between the binder and the supporting solid, lend some support for this view.

For briquetting of "tempered" mixes, presses with horizontal or vertical revolving tables, exemplified, respectively, by the Couffinhal [31] and Yeadon [32] table presses, have been used; but modern operations, mostly designed to manufacture small "ovoid" or "cushion" briquettes for domestic markets, almost always employ either double-roll (e.g., Komareck–Greaves [33]) or ring-roll (e.g., Apfelbeck [34]; Piersol [35]; or Krupp–Herglotz [36]) presses.

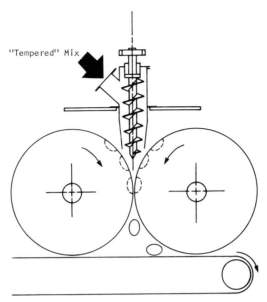

Fig. 8.3.2 Schematic illustrating the working principles of a double-roll briquetting press.

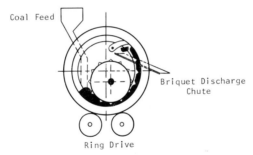

Fig. 8.3.3 Schematic illustrating the working principles of a ring-roll briquetting press. (By permission of J. Wiley and Sons Inc.)

The working principles of these machines are illustrated in Figs. 8.3.2 and 8.3.3. Although capable of attaining much higher pressures, most are routinely set for briquetting under ~ 1 ton/in.^{2} (150 kg/cm^2).

The quality of briquettes, whether made with or without binders, is judged by their crushing strengths and/or abrasion resistance, the latter measured by tumbler tests, such as prescribed by ASTM D 3402-72; and where the briquettes are intended for use as a fuel (i.e., without further processing by carbonization; see Sections 11.1 and 11.6), considerable importance also attaches to their water resistance and their ability to withstand weathering.

Notes

A. Cleaning, drying, and screening—the latter to meet demands for particular size cuts (see, e.g., Section 10.3)—are sometimes collectively referred to as coal *preparation*.

B. In passing it should be observed that the presence of a high proportion of colloidally disseminated inorganic matter in coal constitutes prima facie evidence of *allochthonous* development.

C. It is appropriate to note here that heavy media cells cannot be used if the coal contains substantial amounts of clay minerals that will disperse in the medium.

D. Alternatively, the effect of moisture can be associated with a reduction of the free surface energy F of the coal. Thermodynamically, such a reduction is equivalent to a two-dimensional surface pressure, expressed in dynes per centimeter (20); and in the case of a rod with a periphery $= 2\pi r$, a reduction of $-\Delta F$ would be equal to a pull of $2\pi r\Delta F$ at each end of the rod [21]. It might be observed that this phenomenon is, in part, also responsible for swelling when coal sorbs vapors.

References

1. F. W. Mayer, *Glückauf* **86,** 498 (1950).
2. K. F. Tromp, *Glückauf* **73,** 121 (1937).
3. R. Symington, G. H. Higginbotham, and F. Armstong, *Proc. Symp. Coal Prep., 2nd Univ. of Leeds* (October 1957).
4. B. M. Bird and D. R. Mitchell, *in* "Coal Preparation" (D. R. Mitchell, ed.), Chapter 12. American Institute Mining Engineers, New York, 1950.
5. H. F. Yancey and M. R. Geer, U. S. Bureau Mines Rep. Invest. No. 5308 (1957); No. 5412 (1958).
6. E. H. Citron, *Min. Eng.* **10,** 488 (1958).
7. A. P. Massmann, *Mechanization* **20** (5), 87 (1956).
8. M. R. Geer, M. Sokaski, J. M. West, and H. F. Yancey, U. S. Bureau Mines Rep. Invest. No. 5354 (1957).
9. R. L. Whitmore, *Colliery Eng.* **36,** 151 (1959).
10. H. B. Charmbury and D. R. Mitchell, *Mechanization* **20** (9), 79 (1956).
11. F. S. Ambrose and D. H. Davis, *Min. Congr. J.* **43** (11), 41 (1957).
12. J. Visman, Canadian Patent 686,112 (1964); 988,460 (1974).
13. J. Visman and D. Riva, *Proc. Int. Coal Prep. Congr., 6th Paris* (March 1972).
14. H. M. Smith and I. E. Puddington, *Can. J. Chem. Eng.* **38,** 1911 (1960); J. R. Farnand, H. M. Smith, and I. E. Puddington, *Can. J. Chem. Eng.* **39.** 94 (1961); A. F. Sirianni, C. E. Capes, and I. E. Puddington, *Can. J. Chem. Eng.* **47,** 166 (1969); C. E. Capes, A. E. Smith, and I. E. Puddington, *Bull. Can. Inst. Min. Metall., July 1974.*
15. T. Fraser and H. F. Yancey, *Trans. Am. Inst. Min. Metall. Eng.* **69,** 447 (1923).
16. G. Franke, *in* "Handbuch der Brikettierung," Vol. 1, p. 2. Enke, Stuttgart, 1930; A. Haacke and A. Meyer, *ibid.,* Vol. 1, p. 57; G. A. H. Meyer, *in* "Der Deutsche Steinkohlenbergbau," Vol. 3, p. 90. Glückauf, Essen 1958.
17. G. E. Baragwanath and W. H. Finlayson, BIOS Rep. No. 1760, 60 (1947); G. E. Baragwanath, *in* "Brown Coal: Its Mining and Utilization" (P. L. Henderson, ed.), p. 150. Melbourne Univ. Press, Melbourne, Australia, 1953.
18. O. Werner, *in* "Leitfaden der Brennstoff-Brikettierung," p. 152. Enke, Stuttgart, 1953.

19. N. Berkowitz, *Proc. Biennial Briquetting Conf.*, *3rd* Natural Resources Research Institute, Univ. of Wyoming, 1953.
20. N. K. Adams, "The Physics and Chemistry of Surfaces." Oxford Univ. Press (Clarendon), London and New York, 1941.
21. D. H. Bangham, *Proc. Conf. Ultrafine Struct. Coals Cokes* p. 18. BCURA, London, 1944.
22. H. Herman, *in* "Brown Coal," p. 192. State Electricity Comm., Victoria, Australia, 1952.
23. M. Kaiser, *Braunkohle* **50,** 333 (1951).
24. H. Mayer, *Bergbautechnik*. **2,** 213 (1952).
25. E. R. Sutcliffe, British Patent 102,918 (1916); U. S. Patent 1,267,711 (1918); 1,335,303 (1920); N. Y. Ten Bosch Octrovien, British Patent 406,492 (1934); 440,811 (1936); 450,633 (1936); 456,114 (1936); 512,374 (1940); U. S. Patent 2,133,675 (1938); W. Idris Jones and D. C. Rhys Jones, British Patent 616,857 (1949).
26. P. D. Haigh, Natural Resources Research Institute, Univ. of Wyoming, Information Circ. No. 3 (1949).
27. L. J. Wood, Coal Tar Research Assoc. Rep. No. 0140 (1955).
28. R. Broadbent, *J. Inst. Fuel* **28,** 3 (1955).
29. E. Swartzman, Natural Resources Research Institute, Univ. of Wyoming, Information Circ. No. 3, p. 870 (1949).
30. D. P. Agrawal and N. Berkowitz, *Proc. Biennial Briquetting Conf.*, *9th.* Natural Resources Research Institute, Univ. of Wyoming, p. 104 (1965).
31. C. Berthelot, *Génie Civ.* **112,** 221 (1938).
32. Yeadon and Co. Ltd., Leeds, England, Company information literature.
33. P. D. Haigh, Natural Resources Research Institute, Univ. of Wyoming, Information Circ. No. 5 (1951).
34. H. Apfelbeck, Czechoslovak Patent 1,376,096 (1921); W. Ronge, *Berg. Hüttenmänn. Monatsh. Hochsch. Leoben* **3,** 59 (1950); **8,** 112 (1955).
35. R. J. Piersol, U. S. Patent 2,021,020 (1934); 2,119,243 (1938); 2,321,238 (1941).
36. A. Streppel, German Patent 713,849 (1941); E. C. R. Spooner, BIOS, Britain, Final Rep. 1830 (1947).

CHAPTER 9

TRANSPORTATION AND STORAGE

When exposed to air for an extended period of time, freshly mined coal will lose moisture and oxidize, and as a result of such "weathering" progressively deteriorate. In general, this deterioration will be the more rapid and extensive the lower the rank of the coal, but even mature bituminous coals will, on occasion, quickly lose their commercial value through, for instance, oxidative destruction of their caking properties.

Aside from their direct costs and the limitations that such costs impose on marketing, long-distance transport and storage of coal can consequently pose serious technical and indirect economic difficulties.

9.1 Transportation

A major advance in long-distance bulk haulage of coal came with the introduction of so-called unit trains in the late 1950s. Typically made up of

100 to 120 cars, each with a 100–110 ton carrying capacity, these trains achieve substantial economies through

(a)   being dedicated to haulage of a single commodity,
(b)   moving directly from a mine or preparation plant to the consumer, and
(c)   operating in conjunction with specialized loading and unloading facilities that promote fast turnabout.

Charging and discharging is thus frequently accomplished while the cars are still in motion along the track; and to speed discharge, some cars are designed for bottom dumping. Others not so equipped are quickly emptied by being turned upside down as they pass through a rotary dumper at the delivery terminal.

In the case of mature coal, the principal problem associated with this form of transport arises from "windage" losses, i.e., losses of small coal and coal fines which become airborne and are carried out of the cars. Experience in Canada and the western United States indicates that in the absence of preventive measures, such losses will run near 0.2 wt %/100 miles (160 km) of travel. To avoid the economic and environmental penalties thereby incurred, some cars are now being built or retrofitted with hinged covers; and where open cars are used, it has proved helpful to top each nearly filled car with *coarse* coal or to spray-coat the exposed coal with heavy hydrocarbons. Several specially formulated compositions are now being marketed for this purpose.

Long-distance haulage of lignites and subbituminous coals in open cars may, however, also pose hazards from autogenous heating and spontaneous ignition (see Section 9.2); and whether these can be substantially eliminated by application of a spray-coat is doubtful.

The technical problems associated with rail movement of coal are skirted—although, to some extent, replaced by others—when coal is pipelined as a slurry, i.e., a suspension in water. This technique has long been used to convey a variety of solids (including coal) across short distances, e.g., from a mine to a preparation plant; but since it is not physically limited to short hauls, it has since the early 1950s attracted growing attention, particularly in Canada, the United States, and the Soviet Union, as a potentially cheaper transportation mode than rail haulage.

The principal features of slurry flow in a pipeline have been discussed by Durand and Condolios [1], Govier and Aziz [2], and Wasp *et al.* [3] and are illustrated in Fig. 9.1.1 as plots of pressure gradients ($\Delta P/\Delta L$) versus flow velocity. At low velocities, to the left of AA', when laminar flow regimes prevail, the solid will tend to settle out of suspension, and this region of the diagram is mainly of interest because it sets a lower velocity limit for slurry

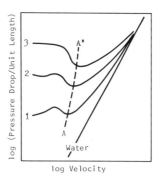

Fig. 9.1.1 The relationship between pressure gradients and flow velocity in a pipe. Curves 1, 2, and 3 refer to slurries with increasing solids contents. (After Durand and Condolios [1].)

flow. However, beyond a characteristic "threshold" velocity, which is identified by the intersection of AA' with the three illustrative concentration curves and usually lies between 2 and 3 ft/sec ($\sim$60 and 90 cm/sec), the slurry behaves like a homogeneous fluid. And as the flow velocities increase and flow regimes become increasingly turbulent, the energy requirements for slurry movement progressively approach the energy expenditures incurred when pumping water itself.

In practice, it is, of course, generally desirable to maximize the solids content of the slurry and to hold pumping costs to economically acceptable levels by operating at low velocities. For design purposes, it is therefore important to note that neither the (internal) pipe diameter (D) nor the density of the suspended solid (d_s) exerts significant effect on the threshold flow velocity (V_{lim}) unless the suspended particles are smaller than 1–2 mm. For larger sizes, the function

$$V_{lim}/[2gD(d_s - d_l)/d_l]^{1/2}$$

in which g is the gravitational acceleration (in ft/sec^2) and d_l the density of the carrier fluid (water), assumes a nearly constant value of 1.34 [1]. But pressure gradients at flow velocities greater than V_{lim} depend on D, as well as on the solids content (c) of the slurry; i.e.,

$$\Delta P/\Delta L = f[(H_{s/w} - H_w)/H_w c, \; V/(gD)^{1/2}]$$

and the general slurry flow equation has therefore been written [1] in the form

$$(H_{s/w} - H_w)/H_w c = K\{V^2 X^{1/2}/[gD(d_s - d_l)/d_l]\}^{1/2}$$

In these expressions, $H_{s/w}$ and H_w are the head losses of the slurry and of water at the flow velocity V, both expressed in feet of water per foot of pipe length; X is the drag coefficient; and K is a dimensionless proportionality constant. Both X and K have been experimentally evaluated by tests in model pipelines [4–6], and some of these studies [4], as well as others [7, 8], have also provided information about size degradation of coal during slurry movement.

Development work in connection with the commissioning and operation of Consolidation Coal Company's 108-mile (175-km) long Cadiz, Ohio, pipeline [9] and, subsequently, for the 273-mile (435-km) long Black Mesa, Nevada, line [10] has shown that optimization of coal slurry transport requires considerable care in sizing the coal and preparing the slurry; and some problems can also be posed by slurry dewatering at the delivery end. Nevertheless, while Consol's line had to be abandoned in 1963, when a (transient) reduction of tariffs for coal haulage on parallel rail trackage made it uncompetitive, both it and the Black Mesa facility have clearly demonstrated the practicality of long-distance slurry pipelining.

Consol's was a 10-in. (25.4-cm) -i.d. line which was commissioned in 1957 and, when fully operative, moved some 1.25 million tons of coal per year from the mine near Cadiz to Cleveland's Eastlake generating station. The slurry was initially made up to consist of 50 wt % <1.2-mm coal in water, but later improvements allowed conveyance of <4-mm coal at 60 wt % concentration. Smooth flow at ~ 4.9 ft/sec (1.5 m/sec) was maintained by three pumping stations equipped with 450-hp high-pressure reciprocating pumps. At the delivery end, the slurry was centrifugally pumped from short-term holding tanks to vacuum filters, which yielded filter cakes with 20% moisture, and these were then sent to flash dryers in which residual water was removed by hot ($1600°F = 870°C$) combustion gas from coal-fired furnaces.

The Black Mesa installation, which began operating in 1971, is a substantially similar, though larger, line and moves some 4.8 million tons of coal per year in an 18-in. (45.7-cm) -i.d. pipe from a mine in Arizona to the Mohave power plant in southern Nevada. The slurry contains 50 wt % of -8-mesh (<2.4-mm) coal and flows at 5.8 ft/sec (~ 1.8 m/sec). Three of the four on-line pumping stations (of which one is on stand-by) consist of three piston pumps each, working at 1000 psi, while the fourth is comprised of four such units, operating at 1500 psi. At the power plant site, the slurry is taken to large storage tanks and then withdrawn from there to centrifuges which remove 75% of the water. Final drying and further pulverization to make the coal suitable for use in the pulverized fuel firing system is accomplished in bowl mills from which it is pneumatically conveyed to the boilers.

Some attention has also been given to the possibility of conveying *coal-in-oil* slurries [11]; and tests in a 1-in.- (2.54-cm-) i.d. experimental pipeline, in which the flow properties of bituminous and subbituminous coals in two dissimilar oils were investigated, have in fact yielded results that corresponded closely to those obtained with equivalent water slurries. At velocities above ~ 3 ft/sec (0.9 m/sec) and coal concentrations less than 50 wt %, pressure gradients conformed with conventional friction factor/Reynolds number plots, which generally allow better prediction of $\Delta P/\Delta L$ than the Durand–Condolios equation; but as expected, no satisfactory correlations could be established at $V < 2$ ft/sec (0.6 m/sec), when flow is affected by particle settling, or at coal loadings above 50%, when coal particles tended to concentrate near the pipe axis and to flow as a distinct core.

In practice, coal-in-oil slurry pipelining is only likely to prove feasible where coal and oil occur in close proximity and would otherwise, in order to meet demand, have to be separately transported to a common distant market area.[1] In such cases, it could conceivably bring about significant further improvements in transportation economics through the fact that the slurry line would always, regardless of the extent to which the oil is loaded with coal, carry a 100% payload (Note A).

However, where conditions favor coal-in-oil slurry pipelining, equal or greater benefits could, possibly, also accrue from a more novel concept that envisages forming comminuted (say, -10-mesh) coal into a stiff paste with water, extruding the paste into cylindrical segments, and then injecting these into an oil pipeline [12]. This procedure would reduce the effective payload by an amount equal to the water content of the paste but in return greatly simplify solids retrieval at the delivery terminal. Experiments in a 70-ft (21.3-m) -long, 1-in.-diam closed loop, through which paste slugs could be propelled at velocities up to ~ 11 ft/sec (3.3 m/sec), showed that the slugs retained their integrity almost indefinitely and moved, either singly or as long trains, at approximately the same velocity as the carrier oil. Optimum slug : pipeline diameter ratios were found to lie between 0.7 and 0.9; optimum slug length : diameter ratios were 5–7; and suitable pastes could be made by mixing the coal with approximately 30 wt % of water. Even after several days' travel in the test loop, no oil had penetrated into the slugs, and for retrieval at termination of a run, it sufficed to direct the pipe effluent onto a draining screen. As a rule, a 5-min draining period proved long enough to reduce residual oil on the slugs (all adhering to their external surfaces) to 2–3 wt %.

Because they flow as large discrete (capsulelike) bodies, paste slugs also generate much lower pressure gradients than slurries with equivalent solids contents. However, whereas slurries can be accelerated by passage *through* pumps, slugs must be temporarily isolated from the liquid carrier at pump stations along the pipeline route; and while several such bypass systems have been described in patent literature [13] (and a number of prototypes have been successfully operated on experimental paste slug [12] pipelines), a choice between slurry and paste slug transportation would therefore, even

[1] Two instances are movement of coal and oil from (a) Alberta to central Canada and (b) Siberia to the western USSR and beyond.

if both techniques eventually prove to be equally feasible, involve some economic trade-offs.

In a similar comparison between rail haulage and pipeline conveyance, it must also be borne in mind that slurry pipelining may not be appropriate for *all* types of coal. Recent laboratory simulation experiments show that some bituminous coals will suffer disastrous losses of caking properties when so transported [14, 15]. This deterioration has been attributed to mechanical degradation [15]; but since it occurs even when there is no significant change in the size consist of the coal, it may be more correct to associate it with chemical effects induced by cavitation in the liquid carrier during turbulent flow [14]. Unfortunately, the nature of these changes and the reasons why only some coals suffer them are still unknown, and whether or not deterioration will occur in a particular case cannot be predicted without specific testing. Nor has it so far been established whether the effect is confined to slurries or will also occur in paste slugs.

9.2   Quality Deterioration and Autogenous Heating

As already noted, exposure of a freshly mined coal to air causes it to lose moisture and to oxidize. These events will not only gradually destroy any caking properties that the coal may possess but will also lower its calorific value and tend to make it fall apart. In some instances they may even promote autogenous heating and eventual "spontaneous" ignition of the coal.

Modern transportation and storage methods (see Section 9.3) are much influenced by the need to minimize these hazards.

For mature hvb and higher-rank coals, which hold relatively little moisture at saturation, the loss of water (to equilibrium with the ambient atmosphere) is, as a rule, of no particular significance. But in the case of lignites and subbituminous coals, which may lose as much as 25–30% of their original weight during air-drying, such dehydration is accompanied by extensive, partially irreversible shrinkage, and the internal stresses set up thereby quickly cause the coal to lose its cohesion and to disintegrate into progressively smaller pieces. Commonly referred to as *decrepitation* or *slacking*, and analogous to the spalling of rocks exposed to alternate freezing and thawing, this process begins almost immediately after the coal has been mined and can be far advanced within as little as 24–48 hr.

Empirical laboratory tests [16, 17] confirm industrial experience that slacking tends to be confined to coals with capacity moisture contents greater than 10% and becomes increasingly pronounced beyond this limit. Figures 9.2.1 and 9.2.2 (the latter reflecting the fact that swelling of coal in water

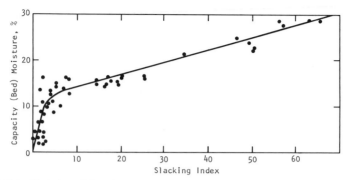

Fig. 9.2.1 Variation of the "slacking index" with capacity moisture contents of coals. (After Yancey *et al.* [16].)

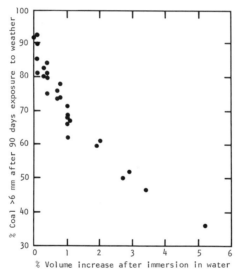

Fig. 9.2.2 The variation of "slacking" (or decrepitation) with swelling of coal in water. (After Roga and Pampuch [17]; by permission of J. Wiley and Sons, Inc.)

also varies with capacity moisture contents beyond ~10–12%) illustrate typical behavior.

The effects of oxidation likewise vary with coal rank and are most immediately and importantly reflected in

(a)   progressive deterioration of free-swelling indices and related plastic properties (see Section 6.3), and

(b)   steady lowering of the calorific value of the coal.

How fast this deterioration will occur in any particular case appears to depend not only on coal rank and ambient conditions, but also on the specific chemical composition and reactivity of the coal and is difficult to predict with any degree of confidence. For example, while the free-swelling indices of some (stockpiled) bituminous coals have been found to fall to less than half their initial values in a matter of days, other seemingly very similar coals can be stored in the open for weeks and even months without being seriously affected thereby. For practical purposes it is therefore necessary to monitor possible adverse effects of extended transportation and/or storage by regular coal sampling and testing programs.

However, also directly linked to (strongly exothermic) coal oxidation,[2] and always of overriding concern where *low-rank* coals are transported or stored in the open, is the notorious liability of such coals to heat up and "spontaneously" ignite [18].

This phenomenon typically involves the development of "hot spots" in a coal stockpile, which, unless killed by heat dissipation, eventually reach temperatures at which the coal begins to burn, and has long been regarded as triggered by air-oxidation of pyrites or by surface peroxide complexes in the coal. A specific mechanism has been advanced by Mapstone [19], who postulated oxidation of FeS to pyrophoric ferrous sulfide as well as to Fe^{2+} and Fe^{3+} sulfates. These reactions would ultimately release some 840 kJ/mole and create regions near coal-to-pyrite interfaces at which coal oxidation could very easily become self-accelerating. But the fact that substantially sulfur-free lignites and subbituminous coals are no more immune to autogenous heating than those containing massive amounts of pyrite makes this hypothesis unsatisfactory except insofar as it points to a special *contributory* cause; and a more general trigger of sustained, self-accelerating coal oxidation has been seen in heat releases that accompany wetting of partly dry coal [20]. Laboratory measurements have shown that such heats of wetting can range as high as 85–105 J/gm—sufficient to raise the temperature of the coal by 25–30°C and, consequently, increase oxidation rates six- to eightfold.

Qualitative support for this mechanism is afforded by repeated observation that stockpiled coal most often tends to heat up when exposed to rain after a period of dry, sunny weather [21] or when wet coal is placed on a dry pile [22]. In the latter case, ignition occurs generally at the interface between dry and wet coal. Other, more recent, confirmation has been provided by, inter alios, Shea and Hsu [23], who have also described a simple laboratory procedure that may be useful for assessing the liability of coal and other hydrocarbons to autogenous heating.

---

[2] The rate of this process (in effect, $C + O_2 \rightarrow CO_2$) will approximately double with each 10°C rise in temperature; if going to completion, the reaction will release 395 kJ/mole.

Since heat releases associated with wetting are proportional to the extent of the wetted surface and therefore, like oxidation rates, functions of the porosity (or capacity moisture content) of the coal (see Section 4.1), the hypothesis also predicts, in agreement with practical experience, that autogenous heating is only likely to be observed in hvCb and lower-rank coals with porosities corresponding to capacity moisture contents $> 10-12\%$ and to pose increasingly greater hazards as capacity moisture contents increase above this limit.

9.3   Storage Methods

In order to curtail slacking and oxidative deterioration of coal during transport and storage and to reduce the risk of autogenous heating, current techniques seek to control loss of moisture *from* coal and access of water and oxygen *to* it by mechanically restricting airflow through the coal. Attempts to inhibit oxidation by impregnating the coal with antioxidants, notably hydroquinones and amines, have generally yielded disappointing results [24] or only encountered some success when high application rates were used. Hydroquinone thus had to be applied as a 3% aqueous solution and even then provided protection for little more than three weeks [25] (Note B).

For protection of large storage piles, such as those often maintained at generating stations or port facilities, the preferred procedures accordingly all involve compaction of the coal; and to this end, several distinctive pile construction techniques have been developed. Typically, these require [27] that the coal be deposited in successive layers, each not thicker than 10–12 in (25–30 cm); that each layer be thoroughly compacted before the next is placed on it; and that the side slopes are raised at less than $14°$ (in order to prevent segregation of sizes). Built in this manner, stockpiles holding as much as 500,000 to 600,000 tons at approximately 80 lb/ft^3 (1.26 tonne/m^3) have remained "safe" for over three years [28].

As a further precaution against slacking and autogenous heating, some large stockpiles have been capped with asphalt [27], but this is now thought to be superfluous if the top of the pile is slightly crowned (to facilitate drainage) and smoothed (to reduce windrowing), and if care is taken to ensure that water runoff does not scour the sides of the pile.

Pile temperatures are usually monitored by means of thermocouples, which are inserted through tubes that penetrate the coal *horizontally* and so largely eliminate "chimney" effects created by vertical entry; and reclaiming is done by stripping across the top and, where necessary, recompacting the freshly exposed coal.

Where slacking is of particular concern and it is impractical to store the coal in a high-humidity environment, it is sometimes covered with several inches of tightly packed coal fines, or drying out is slowed by regular and frequent watering of the pile. However, neither of these procedures is entirely satisfactory, and both can, indeed, prove hazardous. In the former case, care must be exercised to prevent hot spots under the cover from igniting when the coal is being reclaimed; and in the latter, an improperly regulated watering program or a program made inappropriate by a period of unexpectedly dry, warm weather will actually promote autogenous heating. The same is true of the suggestion [29] that excessive heating in storage piles can be combatted by placing wet straw over them.

Where temperatures rise to dangerous levels and no lasting improvement can be effected by thorough dowsing with water, the only safe (though necessarily tedious) method is to excavate the hot spots and to carefully seal all entries.

Coal in small stockpiles, which cannot be economically compacted, can be protected against deterioration by spraying with petroleum products (or aqueous emulsions of hydrocarbons), soaps, fatty acids, or milk of lime [30]; and this method can, in principle, also be used to protect coal in transit. But during transportation, difficulties can arise from disturbances of the coal by the motion of the train (or ship); and whether a seal coat remains sufficiently intact to offer an effective barrier against weathering remains to be demonstrated.

A better solution may therefore lie in processing low-rank coal before shipment and/or storage in such a manner as to permanently protect it

Table 9.3.1

Effects of immersion carbonization:[a] 5-min treatment of $-\frac{1}{2}+\frac{1}{4}$-in. ($-1.25 +0.64$ cm) coal[b]

Temperature (°C)	Coal #1		Coal #2		Coal #3	
	Capacity moisture (%)	Caloric value (Btu/lb)	Capacity moisture (%)	Caloric value (Btu/lb)	Capacity moisture (%)	Caloric value (Btu/lb)
—	32.0	7,710	18.6	9,430	15.5	10,450
350	18.1	9,490	12.2	10,240	6.3	11,620
366	17.1	9,690	12.0	10,340	5.9	11,740
385	12.6	10,550	9.2	10,920	4.7	12,190
410	11.8	10,610	8.6	11,540	3.6	12,490

[a] After Berkowitz and Speight [31].

[b] All calorific values relate to the product coal as tested with its capacity moisture content: #1, lignite (Saskatchewan, Canada) with 5.1% ash; #2, subbituminous B coal (central Alberta, Canada) with 4.4% ash; #3, subbituminous A coal (southern Alberta, Canada) with 4.1% ash.

against weathering. One such method, termed *immersion carbonization*, has recently been described [31]. This makes use of the fact that tar vapors, if prevented from escaping from the interior of a coal particle and condensed instead by cooling, will plug the pores of the coal and so block subsequent reentry of moisture and oxygen. It therefore involves immersing the coal in hot oil at a temperature slightly above the decomposition temperature (T_d) of the coal (see Chapter 6) and then quickly quenching it. The permanent reductions of porosity (and hence, of the liability to autogenous heating) that can be achieved in this manner are illustrated by capacity moisture data in Table 9.3.1. The simultaneous improvement in heating value is a consequence of the limited pyrolysis that generates the tar vapors.

Notes

A.   In contrast, since no credit can be taken for water, the payload of a water slurry line is solely represented by coal and, hence, is less than 50%. However, some of the economic gain that accrues from substituting oil for water is likely to be offset by the greater cost of separating coal from oil and ensuring that the recovered oil meets refinery specifications relating to particulate matter in hydrocarbon feedstocks.

B.   The only exception to the generally unsatisfactory results so far obtained with antioxidants is the observation that spraying a lignite with 0.165% aqueous calcium bicarbonate prevented autogenous heating in 1000-ton stockpiles for 18 months [26].

References

1. R. Durand and E. Condolios, *Proc. Colloq. Hydraul. Transport Coal, November* p. 39. National Coal Board, Great Britain (1952).
2. G. W. Govier and K. Aziz, "Flow of Complex Mixtures in Pipes." Van Nostrand-Reinhold, Princeton, New Jersey, 1972.
3. E. J. Wasp, J. P. Kenny, and R. L. Gandhi, "Series on Bulk Materials Handling," Vol. 1, No. 4. Trans. Tech. Publ., Clausthal, 1977.
4. R. C. Worster, *Proc. Colloq. Hydraul. Transport Coal,* November, p. 5. National Coal Board, Great Britain (1952).
5. D. M. Newitt, J. F. Richardson, M. Abbott, and R. B. Turtle, *Trans. Inst. Chem. Eng.* **33,** 93 (1955).
6. R. A. Smith, *Trans. Inst. Chem. Eng.* **33,** 85 (1955).
7. R. C. Worster and D. F. Denny, *Proc. Inst. Mech. Eng. London* **169,** 563 (1955).
8. R. K. Bond, *Mech. Eng.* **79,** 1171 (1957).
9. G. H. Walker and E. J. Wasp, *Proc. World Power Conf., 6th, Melbourne, October* Paper No. 81 111.31/4 (1962).
10. J. G. Montford, Pipe Line News, p. 10 (May 1972); M. L. Dina, *Inter. Conf. Slurry Transport, Battelle Memorial Institute, Columbus, Ohio, Feb. 1976.*
11. N. Berkowitz, C. Moreland, and G. F. Round, *Can. J. Chem. Eng.* **41,** 116 (1963); N. Berkowitz and C. Moreland, Canadian Patent 702,627 (1965).
12. N. Berkowitz, R. A. S. Brown, and E. J. Jensen, *Can. J. Chem. Eng.* **43,** 280 (1965); N. Berkowitz and E. J. Jensen, Canadian Patent 735,760 (1966); U. S. Patent 3,190,701 (1965).

13. N. Berkowitz, R. A. S. Brown, C. De Zeeuw, and E. J. Jensen, U. S. Patent 3,339,984 (1967); Canadian Patent 840,880 (1970); 840,881 (1970); 841,486 (1970).
14. S. Parkash, A. J. Szladow, and N. Berkowitz, *Bull. Can. Inst. Min. Metall.* November (1971).
15. V. A. Korshunov, I. M. Lazovskii, L. G. Olshanetskii, and A. K. Vetrova, *Koks Khim.* No. 8, 6 (1974).
16. H. F. Yancey, N. J. F. Johnson, and W. A. Selvig, U. S. Bureau Mines Tech. Paper No. 512 (1932).
17. B. Roga and P. Pampuch, *Pr. Glownego Inst. Gornictawa, Komun.* 189 (1956).
18. G. R. Yohe, Illinois State Geol. Survey Rep. Invest. No. 207 (1958).
19. G. E. Mapstone, *Chem. Ind. (London)* 658 (1954).
20. N. Berkowitz and H. G. Schein, *Fuel* **30,** 94 (1951).
21. E. Erdtmann and H. Stoltzenberg, *Braunkohle* **7,** 69 (1908); R. Threllfall, *J. Soc. Chem. Ind. London* **28,** 759 (1909).
22. A. J. Hoskin, Purdue Univ. Experimental Station, Bull. No. 30, p. 45 (1928); W. Francis, *Fuel* **17,** 363 (1938).
23. F. L. Shea, Jr. and H. L. Hsu, *Ind. Eng. Chem. Prod. Res. Dev.* **11** (2), 184 (1972); see also, J. F. Cudmore, *Chem. Ind.* **41,** 1720 (1964); M. Guney and D. J. Hodges, *Colliery Guardian* **217,** 176 (1969).
24. G. R. Yohe, R. H. Organist, and M. W. Lansford, Illinois State Geol. Survey Circ. No. 201, GRY-2 (1951–1952).
25. A. Gillet and M. L. Fastré, *Bull. Soc. Chim. Belges* **53,** 83 (1944).
26. A. I. Khrisanfova and A. K. Shubnikov, *Khim. Tecknol. Masel* **3,** 21 (1958); *Chem. Abstr.* **52,** 16721 (1958).
27. R. R. Allen and V. F. Parry, U. S. Bureau Mines Rep. Invest. No. 5034 (1954).
28. J. N. Ewart, *Trans. ASME* **72,** 435 (1950).
29. L. Hegedus, *Mezogazd. Ipar* **3** (5), 11 (1949).
30. E. A. Terpogosova, *Izv. Akad. Nauk USSR, Otdel. Tekh. Nauk* **4,** 147 (1954); *Chem. Abstr.* **49,** 4968 (1955).
31. N. Berkowitz and J. G. Speight, *Bull. Can. Inst. Min. Metall.* August (1973); Canadian Patent 959,783 (1974).

CHAPTER 10

COMBUSTION

Authentic documented evidence places the first use of coal as heating fuel in late twelfth-century England—but also records that combustion of coal soon aroused considerable hostility. In the reign of Edward I (1239–1307), the sulfurous fumes and soot emitted when "sea coal" was burned (Note A) so discomfited the nobles of the day that it was made an offence punishable by death. And even 300 years later, during the reign of Elizabeth I (1533–1603), when attitudes had moderated, coal was not allowed to be burned in London whenever Parliament was in session. Widespread use of coal as a heating and steam-raising fuel and development of coal combustion methods that mitigated its environmental impact with varying degrees of success began, therefore, only in the eighteenth century, with the advent of the industrial revolution; and combustion appliances of the kind that subsequently evolved into the sophisticated stoker and boiler systems now in use did not make an appearance until the late nineteenth century.

10.1   The Chemistry of Combustion

Given sufficient air, combustion of coal proceeds sequentially through vapor-phase oxidation and ignition of volatile matter released from the coal to eventual ignition of the residual char. The first of these two quite distinct, though partly overlapping, stages is generally only of interest because the relative amounts of volatile matter may affect flame stability and the temperature at which the char will ignite (see Section 10.2). But the second involves a complex reaction mechanism that has much to do with how fast and efficiently combustion progresses.

Detailed theoretical and experimental studies (e.g. [1–4]) have shown that combustion begins with chemisorption of oxygen at "active" sites on char surfaces, and that decomposition of the resultant surface oxides (which continuously exposes fresh reaction sites) mainly generates CO, which is oxidized to CO_2 in a gaseous "boundary zone" around the char particle [5, 6]. CO_2 then either escapes into the off-gas stream or is again reduced to CO if it impinges on the char. Overall, therefore, combustion of char involves at least four carbon–oxygen interactions, viz.,

$$C + \tfrac{1}{2}O_2 \rightarrow CO$$
$$CO + \tfrac{1}{2}O_2 \rightarrow CO_2$$
$$CO_2 + C \rightarrow 2CO$$
$$C + O_2 \rightarrow CO_2$$

as well as concurrent, though quantitatively less important, oxidation of non-C atoms, notably

$$S + O_2 \rightarrow SO_2$$
$$H_2 + \tfrac{1}{2}O_2 \rightarrow H_2O$$

which may be followed by

$$H_2O + C \rightarrow CO + H_2$$
$$CO + H_2O \rightarrow CO_2 + H_2$$

Since the rate at which the carbon of the char is consumed thus depends on availability of oxygen at the char surface (where $C \rightarrow CO$), an overall combustion rate R can be written as

$$R = k_s p_{O(s)}$$

where $p_{O(s)}$ is the oxygen partial pressure at the surface and k_s the velocity constant of the primary reaction. The temperature coefficient of combustion then takes the conventional form

$$k_s = Ae^{-E/RT}$$

where A is a frequency factor and E the effective activation energy. But since

(a) $p_{0(s)}$ is obviously a "resultant" which is critically affected by oxygen transport to the char surface and cannot be directly determined from the nominal oxygen partial pressure of the combustion system, and

(b) R depends on the rate of $CO_2 + C \rightarrow 2CO$, as well as on $CO + \frac{1}{2}O_2 \rightarrow CO$, and is consequently sensitive to the CO/CO_2 ratio in the boundary zone,

knowledge of R itself contributes little toward an understanding of the *mechanism* of combustion, and more has been gained from studies that sought to define and evaluate the components of k_s.

A common point of departure in theoretical analyses of combustion processes is the assumption that free movement of oxygen toward the char surface is impeded by nitrogen and combustion products (i.e., CO, CO_2, H_2O, etc.) in the boundary zone (see Fig. 10.1.1); and because chars are porous solids, it is also supposed that some of the oxygen reaching the char

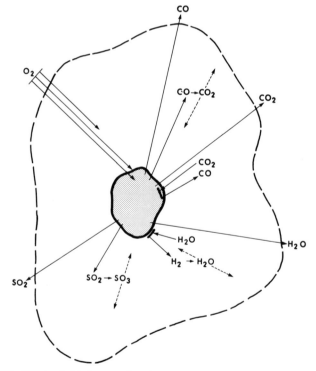

Fig. 10.1.1   Oxidation reactions in and near a burning char particle.

surface migrates into pores and thereby sustains limited "internal" combustion of the char particle. Proceeding from such considerations, the kinetics of combustion have been couched in terms of diffusion-controlled (Langmuir-type) sorption/desorption processes, and theoretical rate equations formulated on this basis show that effective burning rates depend on carbon configuration and density as well as on temperature and oxygen partial pressure [7].

While some of the terms in the theoretical relationships are difficult to measure, laboratory studies of combustion of freely suspended carbon spheres, carbon tubes, and coal particle "clouds" under closely controlled conditions have broadly validated predictions from theory and, in some instances, also further refined theoretical analyses. A case in point is the recognition [8] that diffusional processes only impose constraints on the burning rates of comparatively large (> 50–100 μm) coal particles.

Not surprisingly, in view of its complexity, several details of the combustion mechanism are still contentious, and there is indeed even still some disagreement about whether decomposition of surface oxides formed by initial chemisorption of oxygen on the char delivers only CO [9] or also generates CO_2 [6]. Despite such unresolved questions, however, basic data now at hand allow mathematical modeling of combustion systems. And some of these models make it possible to compute design and operating parameters, e.g., primary and secondary oxygen demands, optimum burner configurations, particle burning times, flame temperatures, and heat-flow patterns, which are increasingly influencing the design of modern boilers.[1]

10.2 Combustion Modes

The manner in which coal is burned and the devices in which it is burned are primarily determined by the required heat rate, which is expressed as an hourly Btu (or kJ) release per unit grate surface or combustion chamber volume, or stated in terms of an hourly steam demand (Note B).

Fixed-bed combustion appliances, which run the gamut from domestic space heaters to fully automatic stokers (see Section 10.3), can be used for heat rates up to 10^6 Btu/hr ft^2 of grate ($\sim 11 \times 10^6$ kJ/hr m^2) or up to 300,000 lb (~ 135 tonnes) steam/hr. These appliances burn relatively coarse coal; but except for domestic grates and ovens, which can accept (and often actually prefer) run-of-mine coal, large lumps, or briquettes (see Section

---

[1] Even where simplified by nomograms, these computations are usually quite lengthy and, for this reason, not illustrated here. Details can be found in the cited references, especially Field *et al.* [10] who gives a comprehensive treatment of pulverized coal combustion, and in other literature on solid fuel combustion.

Table 10.2.1

Coal size designations and size limits[a]

	Top size		Bottom size	
Size designation	in.	cm[b]	in.	cm[b]
Run-of-mine	Variable		0	0
Large lump	Variable		4	10
Lump	Variable		1	2.5
Cobble, egg, or stove	6	15.2	2	5
Nut	2	5	3/4	2
Prepared stoker				
Large	2	5	1/4	0.6
Intermediate	1	2.5	1/8	0.3
Small	3/4	2	1/16	0.2
Nut slack	2	5	0	0
Slack	1 1/4	3	0	0
Fines	1/2	1.3	0	0

[a] As determined with round hole screens.
[b] Approximate metric equivalents.

8.3), most require a smaller, sized fuel. Table 10.2.1 lists the more commonly used size designations and their respective size limits.

For steam rates greater than 300,000 lb/hr,[2] recourse is made to *suspension firing* in pulverized fuel systems or cyclone furnaces (see Section 10.4). Two other suspension-fired systems, based on fluidized bed combustion and combustion in magnetohydrodynamic (MHD) devices, are noted in Sections 10.5 and 10.6 but are not yet used commercially.

What kind of coal is burned in any particular combustion appliance is ultimately decided by availability and cost per million Btu, but choices among potential alternatives are often restricted by

(a) the inability of some types of combustors to satisfactorily burn strongly caking coals or coals of higher than mvb rank, and

(b) chemical compositions or properties that may make the use of some coals undesirable.

With respect to (b), particular attention must be given to sulfur contents, which may be too high to meet SO_2 emission standards (see Section 15.1); to ash compositions, which may create serious corrosion problems (see Section 10.7); and to ash fusion temperatures, which, if low, could render

[2] In some instances, suspension firing is also used to meet steam demands down to 100,000 lb/hr.

the coal unsuitable for burning in "dry-bottom" ash discharge equipment (Note C) or, if high, prevent its use in combustors that remove ash as a molten slag (see Section 2.4). If the coal is destined for suspension firing, some economic importance also attaches to a sufficiently low grindability index (see Section 4.5). And where it must be stockpiled as a safeguard against interrupted supply, its storage characteristics must be taken into account.

While new installations can usually be designed to accommodate the most economically available coal, these technical matters may override direct fuel costs where an existing plant is obliged to change its sources of supply. Particular difficulties arise in such cases if bituminous coal must be replaced by subbituminous coal. Although, other factors being equal, the latter can offer some operational advantages through lower char ignition temperatures and greater flame stabilities, their lower calorific values will reduce steam output and may consequently force some "derating" of the installation (Note D).

10.3   Fixed-Bed Combustion

Aside from domestic space heaters, modern versions of which are exemplified by hand-fired, "smokeless," warm-air furnaces and residential, "packaged" automatic boilers (see below), the most important fixed-bed combustion devices are industrial stokers. These appliances are generally engineered for steam raising, but because they employ different techniques for feeding coal to the grate and removing ash to a disposal pit, a distinction must be made between (a) underfeed stokers, (b) spreader (or overfeed) stokers, and (c) cross-fed traveling (or vibrating) grate stokers. The operating principles of these basic designs are illustrated in Fig. 10.3.1.

Each stoker type tends to have a fairly specific fuel and/or fuel size requirement (see Table 10.3.1), and practical restrictions on grate size also limit its capacity (see Table 10.3.2). When selecting a stoker for a particular application, projected heat demands and demand fluctuations, as well as availability of suitable fuels, must therefore be taken into account. However, in most instances there is considerable freedom to vary design details and thereby change the performance characteristics of a stoker. A case in point are traveling or inclined vibrating grate stokers with spreader-type coal feeding [11]. Such units can handle almost all types of coal, accept a wider fuel size range, and respond more quickly to load changes.

It has also been demonstrated that overfire air jets (which are now commonly installed in large stokers in order to increase gas turbulence and thereby maintain higher grate temperatures) ensure more uniform ignition of the incoming coal [12], and that problems arising from buildup of coke during combustion of caking coals on cross-fed traveling or vibrating grates

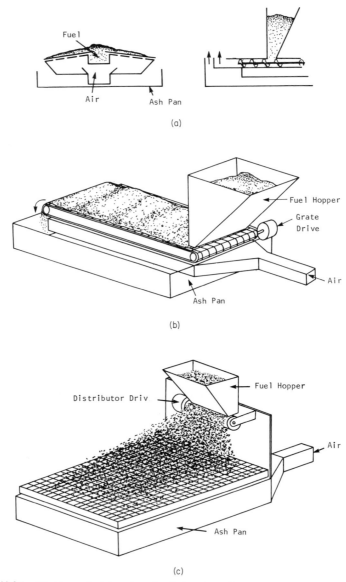

Fig. 10.3.1   Working principles of mechanical stokers. (a) Underfeed stoker; (b) traveling grate stoker; (c) spreader stoker.

Table 10.3.1

Stoker fuel requirements

	Coal types	Coal sizes[a]
Underfeed stokers	Bituminous coals or anthracite	Nut or prepared stoker (large); smaller sizes acceptable if $<50\%$ $-1/4$ in. (~ 6 mm)
Spreader stokers	All types	Prepared stoker (small), $<30\%$ $-1/4$ in. (~ 6 mm)
Traveling and vibrating grate stokers	Noncaking bituminous coals, anthracite or coke breeze	Nut or prepared stoker (large or intermediate), $<20\%$ $-1/4$ in. (~ 6 mm)

[a] See Table 10.2.1 for size definitions.

Table 10.3.2

Capacities of industrial stokers

	Maximum heat release, (10^3 Btu/hr ft^2 of grate)	Steam generation, (10^3 lb/hr)
Underfeed stokers		
Single retort	200	5– 50
Multiple retort	300	40–300
Spreader stokers	1000	10–300
Traveling or vibrating grate stokers	300	10–300

can be alleviated by using more closely spaced grate bars [13] or "tempering" the coal with addition of water [14, 15].

The design principles governing industrial stokers have been successfully applied to smaller (200–2500 lb steam/hr) automatic "packaged" boilers suitable for residential (steam or hot water) heating. These units normally employ underfeed stokers, but some have been built with vibrating grates that can handle all types of bituminous coal, including strongly caking varieties [16]. Maximum coal-burning rates during continuous operation run to ~ 25 lb/hr ft^2 of grate. Also developed since the 1940s were several types of more efficient "smokeless," warm-air furnaces and space heaters (see, e.g. [17]). Waning interest in coal as a domestic fuel during the 1950s and 1960s has precluded significant commercialization of these appliances, but it seems likely that updated versions, adapted for heating high-rise apartments and similar buildings, will command growing attention in the future.

10.4 Suspension Firing

Where, as in central electricity-generating stations, steam rates beyond the capacity of industrial stokers are required, use is now mostly made of pulverized fuel firing, i.e., combustion of air-entrained powdered coal as it passes through a firebox. Progressively refined since first introduced into commercial practice, this technique obviates the need for a supporting grate and thereby eliminates restrictions on equipment size, releases substantially more heat per unit volume of combustion space than stoker firing, and allows satisfactory combustion of virtually any kind of *noncaking* coal. (*Caking* coals pose the obvious hazard of becoming plastic before entering the combustion space, thereby fouling the burner nozzles. Also, some coke may be transiently deposited on firebox walls and boiler tubes and lower the heat-transfer efficiency.) However, in order to maintain a stable intense flame and avoid flashback through the burners, the coal must be injected into the firebox at relatively high velocities, typically, 50 ft/sec (15 m/sec); and efficient combustion with no significant loss of unburned carbon to the stack therefore demands some care in matching burner configurations and firebox dimensions to the coal used in any particular instance.

The nominal coal size for pulverized fuel firing is -200 mesh (<74 μm), but a more definitive size distribution has been cited [18] as $<2\% >50$ mesh (300 μm), with the proportion of -200-mesh material increasing from 65–70% in the case of easily ignited lignites and subbituminous coals to 80–85% for bituminous coals. At this degree of comminution, the average burning time of an individual coal particle is in the order of 0.25 sec [10, 19] or a little over one-tenth of its average residence time in the combustion zone. Pulverization (which, depending on the grindability of the coal, requires an energy expenditure varying from <20 to >40 kWh/ton) is usually achieved in air-swept impact or attrition mills that also "classify" the comminuted coal and prevent egress of oversize particles; and sufficiently fine coal is pneumatically conveyed to burners mounted in the walls of the firebox.[3]

Typical burner arrangements are illustrated in Fig. 10.4.1; these burn the coal between the lower boiler tubes (as in horizontal firing) or create a a turbulent vortex flame (as in opposed or tangential firing). In either case, "secondary" air, i.e., all combustion air other than that used to transport the coal, is injected around each burner in order to promote fast ignition close to the burner tip.

The vertical firing mode depicted in Fig. 10.4.1 has been mainly used in

[3] As a rule, between 10 and 20% of the total combustion air, which in turn amounts to 110–130% of the stoichiometrically required volume, is used for this purpose. (The larger amounts of "excess" air are needed when burning bituminous coal.)

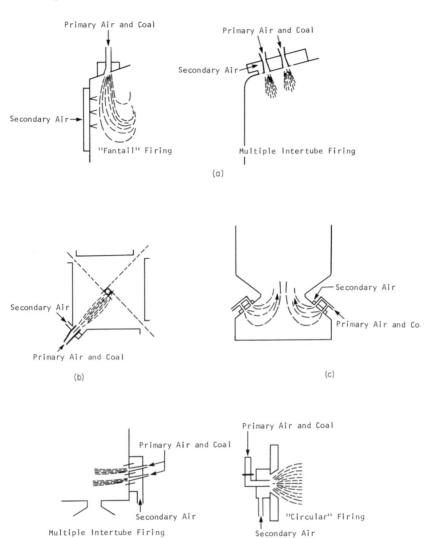

Fig. 10.4.1 Burner arrangements for pulverized fuel firing of boilers. (a) Vertical firing; (b) tangential firing (plane view); (c) opposed inclined firing; (d) horizontal firing.

slag-tap systems (from which ash is removed in its molten state). However, since such systems do not, as was hoped, greatly reduce fly-ash discharge [20] and, additionally, limit the range of usable coals to those with low ash fusion temperatures, most pulverized fuel installations built after the mid-1940s have been designed for horizontal or tangential firing and dry-bottom ash discharge (with 60–90% of the total ash taken off as fly ash; see Section 15.2).

Technically, some important gains over pulverized coal combustion have actually been found to accrue from an alternative suspension-firing method, which entails injecting air-entrained coal at high speed tangentially into a cylindrical horizontal "cyclone furnace" (see Fig. 10.4.2) and burning it while it spirals toward an exit at the opposite end ([21, 22]; Note E). Under appropriate aerodynamic conditions [23], such cyclone-firing generates heat releases up to 500,000 Btu/hr ft³ of combustion space (as compared with at best 150,000 and 400,000 Btu in dry-bottom and slag-tap pulverized fuel systems, respectively); the attendent high flame temperatures ($\sim 3000°F$) make it possible to discharge up to 90% of the ash as a molten slag; and since the slag tends to coat the interior furnace walls with a sticky melt that retains any still unburned coal particles, even relatively coarse ($-1/4$-in.) coal can be efficiently burned.

These features and the ability of cyclone furnaces to accept slurried fuels, e.g., partly dewatered washery refuse streams (see Section 8.1), have prompted the fitting of several large utility boilers with such furnaces. But although operating experiences from these installations have generally been good, cyclone-firing appears now to be falling out of favor because of its tendency to generate excessive amounts of NO_x (see Section 15.2), and a more attractive alternative to pulverized coal combustion is seen in fluidized bed combustion.

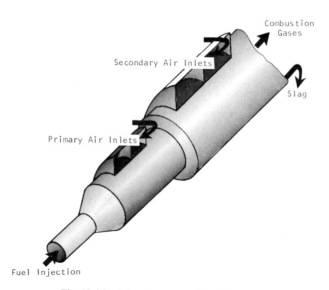

Fig. 10.4.2 Schematic of a cyclone furnace.

10.5 Combustion in Fluidized Beds

When expanded to a *fluidized* state [25], a bed of solid particles behaves in many respects like a liquid and consequently becomes a much more efficient system for heat and mass transfer than it would otherwise be. Since the 1940s fluidized beds, e.g., of catalysts, have therefore become increasingly important components of chemical-processing technology, and more recently, they have also been shown to offer major advantages when used for combustion of coal and coal-fired steam raising.

While the detailed design of an efficient fluidized bed combustor may on occasion pose problems, the basic structure and operating principles of such units are inherently simple. As illustrated in Fig. 10.5.1, the combustion chamber is a vertical cylindrical or squared shell in which a bed of crushed coal, typically $-1/16$ in. (1.5 mm), is supported on a (sometimes water-cooled) perforated distributor plate and fluidized by combustion air pumped through the plate at the required rate (Note F). Ash particles that accumulate near the top of the fuel bed as the coal particles burn out while moving up are removed by being allowed to overflow through an appropriately placed off-take pipe; fresh fuel is continuously, though not necessarily at constant rate, admitted through an inlet just above the distributor plate; and where

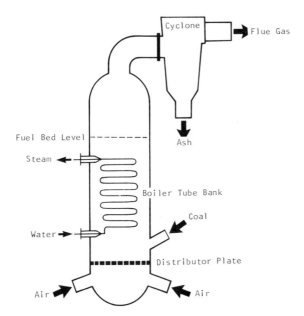

Fig. 10.5.1 Schematic illustrating the principles of fluidized bed combustion.

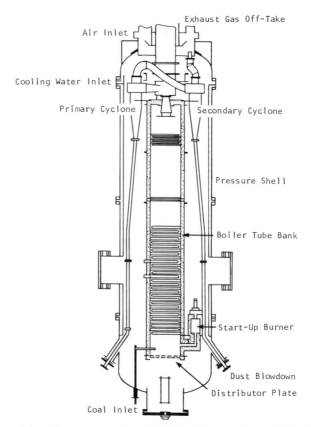

Fig. 10.5.2   Schematic of an experimental fluidized bed combustor. (After Westinghouse Research Laboratories [28].)

combustion serves to provide heat for steam generation, the boiler tubes can be directly submerged in the fuel bed.

Engineering studies of this mode of coal combustion began in Britain in the early 1950s [26] and were subsequently expanded to include investigation of fluidized bed combustion at elevated pressures. A recent form of an experimental combustor, capable of accommodating an 8-ft ($\sim$2.4-m) fuel bed and operating at pressures up to 1.5 MPa, is shown in Fig. 10.5.2. The combustion chamber is in this case a refractory-lined, water-cooled casing mounted inside a 6-ft (1.8-m)-diam pressure shell, which near its top also contains two cyclones for freeing the flue gas of particulate matter [27]. Development work, primarily directed to testing materials of construction under simulated commercial operating conditions and defining the potential of pressurized fluid-bed combustion as the first stage of (two-stage) "combined-cycle" generation of electricity,[4] is now being actively carried forward by Britain's National Research Development Corporation (under sponsorship by the US Energy R & D Administration) and Westinghouse Research Laboratories (under contract with the US Environment Protection Agency). Other pilot plants are operated by, inter alia, Exxon, Argonne National Laboratories, and Combustion Power Corporation; and in the US fluidized bed coal combustion is already projected as contributing up to 40,000 MW of installed generating capacity by the end of the century. A recent appraisal of the status of the technology [28] notes that no problems that would preclude commercialization or make it difficult to meet environmental standards could be identified.

Among the several advantages seen to attach to fluidized bed combustion are (a) significantly lower capital and operating costs, (b) exceptionally high (up to 100 Btu/hr ft^2 °F) heat transfer to boiler tubes, and (c) because of low flame temperatures ($1400-1750°F \equiv 760-950°C$), greatly reduced NO_x formation (Note G). Where high-sulfur coal is burned and it is necessary to reduce SO_2 emissions, it is also easier to use limestone injection (see Section 15.2), which serves to abstract SO_2 from the combustion gas via

$$CaCO_3 \rightarrow CaO + CO_2$$
$$CaO + SO_2 + \tfrac{1}{2}O_2 \rightarrow CaSO_4$$

Capital savings over alternative coal-firing systems are expected to amount to at least 10%, and *pressurized* fluid-bed combustion would, additionally, occupy much less space. A 500-MW facility, operating at 1.5 MPa, would be one-twentyfifth the size of an equivalent conventional thermal plant designed for pulverized fuel firing.

10.6 Magnetohydrodynamic Generation of Electricity

Like the rotor of a turbogenerator, a conducting gas, e.g., a high-temperature combustion gas "seeded" with an easily ionizable metal (such as K or Ce), creates an electric current when flowing at high speed in a magnetic field. This effect, now known as magnetohydrodynamic (MHD) generation, was first described by Faraday over 100 years ago but, because of technical difficulties associated with manipulation of very hot gases, has only recently become a potentially practical generation technique.

Like other generating methods, MHD provides for several alternative

---

[4] In such generation, the hot flue gas exiting from the combustor system is used to drive gas turbines, and the overall electric efficiency increases from a maximum of $\sim 37\%$ (in the case of a large thermal installation) to 43–45%. However, combined-cycle operations are also possible with other firing methods (see, e.g., Section 10.6).

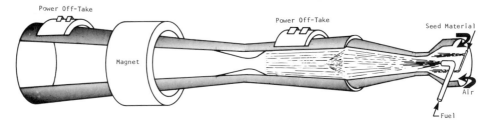

Fig. 10.6.1 Schematic illustrating the working principles of a venturi-type MHD generator.

generator configurations and operating procedures (see, e.g. [30]); and technically, it can also use media other than a combustion gas. In a closed-cycle system, in which the working fluid would be recycled after reheating upon completion of a passage through the MHD channel, it could be a molten metal or metal vapor. But economic considerations are increasingly shifting attention to open-cycle operation of coal-fired venturi-type generators (see Fig. 10.6.1), and Fig. 10.6.2 shows how these would fit into a commercial generating facility.

MHD generation and a variety of related topics (e.g., suitable gas-seeding compounds, seed recovery, and spent gas cleanup, which are all essential for economically and environmentally acceptable operation of open-cycle systems) are being actively investigated in the United States (where Avco's Everett Research Laboratories near Boston, Stanford University, and the University of Tennessee Space Institute are among the leaders), as well as in the USSR; and smaller programs are also being pursued in West Germany, Britain, France, Poland, and Japan [31, 32]. Furthest advanced is the Soviet Union, where a generator designed for 20–25 MW power output was

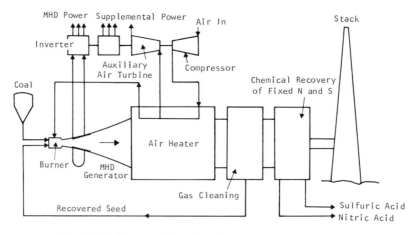

Fig. 10.6.2 Conceptual flow sheet for an MHD generating facility.

Table 10.6.1

Design and preliminary performance data for the U-25 MHD installation, USSR[a]

	Design	Actual
MHD power, MW	20–25	6[b]
Fuel	Natural gas	Natural gas
Oxidant	Air + 40% O_2	Air + 40% O_2
Combustion air temperature, °C	1200	1200–1250
Flame temperature, °C	2600	2500–2550
Seeding	Up to 1% K	Up to 1% K
Flow rate of gas, kg/sec	50	40–43
Pressure at inlet nozzle, atm	2.75	2–2.4
Pressure at outlet, atm	1.07	1–1.1
Gas velocity at nozzle, m/sec	850	1000
Magnet power, MW	2.6	2.4

[a] After Kirillin and Scheindlin [33].
[b] From channel no. 1 with design power of 7–9 MW.

completed in 1971. The design parameters of this installation, referred to as the U-25 plant, and some recent performance data [33] are summarized in Table 10.6.1.

Because of the very high temperatures at which MHD generators must be operated, formation of NO_x from nitrogen in the combustion air may prove a major obstacle to eventual large-scale deployment of such devices. Economic appraisals of MHD technology, when they include downstream components at all, generally merely *assume* that viable gas cleanup methods will be available when needed.[5] However, should this assumption prove correct, MHD generation could conceivably develop into an important and highly efficient alternative to pulverized fuel firing and fluidized bed combustion under utility boilers. The Soviet Union is, in fact, reported to have an 800-MW coal-fired MHD plant already under construction [34]. The coal will be sized to -50 mesh (300 μm), and molten coal ash, which is also expected to contribute some seed potassium to the combustion gas, will be relied on to form a 1-cm-thick layer on the water-cooled generator walls and thereby protect the generator against possible corrosion damage (see Section 10.7).

10.7   Effects of Coal Ash in Combustion Processes

As well as contributing heavily to objectionable stack emissions (see Section 15.1), coal ash and inorganic volatile material generated by thermal

---

[5] Current NO_x removal methods are discussed in Section 15.2.

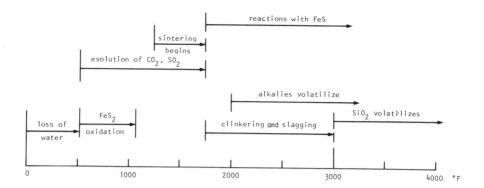

Fig. 10.7.1   Effects of heat on mineral matter in coal. (After Reid [35].)

alteration of mineral matter in coal (see Fig. 10.7.1) will not only adversely affect heat-transfer processes by "fouling" heat-absorbing and -radiating surfaces, but also threaten the integrity of the combustion system by causing corrosion and, in certain circumstances, erosion. Especially in large stoker- and pulverized-fuel-fired installations, the design, materials of construction, and operating procedures must therefore provide for effective countering of these hazards.

In suspension-fired systems, *fouling* is usually limited to

(a)   loosely adhering powdery low-temperature deposits that tend to form in cooler parts of the system (where moisture can condense and then trap fly ash), and

(b)   tightly bonded or sintered high-temperature deposits that develop quite specifically in the superheater and reheater sections of a boiler and typically consist of a strongly adhering dense core surrounded by a fly-ash-like layer. The most common forms are bonded by alkali metal compounds, notably sulfates and pyrosulfates, but calcium- and phosphate-bonded deposits have also been reported from combustion of coals rich in Ca or P.

The mechanism by which high-temperature deposits build up is still uncertain but is evidently connected with the volatility of alkali metal chlorides and sulfates. Bonded HT deposits thus tend to be particularly troublesome when coals with more than 0.5–0.6% chlorine are burned.

In slag-tap systems and stoker-fired equipment in which high grate temperatures are generated, fouling occurs principally through development of fused slag deposits, which form by impaction of superficially sticky gas-entrained particles on solid surfaces.

Impaction of particulate matter, whether partly fused or not, can also cause extensive *erosion*. The severity of this problem depends mostly on the size, shape, hardness, and velocity of the particles but also appears to vary with the angle of impact. According to Stoker [36], erosion rates under otherwise very similar conditions reach a maximum where particles impinge on a surface at $\sim 30°$.

Corrosion is mainly caused by oxides of sulfur; but in certain parts of a combustion system—specifically on furnace wall tubes with metal temperatures of 550 to 800°F (~ 290 to 425°C) and superheater or reheater tubes with metal temperatures in the range 1100–1300°F (~ 600–700°C)—corrosion can be induced by tube deposits that destroy protective surface oxide coatings.

Corrosion damage nominally ascribed to sulfur is actually caused by sulfuric acid, which is generated from organic and inorganic sulfur-bearing compounds via

$$SO_2 \rightarrow SO_3$$
$$SO_3 + H_2O \rightarrow H_2SO_4$$

Oxidation of the dioxide to the trioxide occurs mostly in flames where (transient) atomic oxygen species are thought to form [37] by such reactions as

$$H + O_2 \rightarrow OH + O$$
$$CO + O_2 \rightarrow CO_2 + O$$

It can, however, also be catalyzed by ferric oxides [38], which commonly exist on boiler tube surfaces and display maximum catalytic activity for SO_2 oxidation at $\sim 1100°F$ ($\sim 600°C$), i.e., at metal temperatures obtaining, e.g., in the superheater section of a boiler.

Under normal combustion conditions, less than 3% of SO_2 is oxidized to SO_3, and concentrations of sulfuric acid vapor in flue gas are therefore generally low (10–50 ppm). But even such levels will raise the so-called dew point (the lowest temperature at which a vapor can begin to condense on a solid surface) from a water dew point of ~ 100 to 120°F (~ 38 to 49°C) to an acid dew point of 250 to 350°F (~ 120 to 175°C); and relatively strong (70–90%) H_2SO_4 can then condense on and seriously corrode any boiler section (in particular, the air heater and economizer) in which these metal temperatures obtain. H_2SO_4 can also interact with fly ash in other sections to form sulfates of Na, K, Al, and/or Fe, which tend to promote corrosion on hotter surfaces (Note H).

To combat fouling, modern combustion equipment is designed to ensure that particles are cooled to well below their fusion temperatures before they can reach the banks of closely spaced tubes in the upper regions of the boiler; and in pulverized fuel systems, provision is usually made for tilting the burners and thereby periodically altering the heat regime. In addition, tube deposits are routinely dislodged by frequent "soot blowing," i.e., by inserting perforated lances through which jets of high-pressure air or steam can be sent between boiler tubes [41].

For corrosion control and an acceptable service life of the combustion system, reliance is, first and foremost, placed on using constructional materials that can resist attack by H_2SO_4 and corrosive sulfates; and metal temperatures are controlled by various flue gas recirculation techniques. Careful attention is also given to optimizing air distribution, so that excess air, which would encourage formation of SO_3, can be held to a minimum. And whenever possible, further retardation of corrosion (as well as compliance with modern environmental standards; see Section 15.2) is sought by burning low-sulfur coals. Where such coals are not economically available, additives that abstract SO_2 or inhibit its oxidation to SO_3 are employed. The most widely used material for this purpose is dolomite, but others include ammonia, magnesium oxide, powdered zinc, and Mg or Zn naphthenates.[6]

Notes

A.   The meaning of the term *sea coal* is obscure, but the term suggests that such coal, evidently sometimes washed up on beaches, was gathered rather than mined.

B.   This latter convention reflects the fact that in most industrial operations coal combustion serves to generate steam. Major exceptions are processes (e.g., coal gasification; see Chapter 12) in which combustion is used to raise reactor temperatures to the required levels.

C.   Coals with low ash fusion temperatures can, for example, prove troublesome when burned on chain grates or vibrating grates because the ash may clog air ports.

D.   Other factors that may require derating of a boiler installation when a low-rank coal is substituted for a more mature one are insufficient air blower capacity, insufficient capacity of the coal delivery system, excessive flame lengths (due to high moisture contents), and/or larger amounts of fly ash passing through the superheater section.

E.   An interesting historical account of cyclone-firing and a description of the first commercial cyclone furnaces put into operation in the late 1940s have been presented by Grunert *et al.* [24].

F.   Where the critical gas-flow rate for full fluidization exceeds the optimum air rate, the combustion air can be augmented by recycling part of the flue gas. More commonly, however, combustion air rates exceed fluidization requirements and consequently tend to carry entrained solids out of the fuel bed. To minimize this, the upper part of the combustor is sometimes widened in order to reduce the gas velocity.

G.   At temperatures below $\sim 980°C$ (1800°F), there is little, if any, fixation of atmospheric nitrogen by the Zeldovich mechanism [29], and NO_x forms only from bound nitrogen in the coal. Typical NO_x emissions from experimental fluidized bed combustors have been reported as 0.3–0.4 lb/10^6 Btu (0.13–0.17 kg/GJ) or about one-half of the current US emission standard [28].

---

[6] Further reference to the use of some of these additives is made in Section 15.2.

H. A good review of corrosion and fouling in coal-fired combustion systems has been written by Lyons [39]. Britain's Boiler Availability Committee has also published several annotated bibliographies on these topics [40].

References

1. H. C. Hottel and I. McC. Stewart, *Ind. Eng. Chem.* **32,** 719 (1940).
2. D. A. Frank-Kamenetskii, *in* "Diffusion and Heat Exchange in Chemical Kinetics" (English Transl. by N. Thon). Princeton Univ. Press, Princeton, New Jersey, 1955.
3. D. B. Spalding, *Proc. Inst. Mech. Eng. London* **168,** 545 (1954).
4. K. G. Denbigh, *in* "Chemical Reactor Theory." Cambridge Univ. Press, London and New York, 1965.
5. L. Meyer, *Z. Phys. Chem. Abt. B* **17,** 385 (1932).
6. J. R. Arthur, *Nature (London)* **157,** 732 (1946); J. R. Arthur and D. H. Bangham, *J. Chim. Phys.* **47,** 559 (1950); J. R. Arthur, *Trans. Faraday Soc.* **47,** 164 (1951).
7. M. W. Thring and R. H. Essenhigh, *in* "Chemistry of Coal Utilization" (H. H. Lowry, ed.), Suppl. Vol. Wiley, New York, 1963.
8. R. H. Essenhigh, *Conf. Pulv. Fuel, 2nd* Paper B-1. Institute Fuel, London, 1957.
9. J. D. Blackwood and F. K. McTaggart, *Aust. J. Chem.* **12,** 114 (1959).
10. M. A. Field, D. W. Gill, B. B. Morgan, and P. G. W. Hawksley, "Combustion of Pulverized Coal," British Coal Utilization Research Association, 1967.
11. J. K. L. Mignacca, *Power Eng.* **62** (6), 76 (1958).
12. O. De Lorenzi, "Combustion Engineering." Combustion Engineering Co., New York, 1952.
13. J. J. Solari, *J. Inst. Fuel* **26,** 243 (1953).
14. E. P. Carman and W. T. Reid, U. S. Bureau Mines Bull. No 563 (1957).
15. E. J. McDonald and M. V. Murray, *J. Inst. Fuel* **28,** 479 (1955).
16. P. O. Koch, *Mech. Eng.* **80** (5), 108 (1958).
17. B. A. Landry and R. A. Sherman, *Trans. ASME* **72,** 9 (1950).
18. A. F. Duzy, *Mech. Eng.* **82** (4), 109 (1960).
19. H. K. Griffin, J. R. Adams, and D. F. Smith, *Ind. Eng. Chem.* **21,** 808 (1929); I. P. Ivanova and V. L. Babii, *Teploenergetika* **13** (4), 54 (1966).
20. G. W. Kessler, *Proc. Am. Power Conf.* **16,** 78 (1954).
21. W. H. Rowand and E. G. Kispert, *Mech. Eng.* **82** (4), 103 (1960).
22. A. G. Roberts, *BCURA Bull.* **24,** 206 (1960).
23. M. Ledinegg, *Z. Ver. Dtsch. Ing.* **94,** 921 (1951).
24. A. E. Grunert, L. Skog, and L. S. Wilcoxson, *Trans. ASME* **69,** 613 (1947).
25. "Fluidization Technology" (D. E. Keairns *et al.*, eds), Vols. 1 and 2. Hemisphere Publ. Corp., Washington, D.C., 1976.
26. J. McLaren and D. F. Williams, *J. Inst. Fuel* **42,** 303 (1969); H. R. Hoy and J. E. Stanton, *Joint Meeting CIC/ACS, Toronto, May* (1970); R. W. Bryers and W. E. Kramer, *TAPPI Eng. Conf., Houston, Texas, October* (1976).
27. National Research Development Corporation, London, Quarterly Tech. Progr. Rep., January–March 1976, ERDA, Washington, D.C., FE-1511-23.
28. Westinghouse Research Laboratories, NTIS PB-246 116, U. S. Dept. of Commerce, Washington, D.C. (1975).
29. Y. Zeldovish and Y. P. Raizer, *in* "Physics of Shock Waves and High Temperature Hydrodynamic Phenomena," Academic Press, New York, 1966, pp. 375–78; B. R. Bowman, D. T. Pratt and C. T. Crowe, *Proc. Inter. Symp. on Combustion, 14th, Pennsylvania State University, August 1972;* F. C. Goulain, *Combust. Sci. Technol.* **7,** 33 (1973).

30. W. D. Jackson, M. Petrick, and J. E. Klepeis, *ASME Winter Meeting, Los Angeles, California, November* Paper No. 69-WA/PWR-12 (1969).

31. J. B. Dicks, S. Way, T. R. Brogan, and M. S. Jones Jr., *Mech. Eng.*, August, 18 (1969).

32. J. B. Dicks, *Mech. Eng.* May, 14 (1972).

33. V. A. Kirillin and A. E. Scheindlin, *Teplofiz. Vys. Temp.* **12** (2), 372 (1974) [English transl. UDC 621.313.12:538.4. Consultants Bureau, New York, 1974].

34. F. D. Johnson, *Design Eng.* January, 36 (1976).

35. W. T. Reid, *in* "External Corrosion and Deposits: Boilers and Gas Turbines." Elsevier, New York, 1971.

36. R. L. Stoker, *Ind. Eng. Chem.* **41,** 1196 (1949).

37. A. Dooley and G. Whittingham, *Trans. Faraday Soc.* **42,** 354 (1946); G. Whittingham, *ibid.* **44,** 141 (1948).

38. W. F. Harlow, *Proc. Inst. Mech. Eng. London* **151,** 293 (1944); **160,** 359 (1949); *J. Inst. Fuel* **32,** 126 (1959).

39. C. J. Lyons, *in* "Review and Analysis of Information on External Deposits and Corrosion in Coal-Fired Boilers and Gas Turbines." Battelle Memorial Inst., Columbus, Ohio, 1956; see also H. H. Krause, *in* "Corrosion and Deposits in Coal- and Oil-Fired Boilers and Gas Turbines." Battelle Memorial Inst.; American Society Mechanical Engineers, New York, 1959.

40. Boiler Availability Committee, Bull. MC/235 (February 1953); Bull. MC/236, (February 1953); Bull. MC/299 (March 1959).

41. Boiler Availability Committee, Bull. MC/286 (August 1957).

CHAPTER 11

CARBONIZATION

Coal carbonization processes are customarily regarded as low-temperature (LT) operations if conducted below 700°C ($\sim$1300°F) or as high-temperature (HT) ones if carried out at or above 900°C ($\sim$1650°F). Although somewhat arbitrary,[1] this distinction reflects the particularly pronounced physical changes that coal undergoes at temperatures between 600 and 800°C (see Section 6.4) and underscores the fact that commercial development of carbonization, stimulated by the divergent requirements of domestic fuel markets and an increasingly sophisticated iron and steel industry, has indeed created two very different carbonization technologies.

11.1 Low-Temperature Carbonization

LT carbonization was mainly designed to provide *town gas* for residential and street lighting and/or to manufacture substantially devolatilized "smoke-

---

[1] Occasionally, LT processes reached into the 700–900°C range and were then termed medium-temperature processes. For present purposes, however, it suffices to differentiate only between LT and HT operations.

less" solid fuels that were sufficiently reactive to be burned on domestic grates. Great economic importance attached, however, also to the by-product tars, which were often essential feedstocks for the chemical industry or refined to motor gasolines, heating oils, and lubricants (see Section 11.3).

The preferred coals for LT carbonization were, as a rule, lignites or sub-bituminous and high-volatile bituminous coals, which, when pyrolyzed at temperatures between 600° and 700°C, yield porous chars (or so-called semicokes[2]) whose reactivities are typically not very much lower than those of their parent coals. Higher-rank (caking) coals, unless pretreated to destroy their caking properties, were less suitable because they formed residues that tended to stick to the walls of the carbonization chamber and thereby impeded fast discharge of the char. But numerous LT processes were also developed for carbonization of briquettes (see Section 8.3), which were sometimes manufactured from blends containing caking coals.

Commercial LT carbonization evolved and flourished principally in industrialized European countries (whose energy economies were founded on coal) and was even there gradually abandoned after 1945 as oil and natural gas became widely available and demands for solid fuels dwindled. Subsequent rapid escalation of oil and gas prices, environmental constraints on combustion of "raw" coal (especially in urban areas), and growing interest in recovery of gaseous and liquid hydrocarbons from coal before burning it are, however, beginning to redirect attention to LT processing; and the classic pre-World War II practices (which have been reviewed by, inter alia, Gentry [1], the Utah Conservation and Research Foundation [2], and Wilson and Clendenin [3]) offer a body of experiences that are very relevant to the development of modern carbonization and tar-refining technologies.

The numerous technical options for efficient LT carbonization are illustrated by the diversity of the equipment employed in commercial LT installations. Vertical and horizontal retorts have been used for batch, as well as for continuous, operation—with process heat supplied directly (by circulating hot combustion gases through the charge) or indirectly (by passing such gases through external flues between contiguous retorts). Both stationary and revolving horizontal retorts have been successfully operated. And several processes for carbonizing fluidized or gas-entrained coal have been described. Many of these systems are, of course, now only of historical interest. But some, including even such *batch*-type carbonizers as the Brennstoff Technik, Krupp–Lurgi and Parker retorts, merit noting here because of their ability to process an exceptionally wide range of coals without loss of efficiency.

The *Brennstoff Technik* retort [4] was specifically designed for manufac-

---

[2] This term or its German equivalent *Schwelkoks* is commonly used in European literature.

ture of lumpy char and was a narrow (6–12.5 cm wide) cell that resembled a slot-type coke oven (see Section 11.2) and held a ~1-tonne charge. It was built of steel to facilitate rapid heat transfer to the coal and was heated by circulating hot combustion gas through flues in the long sidewalls; and fast discharge of the char to a bottom quench car was ensured by hinging the sidewalls at the upper edges and briefly swinging them out at the end of each processing cycle. Commercial installations consisted of several blocks, each with 12 retorts; and depending on whether briquetted or unbriquetted coal was charged, a full cycle, which carbonized the coal at 650°C, took 2–4 hr. Typical yields per tonne of high-volatile bituminous coal (with ~30% volatile matter) were ~750 kg char, 94 liters tar and light oil, and 127 m^3 gas (with 25.3–29.4 MJ/m^3; Note A). Successful operation of 50- to 60-tonne/day Brennstoff Technik plants in Germany led, in 1944, to start of construction of three 1400-tonne/day installations, but these remained uncompleted when World War II ended and were subsequently abandoned.

Another prominent German LT system employed the *Krupp–Lurgi* retort [5], which had a 3-tonne/day capacity and consisted of six narrow, rectangular cells that were separated by heating chambers through which hot combustion gas circulated. Coal was admitted through the top from a hopper, which moved across the battery of retorts (and could also be used to increase the bulk density of the charge by stamping it into the carbonization cells), and the char was withdrawn through a bottom trapdoor. The largest Krupp–Lurgi plant was a 56-retort installation, built in 1943 at Wanne-Eickel, which produced 225,000 tonnes of char per year. The by-product tar yielded some 1325 tonnes of fuel oil and 2100 tonnes of motor fuels.

These and similar German LT systems, e.g., the *Otto* [6] and *Weber* [5, 6] carbonizers, had their British counterparts in the *Phurnacite* process [6, 7], which carbonized ovoid briquettes in tall slot-type ovens, and in the *Parker* retort [8]. The latter consisted of a cast-iron block which contained twelve 9-ft-long tubes that tapered from $4\frac{1}{2}$-in. (~11.5-cm) diam at the top to $5\frac{1}{2}$ in. (~14 cm) at the bottom; and 40 such retorts, alternating with combustion chambers in which fuel gas was burned, constituted a battery. Carbonization at 650°C was completed in $4\frac{1}{2}$ hr; and at the largest plant—a 19-battery facility at Bolsover, commissioned in 1936—an average of ~1500 lb char, 20 gal tar and light oil, and 4 Mcf gas (at 700 Btu/scf) was obtained from each ton of coal.

Also of the same genre was the British *Rexco* process [9], which carbonized 4- × 1-in. (~10 × 2.5-cm) lump coal at 700°C and produced a hard, dense domestic fuel coke. The Rexco retort was a 25-ft-high, 11-ft-diam mild steel shell lined with firebrick, held a 34-ton charge on a grate support, and was internally heated by combusting part of the charge. A complete operating cycle, including a 7-hr cooling period in the carbonizer, took $16\frac{1}{2}$ hr, but this was later reduced to $13\frac{1}{2}$ hr by improving the design of the retort. The first Rexco plant came on stream in Nottingham in 1936 and produced an average of 1350 lb char, 19 gal tar and light oil, and 21.5 Mcf gas (at 140 Btu/scf) per ton of coal charged to it.

From the mid-1930s on, however, production from batch-type carbonizers was also increasingly augmented by operation of a number of *continuous* retorting processes, which allowed much greater throughput rates than were previously possible. These processes employed rectangular or cylindrical vessels of sufficient height to carbonize the coal while it traveled from top to bottom, and the charge was usually heated by a countercurrent flow of hot combustion gas. Particularly interesting examples, because they incorporate design and operating features reminiscent of the coal gasifiers that made their debut at that time (see Section 12.2), are the *Lurgi-Spülgas* retort [10] and the *Koppers continuous "steaming" oven* [5].

The Lurgi-Spülgas retort, which could process up to 300 tonnes of coal per day, was a 40- to 45-ft-high, 10-ft-diam brick-lined steel vessel which was internally divided into a coal-drying chamber and a 20-ft-high carbonizer. These two chambers were connected to each other by vertical pipes through which the coal moved down, but each was separately heated by injection of externally generated hot combustion gas and provided with its own gas off-take (see Fig. 11.1.1). Coal was continuously admitted from a hopper mounted directly above the retort; carbonized char was removed through a bottom extractor into a revolving cone where it was quenched by water sprays; and by-product gas and tar vapors were passed through the carbonizer gas off-take to a scrubber and electrical precipitator (which freed it of particulate matter and tar) before being chilled (to recover lighter hydrocarbons). The drying chamber was usually operated at temperatures between 150 and 220°C in order to reduce the moisture content of the feed coal to ~0.5% and vented to a stack, and carbonizer temperatures varied between 500 and 600°C. Total residence time of coal in the retort was approximately 20 hr.

Lurgi-Spülgas installations were extensively used in Germany during the 1930s and 1940s to manufacture char for domestic and industrial consumption, and tars accruing in these plants were major sources of motor fuels. Similar facilities were also operated in Japan [11]; and a small Lurgi-Spülgas plant, comprised of two retorts with a combined daily throughput of 280 tons, has been reported from New Zealand [12].

The Koppers continuous "steaming" oven was of more conventional design, consisting essentially of an indirectly heated rectangular chamber which widened from approximately $9\frac{1}{2}$ in. (24 cm) at the top to 13 in. (33 cm) near its base, but merits special mention because it provided for injection of steam into the carbonizing coal charge. This facilitated carbonization

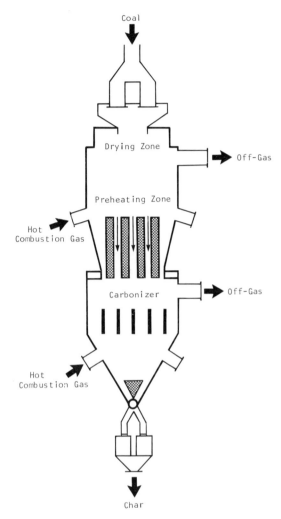

Fig. 11.1.1 Simplified schematic of the Lurgi-Spülgas retort for low-temperature coal carbonization.

through improved heat transfer and also served to increase the total gas yield by reacting with the hot char to form water gas (see Section 12.1). The residence time of coal in the oven was 20–22 hr, and individual ovens were built for throughputs up to 180 tonnes/day.

Like the Lurgi-Spülgas retort, the Koppers continuous "steaming" oven was mainly used in Germany and, later in Japan. Reid [11] has described the more important of these installations and given details of their performance.

British development of continuous LT carbonization in vertical retorts is represented by the *Rochdale* process [13], which, instead of injecting steam, cycled producer gas through the base of the retort in order to improve heat transfer and accelerate carbonization. Since producer gas does not interact chemically with coal, most of the sensible heat acquired by the gas as it passed upward through the char into the still more or less undecomposed charge was effectively transferred to the coal. A similar scheme was employed in the National Fuels Corporation's process, which was used in the United States (at Bethlehem, Pennsylvania) to produce uniformly sized char from briquetted coal, and in several experimental continuous carbonizers [3].

Among horizontal LT carbonization retorts, one of the earliest examples was the externally heated *Shimomura* retort [5], which was used in Japan during the 1920s to produce char suitable for incorporation into coke-oven blends (see Section 11.2). The complete unit consisted of two 12-ft-long, ~2-ft-diam cast-iron cylinders which were mounted one above the other and interconnected at one end by a short vertical drop pipe. Each cylinder was equipped with a coaxial rabble-armed rotating shaft which moved the coal through the cylinder, and the entire assembly was heated in a coke or gas-fired oven. The capacity of the Shimomura retort, which was usually operated at 500°C, was about 8 tons/day.

More recent practices are illustrated by the *Disco* process [14], which has been operated in a major US plant (near Pittsburgh, Pennsylvania) since the 1930s and proved so successful in an otherwise unpromising market area as to warrant subsequent overhaul and expansion of the installation to a rated capacity of 800 tons/day. The Disco process produces closely sized spherical chars from caking coal fines by utilizing the ability of plastic coal to incorporate inert material (see Section 6.3) and, when set into appropriate rolling motion, to form coal balls. It therefore begins with carefully controlled air-oxidation, which reduces the plastic properties of the feed to the desired level, and then introduces the oxidized material into a slightly inclined horizontal retort where it almost immediately becomes plastic, mixes with recycled char fines, and gradually shapes itself into small spheres as it moves through toward the discharge. The retort itself consists of a 125-ft-long, 9-ft-diam revolving steel cylinder in a stationary outer steel shell and contains a number of 8-in. baffles which temporarily hold up the coal balls and thereby extend their residence time in the carbonizer. Process heat is supplied by fast circulation of hot combustion gas through the annulus between the inner and outer cylinders. The carbonizing temperature is normally ~1070°F (575°C) and furnishes char with ~15–17% volatile matter in ~75% yield.

Variants of the Disco retort design are exemplified by the *Hayes* carbonizer [15], in which the charge is moved along by a continuous coaxial screw, and by two Japanese versions, the *Wanishi* and *Mimura* retorts [11].

Directly heated continuous horizontal LT retorts have so far only been tested in pilot plants. In the University of Kentucky retort [16], the heat carrier was combustion gas; in the Humboldt carbonizer [4] it was sand; and Korobchanski [17] has used hollow metal spheres filled with a mixture of sodium and calcium chlorides. Molten aluminum [18], as well as molten lead [19], has also been proposed as a heat carrier.

Since the 1940s some work has also been carried out with experimental conveyor carbonizers, such as the *Storrs* retort [20] in which the coal is moved along on a chrome–molybdenum wire-mesh belt under infrared heating elements, and with "flat bed" (or "tray") retorts [21, 22]. However, because carbonizers of this type require the coal to be spread out in relatively thin layers, their throughput capacities are low, and much greater attention has come to focus, especially in the United States, on pyrolysis of air- or gas-entrained coal and of fluidized coal beds. Such systems are, of course, unsuitable for production of lumpy char and at present mainly command interest as means for efficiently *gasifying* coal.[3] But they could, when required, be easily adapted for manufacture of granular LT chars and concurrent tar recovery. The *Parry* process [23], a prototype of which was used in the mid-1930s by the Texas Power and Light Company to produce and evaluate lignite chars as boiler fuels, illustrates such an application.

11.2 High-Temperature Carbonization

If carbonized at temperatures above ∼650–700°C, LT chars become progressively less reactive through devolatilization and loss of porosity and, as a result, tend to lose the properties that make them attractive domestic and industrial boiler fuels. In practice, HT carbonization is therefore only employed for production of coke (which *acquires* much of its industrial utility through processing at high temperatures); and this "specialization" not only limits the range of useful coals available to HT technology (see Section 6.3), but also determines its format. While some coke is used for manufacture of calcium carbide and electrode carbons, or as reductant in certain ferrous and nonferrous open-hearth operations, by far the largest consumption (accounting for well over 90% of total output) is in blast furnaces, and modern coke-making practices are therefore virtually dictated by the coke quality criteria of this market.

As an art, coke making dates from the late sixteenth century when

---

[3] They are, in fact, already used in two commercial gasification processes and in several "second-generation" gasifiers that are now in advanced phases of development (see Sections 12.2 and 12.3).

English ironmasters, facing Parliamentary restrictions on the use of wood charcoal on which they had traditionally relied, began to direct attention to alternative fuels; and from 1589 on, several patents relating to the manufacture of iron with coal were, in fact, issued. But wide use of coke in place of charcoal came only in the early 1700s, after Abraham Derby and his son showed that coke burned more cleanly and with a hotter flame than coal. This led to various adaptations of conventional wood-charring methods to coking and eventually to the emergence of the beehive oven, which by the mid-1850s had become the most common coking plant.

The beehive oven was a simple domed brick structure into which coal could be charged through an opening at the top and then leveled through a side door to form an approximately 2-ft-thick bed. Process heat was supplied by burning the volatile matter released from the coal, and carbonization consequently progressed from the top down through the charge to yield a characteristic columnar coke which was raked or pushed out from the side. As a rule, some 5–6 tons of coal could be charged, and 48–76 hr were required for carbonization.

With minor improvements and additions of waste heat boilers (which allowed heat recovery from the combustion products), some beehive ovens are still in operation (Note B). But for the most part they have been superseded by wall-heated, horizontal chamber or "slot" ovens, in which higher temperatures can be attained and better control over coke quality can be exercised. Since their introduction in the early 1900s, the basic design of these ovens has remained substantially unchanged, and greater efficiencies have accrued primarily from improved heating and incorporation of a number of labor-saving devices, such as self-sealing doors and more effective door extractors. Ovens now in use or offered by the major constructors differ little, therefore, except in the arrangement and operation of their heating flues and ancillary firing equipment.

Modern slot-type coke ovens are 50- to 55-ft-long, 20- to 22-ft-high chambers whose widths are chosen, before construction, to suit the carbonization behavior of the coals they are intended to process.[4] (The most common widths are 18 and 20 in., but some ovens are as narrow as 12 in. and a few are 22 in. wide.) A number of such chambers (usually 20 or more), alternating with similar cells that accommodate heating flues, are built as a battery over a common firing system through which hot combustion gas is sent to the flues. The flat roof of the battery provides the deck for a mobile,

---

[4] This connection between oven width and coal behavior tends to make a coke-oven installation captive to a coal source and is one reason why coke makers, once decided on their feedstocks, are usually so reluctant to consider substitutes. At the very least, a fairly lengthy test program in the plant is necessary to determine the suitability of the proposed substitute.

electrically powered charging car from which coal enters each oven through three openings along the top and is leveled by a rectractable bar; and finished coke is pushed from the rear of the oven through its opened front door onto a quenching platform or into rail cars that move it through water sprays. By-product gas and tar vapors are led from each end of the oven to collector mains for further processing (see Section 11.3) or in-battery fuel use.

Oven capacities average 25–40 tons (at a bulk density of 50 lb/ft^3), and the operating cycle is approximately equivalent to 1 hr/in. of oven width.

The almost universal adoption of slot-type ovens for coke making attests to their technical and economic efficiency but demands that control over coke quality be exercised by close attention to the composition of the feedstock rather than by manipulation of the processing equipment. Much of what is properly called HT carbonization technology relates therefore to such matters as selection and preparation of oven charges.

Fundamental to quality control on the coke is the choice of suitable coal, and in this coke makers have gained considerable freedom through recourse to *blending*—i.e., by charging binary or ternary mixtures rather than single coals. This practice developed originally from the observation that addition of weakly caking or noncaking coals to an oven charge could eliminate the technical difficulties that were occasionally encountered when carbonizing coals with high FSIs and/or high Gieseler fluidities (see below), but was later increasingly used to stretch supplies of scarce or costly "prime" caking coals and quickly became the principal method for manipulating coke characteristics.

In all cases it remains, of course, axiomatic that at least one component of the blend must possess strongly expressed caking propensities (as evidenced by a high FSI or Roga index), and where that coal lacks sufficient fluidity, it must be combined with another that remedies this defect. Potential blend components and variously proportioned mixtures of two or three such components must therefore be carefully screened by plastometric and dilatometric tests. For practical purposes, however, much reliance is also placed on petrographic analyses from which it is possible to predict the strength and stability of the coke that the blend would yield in a commercial oven and to determine optimum blend compositions. The basic correlations among petrographic composition, blend composition, and coke stability that such predictions require have been empirically worked out by Shapiro, Gray, and Eusner [24], and industrial experience to date seems to have substantially confirmed their validity. But recent reports suggest that they may only be directly applicable to *Carboniferous* bituminous coals. When used with Cretaceous coals, such as western Canadian mvb and lvb coals, which are now beginning to be commercially exploited for coke making, they frequently yield fallacious results [25], and more reliable predictions may become possible from correlations between coke stability and the reactive oxygen contents of the coals from which the coke is made [26].

Also bearing on the suitability of a coal for coke making is its mineral matter (or ash) content and the composition of the ash. As noted in Section 6.3, mineral matter contents greater than ~ 10 or 12% "dilute" the coal sufficiently to diminish its caking properties, and in many instances a coal that would otherwise be rejected by coke makers can be upgraded into an attractive blend component by conventional cleaning (see Section 8.1). Beyond that, however, there is some evidence that high ash contents also adversely affect coke quality per se. Although acceptable blast-furnace cokes can, where better coals are lacking, be made from coals with as much as 20 or 25% ash, full-scale oven tests with washed and unwashed coals [27] have shown that even small reductions in the ash content of the charge result in much superior coke and smaller amounts of coke fines (or so-called breeze, which cannot be used in blast furnaces). And in blast-furnace tests with cokes containing ~ 6 and 10.2% ash, respectively, it was found that the former allowed $\sim 10\%$ greater iron production at $\sim 9\%$ less coke consumption [28]. But cleaning costs partially offset such gains: Studies that measured the improvements in coke quality against the relative costs of raw and variously cleaned coal [29] suggest that reduction of ash contents below 9–11% is economically unjustified.

Limitations that ash *compositions* impose on the use of a coal in coke-oven charges arise from adverse effects of phosphorus, sulfur, and silicon in coke on iron and steel quality. Users of metallurgical coke consequently specify the maximum concentrations of these elements in cokes that they would be prepared to accept (Note C); and since phosphorus, unlike sulfur and silicon, occurs in coal predominantly in organic combinations and cannot, therefore, be removed by cleaning, coals with excessive amounts of phosphorus are eliminated from further consideration.

Aside from selection and preparation of suitable blends, some control over coke quality can also be exercised by adjusting the bulk density of the oven charge in order to produce denser or more homogeneous cokes. Methods adopted for this purpose include (a) variation of particle size distributions and (b) addition of small amounts of water or oil to the blend. Compression of coal in an oven by *charge stamping*—a technique that was fairly widely used in the 1930s—is now rare and, in any event, practically precluded by the type of oven and charging procedures employed in modern installations.

In fixing on the composition, size distribution, and bulk density of a charge, careful attention must, however, be given to the physical behavior of the charge during the coking process, and especially to the expansion pressures that the charge will exert on oven walls. These pressures arise from the fact that the transient plastic coal zone in a slot-type oven moves slowly inward from the two sides and, constrained by still uncarbonized coal in the center, swells against the outer (coked) layers (see Fig. 11.2.1). In some cases, when high-fluidity hvb coals or blends containing such coals were carbonized, the pressures generated in this manner proved so high as to seriously damage and even totally destroy oven walls. It is therefore now common practice, before accepting new blends for commercial use, to test them exhaustively in scaled-down experimental ovens and, when necessary, to modify blend compositions in the light of such tests. Experimental ovens vary in size, holding as little as 30 lb or as much as 850 lb of coal, but all are designed to simulate a commercial slot-type oven and measure expansion pressure by displacement of a counterbalanced movable wall [31] or with load cells built into their sides [32].

Although there is little consensus about what blast furnaces and other coke-consuming operations can tolerate (specifications are often established by agreement between producers and consumers), the quality of a coke is conventionally assessed in terms of its chemical composition, reactivity, and mechanical properties.

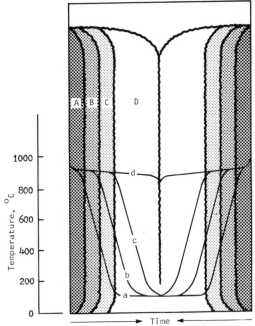

Fig. 11.2.1   Progress of carbonization in a slot-type coke oven. The shaded layers (A, B, C) illustrate how coking progresses with increasing time; and a, b, c, d show the (idealized) temperature gradients across the oven. D, d represent the status just before discharging the coke.

With respect to chemical compositions, which are determined by procedures that are virtually identical with those used for coal (see Section 2.2), practical interest centers primarily on calorific values and on ash, volatile matter, and sulfur contents. Preferred, but by no means exclusively qualified, for metallurgical use are generally cokes with

calorific value	$\geq 14{,}500$ Btu/lb (33,725 MJ/kg)
volatile matter	$< 1.5\%$
ash	$< 10\%$
sulfur	$< 1.0\%$

(all calculated on a dry coke basis).

The reactivity of a coke is loosely defined as its ability to interact with oxygen or steam [33]; but as now understood, reactivity is difficult to measure otherwise than empirically and is therefore determined by standard laboratory tests that express it as an ignition temperature [34], a "critical" air blast [35], or a reaction rate that, in an adiabatic system, will raise the temperature of the coke at some arbitrarily set rate [36]. Comparative studies (e.g. [37]) indicate, however, that reactivities so measured do not correlate well with each other or with the absolute density or porosity (which are theoretically related to reactivity). And it is also still questionable whether such measurements, which are always made on granulated coke samples, really reflect significant aspects of coke performance in industrial operations. Many blast-furnace and cupola operators consequently prefer to evaluate coke reactivity from in-plant material balances or, more simply, by subjective judgments based on experience.

Mechanical properties are, of necessity, similarly assessed by empirical tests but have greater practical significance. Data most frequently sought concern

(a) the size distribution in bulk coke, which is an important factor in determining the permeability of a coke bed and can be obtained from screen analyses; and

(b) coke strength, with respect to which a distinction is made between *hardness* (i.e., resistance to fracture) and *stability* (i.e., resistance to abrasion).

In all cases, test procedures are prescribed by national standards, and several cooperative programs have from time to time been undertaken in efforts to correlate results from different sources [38]. Typical of strength tests are the ASTM drop shatter test (ASTM D 3038-72), which measures the size degradation of a $+2$-in. sample after dropping it from a height of 6 ft onto a steel plate and expresses coke hardness in terms of the resultant $+2$-, $+1\frac{1}{2}$-, $+1$-, and $+\frac{1}{2}$-in. material, and the ASTM tumbler test (ASTM D 3402-72), which equates stability with the percentage of residual $+1$-in. coke when a $+2$-in. sample is subjected to 1400 revolutions in a 36-in.-diam drum at

24 ± 4 rpm. An alternative to the ASTM tumbler method is the Micum drum test [39], which has been recommended by ISO for international use. This yields three indices, viz., $M_{10} = \% -10$ mm, $M_{20} = \% -40- +20$ mm, and $M_{40} = \% +40$ mm, and is considered to discriminate more definitively between initial fracture (due to fissures in coke pieces) and subsequent shatter or attrition. The so-called Micum cohesion is defined by $C_M = (M_{40} + M_{20}) - M_{10}$.

11.3 Coal Tar Processing

Introduction and rapidly expanding use of "coal gas" for street and house lighting in England in the closing years of the eighteenth century generated large quantities of coal tar that during the following 50 years were mostly discarded as troublesome waste. But from the 1850s on, the development of Western Europe's chemical industry lent them increasing economic importance. At first mainly providing precursors for synthesis of textile dyestuffs, coal tars soon established themselves as essential raw materials for the production of solvents, pharmaceuticals, plastics, and synthetic fibers (Note D). And by the early 1930s a coal-tar-processing technology had emerged through which, whenever petroleum was not available, tars could be upgraded to high-quality gasolines and aviation fuels.

Since, additionally, many of the later petroleum refining techniques are also directly applicable to coal tar, the origin of a tar, i.e., whether produced by LT or HT carbonization, therefore now determines only *how* it is most advantageously processed into a useful product slate.

Like crude petroleum, most *high-temperature* tars are first fractionated by distillation into three oil cuts, which are customarily designated as (a) light oil, (b) middle (or tar acid) oil, and (c) heavy (anthracene) oil. This primary separation is carried out in vertical or horizontal (3000–8000 gal capacity) batch stills or in continuous "pipe stills," which heat the tar to a predetermined temperature and release it into a fractionating tower; and by appropriate choice of cut points, one thereby obtains distillate oils that, despite being of different chemical composition than the corresponding petroleum fractions, are effectively interchangeable with them and also afford specific sources of important tar chemicals.

The *light oil* cut[5] (b.p. $<220°C$ or $<430°F$), which consists mostly of benzene (45–72%), toluene (11–19%), xylenes (3–8%), styrene (1–1.5%),

---

[5] In practice, relatively small amounts of light oil are condensed with the tars sent to refineries; most is recovered from gas streams at carbonization plants by passing the crude gas through a series of scrubbers in which light oils are stripped out by a recycled wash oil. Some European producers prefer adsorption on activated carbon and subsequent steaming of the carbon.

and indene (1–1.5%), is either processed into motor gasolines and aviation fuels or fractionated to provide solvents and feedstocks for chemical industries. In either case, upgrading centers mainly on removal of sulfur compounds, nitrogen bases, and undesirable unsaturates, and this is usually accomplished by acid-washing the oils in batch agitators [40] or by hydrogenating them over cobalt–molybdenum or nickel–tungsten catalysts [41]. In the more conventional acid wash, the crude oil cut is intimately mixed with strong sulfuric acid, neutralized with ammoniacal liquor and/or caustic soda, and, after separation of the aqueous phase, steam-distilled or stripped of polymeric matter by centrifuging. Hydrogenation, a much later development, is carried out at 300–400°C under 2–3 MPa pressure and, when first introduced in the 1930s, was primarily designed to saturate olefins in LT tars. But subsequently it proved so attractive for purifying HT oils that many modern refineries prefer it to acid-washing [42].

Where the purified light oils are fractionated into their components rather than sold as motor fuels, the final processing steps involve removal of thiophene and residual saturated hydrocarbons (notably cyclohexane, *n*-heptane, and methyl hexanes) from the distillates; and for this purpose, a variety of techniques (including azeotropic and extractive distillation, selective solvent extraction, and crystallization) is used. The principles of azeotropic distillation have been discussed by Mair *et al.* [43]; extractive distillation is the subject of several US and British patents [44]; and solvent extraction is exemplified by the Udex process [45], which employs mixtures of various polyethylene glycols in water.

Middle oils are usually cut to boil between ~220 and 375°C (~430 and 710°F) and, after sequential extraction of tar acids, tar bases, and naphthalene, processed to meet specifications for diesel fuels, kerosene, or creosote (Note E).

Tar acids, which are mostly comprised of phenols, are recovered by mixing the crude middle oils with a dilute solution of caustic soda, separating the resultant aqueous "carbolate" layer, and passing steam through it in order to remove residual traces of hydrocarbons. The acids are then sprung by treating the carbolate with CO_2 or dilute sulfuric acid and finally fractionated by vacuum distillation. The principal products are phenol, cresols, and xylenols.

Tar bases are isolated in an analogous manner, i.e., by treating the acid-free oils with dilute sulfuric acid, regenerating the bases from the aqueous sulfate solution by adding an excess of caustic soda or lime slurry, and fractionating the mixture. The compounds thus produced are pyridine, picolines, lutidines, and anilines (which can also, though in smaller yields, be obtained from some light oil cuts), as well as quinoline, isoquinoline, and methyl quinolines.

Selective solvent extraction, which has from time to time been proposed as an alternative method for isolating and separating tar acids and bases [46], has so far found little commercial use.

Naphthalene is mostly recovered by fractional distillation of the neutral middle oils and is then purified by recrystallization [46] or by azeotropic distillation with cresols. A very pure product can be obtained by hydrogenating the crude naphthalene over a platinic oxide–alumina catalyst at 500–550°C before fractionating and recrystallizing it [47].

The temperature to which distillation of *heavy* oils is taken depends on what type of (undistilled) pitch residue is desired and lies usually between 450° and 550°C (840 and 1020°F).[6] But in all cases, the distillate is a rich source of higher hydrocarbons, notably anthracene, phenanthrene, carbazole, acenaphthene, fluorene, and chrysene, which are extracted by fractional distillation and thereafter separated and purified by azeotropic distillation and recrystallization. For example, distillation with a monohydric alcohol carries anthracene and phenanthrene over, while leaving carbazole behind [48]; and distillation with diethylene glycol serves to separate anthracene and phenanthrene from carbazole and fluorene [49].

The remaining heavy oils, like corresponding heavy residues from petroleum refining, are marketed as fuel oils or, in part, blended with pitches to meet specifications for various grades of road tar [50] and so-called refined tars that are used in the manufacture of coal tar paints and tar-saturated insulating felts.

The residual coal tar *pitches*, whose softening characteristics depend on the maximum temperature to which tar distillation is carried, are complex mixtures that contain over 5000 separate (mostly polycondensed aromatic) compounds and possess considerable economic importance because of their unique resistance to water and weathering. In addition, they are also extensively used as briquetting binders (see Section 8.3) and as binders in the preparation of all types of carbon electrodes (including "continuous anode" Soderberg electrodes) and other carbon artifacts.

By different means, a substantially similar slate of products can be prepared from *low-temperature* coal tars, especially those obtained by carbonization of subbituminous and bituminous coals.

The most important upgrading technique, which was extensively used during World War II and afterward to convert LT tars into motor gasolines, diesel fuel, heating oils, and waxes, involves hydrogenation and subsequent fractionation of the hydrogenated product mixture. (A good review of

---

[6] The highest boiling compound so far identified in coal tar distillates is fulminene (b.p. 555°C). However, in some instances distillation is continued until a pitch coke residue is produced, and extensive thermal cracking then contributes significant quantities of aromatics to the light oil fraction (which consequently yields improved motor gasolines).

German wartime practices using this approach has been published by Holroyd [51].) Hydrogenation was usually conducted over molybdenum and/or tungsten sulfide catalysts at temperatures between 420 and 500°C and pressures between 14 and 42 MPa; but in some operations, e.g., where less complete conversion to gasoline is required, conditions can be much milder. In one German refinery (at Böhlen), brown coal tar was successfully converted to a light-yellow diesel fuel by hydrogenation at 425–440°C and 1.5–7 MPa over 10% molybdena on γ-alumina [52].

Yields of motor gasolines plus diesel fuel obtained by such pressure-hydrogenation average 80% of the original tar, and hydrogen consumptions are usually modest, averaging 1200–1500 scf per product barrel [53]. However, owing to their lower content of aromatic and naphthenic compounds, *lignite* tars furnish significantly poorer gasolines than tars from more mature coals,[7] and such gasolines must therefore be blended with higher-quality ones. Alternatively, the crude tar or the fraction boiling below 400°C can be subjected to liquid-phase hydrocracking, stripped of tar acids and bases, and aromatized at 5–10 MPa over a molybdena–alumina or chromia–molybdena–alumina catalyst [54].

Recovery of phenols from the hydrogenated tar, which is as important in LT tar refineries as in those processing HT tars, is normally achieved with aqueous caustic soda in continuous-flow equipment [55], but water at 165°C and ~ 1.7 MPa has also been used for that purpose [56].

Where diesel fuel and heating oils rather than gasolines are wanted, hydrogenation or hydrocracking can be replaced by thermal or catalytic cracking. A case in point is provided by the Rositz refinery in Germany [57], where LT lignite tars have been cracked at 430–450°C and 4–6 MPa to yield $\sim 20\%$ diesel fuel, 47% heating oils, 4% gasolines, and 13% fuel gas, the remainder being pitch and coke. In the United States, the success of this operation has prompted several studies of delayed coking, which demonstrated the commercial feasibility of converting bituminous coal LT tars into gasolines and electrode cokes by such means [58]. Of particular interest in this connection is the observation that fresh lignite tar vapors produced at 550°C and rapidly sent through a cracker at 800°C yielded 70 lb methane and 40 lb ethylene as well as 6 lb benzene, 5 lb toluene, and 3 lb naphthalene per ton of dried lignite [59].

And finally: In one large commercial installation, the Sächsische Werke A.G. at Espenhain, it has been shown that gasolines, diesel fuel, heating oils, and waxes can be produced by continuous selective solvent extraction of LT tars with sulfur dioxide, naphtha, and ethylene dichloride [60].

---

[7] Gasolines from lignite tars typically have octane numbers between 60 and 70 while those of gasolines from LT tars of higher-rank coals lie in the range 80–90.

11.4   Flash Pyrolysis

When coal is "flashed" to active decomposition, i.e., heated to $>450°C$ in a fraction of a second rather than over many minutes, its component molecules tend to fragment more extensively than otherwise; and if the resultant radical species are promptly stabilized and prevented from re-combining, the quantities of material then volatilized can exceed the analyti-cally determined volatile matter content of the coal by a wide margin (see Fig. 11.4.1).

This phenomenon has, since the early 1960s, attracted attention as a potentially practical method for obtaining greater yields of useful gaseous and liquid hydrocarbons than accrue from conventional carbonization of coal.

Some of the techniques for flash-pyrolyzing coal involve passage of hydrogen-entrained coal through a preheated pressurized tube reactor [61] or rapid cycling of a hydrogen-fluidized coal bed [62], and are thus variants of what has been termed *hydropyrolysis* [63]. The extent of conversion that can be achieved by such processing is perhaps best illustrated by the observation [64] that hydropyrolysis of entrained bituminous coal at $900°C/5.2$ kPa transformed $\sim 95\%$ of the organic carbon of the coal into methane and other hydrocarbon gases within 1 sec. Other studies [63, 65, 66] have, however, made it evident that yields and product compositions are critically dependent on operating conditions and that high gas yields result from severe secondary hydrocracking reactions. Under milder conditions, which favor formation of liquid hydrocarbons (Note F), conversion is much less extensive and lies

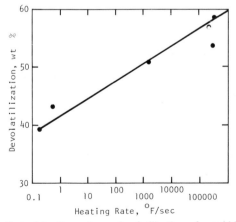

Fig. 11.4.1   The effect of heating rate on devolatilization of a subbituminous coal. The maximum temperature to which the coal was heated in these experiments was 1500°F (815°C). (After Eddinger *et al.* [63].)

Table 11.4.1

Liquid hydrocarbon yields from hydropyrolysis[a]

Reference	Coal	Product	Best yield
Weller et al. [65]	hvC bituminous	Oils	16
Hiteshue et al. [66a]	hvC bituminous	Oils	40
Hiteshue et al. [66b]	Lignite;	Oils	33
	hvC bituminous;	Oils	26
	hvA bituminous	Oils	19
Hiteshue et al. [66c]	Subbituminous	Oils,[b]	40
		C_6 aromatics	6
Albright and Davis [67]	Subbituminous	Oils	28
Squires [68]	Bituminous	Oils[b]	31
Wood and Wiser [69]	Subbituminous	Oils	50
Schroeder [70]	Subbituminous	Oil and tar	70
Steinberg et al. [71]	Lignite	BTX;[c]	10
		benzene	9
Rosen et al. [72]	Subbituminous	Benzene	35

[a] All yields in weight percent of daf coal.

[b] Designated as condensate oils.

[c] This designation is used for a light oil fraction predominantly composed of benzene, toluene, and xylenes.

typically between 30 and 50 wt %. Table 11.4.1 illustrates this with data from recent literature.

Maximum oil yields generally accrue from hydropyrolysis near 550°C [73] and increase almost linearly with hydrogen pressure (see Figs. 11.4.2 and 11.4.3). But if the residence time of the coal in the reactor is very short, the oils resemble LT tars, and significant amounts of light oils (predominantly comprised of benzenoid hydrocarbons) appear only when residence periods are slightly extended. Overall, hydropyrolysis must therefore be considered to proceed in two quite discrete steps, first forming dark, viscous, and partly nondistillable tarlike oils and then hydrocracking these primary liquids in the vapor phase to gas, light oils (BTX), heavy oils, and a solid residue.

Since optimum hydrocracking conditions, which would maximize BTX yields, rarely, if ever, coincide with optimum conditions for oil vapor formation, further development of hydropyrolysis will undoubtedly involve separation of these steps.

In passing, it should be mentioned that oil yields at ~ 500°C can be substantially enhanced by pyrolyzing the coal in the presence of a hydroliquefaction catalyst (such as stannous chloride or molybdenum sulfide [73]). But the effects of such catalysts diminish quickly when temperatures are raised beyond 600°C; and the very high yields (up to 78 wt % within 30 min)

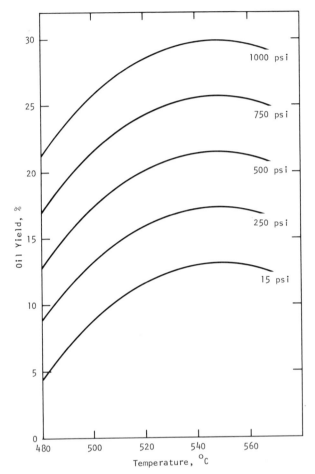

Fig. 11.4.2 The effect of temperature on oil yield from fast hydropyrolysis in a fluidized bed reactor. Pressures refer to the partial pressures of hydrogen used in the tests. (After Holmes *et al.* [73].)

reported by Wald [74] from hydropyrolysis of coal in molten trihalides of antimony, arsenic, or bismuth are almost certainly due to slow hydrogenation of the coal substance rather than to hydropyrolysis per se.

While hydropyrolysis has apparently not yet been taken very far beyond bench-scale study,[8] several variants (all employing the principles of flash pyrolysis but foregoing potential maximum oil production in order to obviate the need for external hydrogen) are undergoing pilot-plant testing. The

[8] The Union Carbide Corporation [67] has reportedly operated a 1250-lb/hr fluidized bed hydropyrolysis unit, but no results from this facility have so far been released.

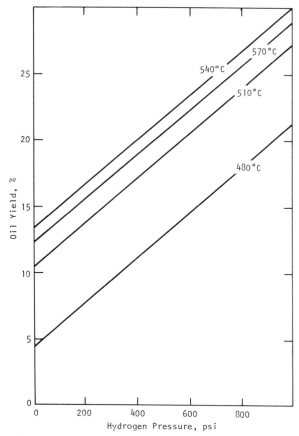

Fig. 11.4.3 The effect of hydrogen pressure on oil yields from fast hydropyrolysis in a fluidized bed reactor. (After Holmes *et al.* [73].)

COGAS and *Garrett* processes, both now sufficiently well developed to be used commercially where economic circumstances warrant it, will illustrate these.

The COGAS process [75][9] combines multistage pyrolysis of gas-entrained coal with downstream gasification of the devolatilized residual char and thus generates a synthetic medium-Btu gas (see Chapter 12), as well as oils. A flow diagram of this scheme is shown in Fig. 11.4.4. Fast pyrolysis is here accomplished by fluidizing previously dried, −1/8-in. (−3.2-mm) coal with a hot (1600°F; ~870°C) countercurrent slipstream of gas from the gasifier and cascading it through a series of reactors, in the last of which it attains a tem-

[9] An earlier version of this process, which contemplated disposing of the devolatilized char as fuel for thermal generation of electric energy, was known as the COED Process.

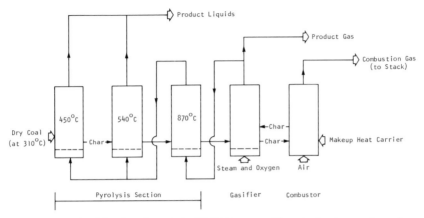

Fig. 11.4.4 Simplified flow sheet of the COGAS process. The retorting temperatures shown in the pyrolysis section are typical but can be varied to suit particular needs.

perature of $\sim 1200°F$ ($\sim 650°C$). The char is then passed to a gasifier where it is partially reacted with steam at 1600°F to form a mixture of hydrogen and carbon monoxide (via $C + H_2O \rightarrow H_2 + CO$; see Section 12.1); and unreacted char exiting from the gasifier is finally burned in the combustor to generate a solid heat carrier,[10] which is recycled to the gasifier.

Typical product distributions from pyrolysis of subbituminous and bituminous coal in a 36-ton/day pilot plant, which has been operated since 1971 near Princeton, New Jersey, are summarized in Table 11.4.2. In favorable

Table 11.4.2

Product yields from COGAS pilot plant[a]

	hvb coal, Illinois	hvb coal, Utah	Subbituminous coal, Wyoming
Char	60.7	59.0	50.0
Oils	18.7	21.9	11.2
Aqueous liquor	5.8	21.5	11.6
Gas[b]	14.6		27.2

[a] After Paige [75], Eddinger and Sacks [75]; all yields in weight percent of dry coal.

[b] Excluding gas from downstream char gasification.

[10] This carrier is either residual unburned char from the combustor or an inert refractory, such as sand, pelletized alumina, or pelletized coal ash. In either case, the "spent" carrier, after having given up its sensible heat to char in the gasifier, is returned to the combustor for regeneration.

cases the total oil yield corresponds to ~1.05–1.1 bbl (barrels) (~0.17 m³) per ton of dry feed coal, and in all instances this oil can be upgraded to a marketable sweet "synthetic crude" by conventional hydrotreating. Recovery of hydrotreated oil is ~93%, with the remainder mostly accounted for in C_1–C_3 gases.

The other scheme, the Garrett Flash Pyrolysis process [76] developed by Occidental Petroleum Company, is designed to produce *either* a hydrocarbon-rich fuel gas (with 600–650 Btu/scf) *or* oil, and a residual char suitable for generation of electric energy. In this process (see Fig. 11.4.5), pulverized coal is fed into a mixing chamber where it is contacted with hot recycle char and then immediately moved into an entrained-flow reactor section in which it is flash-pyrolyzed within 2 sec at ~1600°F (~870°C) if *gas* is to be made, or at ~1050°F (~570°C) if the *oil* yield is to be maximized. (Figure 11.4.6 shows how oil yields from the Garrett process vary with pyrolysis temperatures.) Product gas or oil vapors are then separated from char in a series of cyclones, and some char is returned to the combustor where it is partly burned in air to provide a heat carrier for the pyrolyzer.

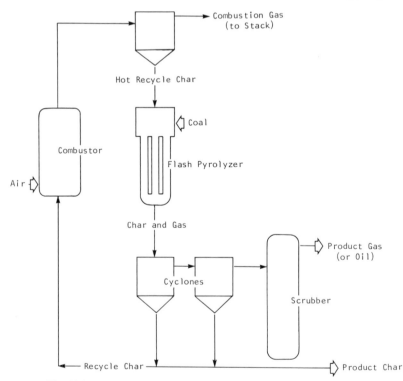

Fig. 11.4.5 Simplified flow sheet of the Garrett flash pyrolysis process.

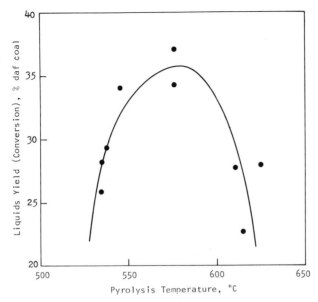

Fig. 11.4.6   Variation of oil yield from Garrett flash pyrolysis with processing temperature. (After Sass [77].)

At optimum reaction temperatures for gas production, a gas yield corresponding to ~ 30 wt% of the dry feed coal, with an average gas composition as shown in Table 11.4.3, can be obtained. In addition, some 10 wt % of a heavy tarlike oil is produced, and this can either be refined or recycled and cracked to gas. The residual char usually has a heating value of $\sim 11,700$ Btu/lb and contains $\sim 10\%$ volatile matter.

Table 11.4.3

Typical composition of gas from Garrett flash pyrolysis[a]

Hydrogen	26.8
Carbon monoxide	30.0
Carbon dioxide	8.5
Methane	22.4
$C_2 +$ hydrocarbons[b]	12.3
	100.0
Calorific value	625 Btu/scf (at 25°C)

[a] After McMath *et al.* [76]; expressed in mole percent of dry, N_2- and H_2S-free gas.
[b] Mainly composed of ethylene.

Table 11.4.4

**Typical product distribution
from Garrett flash pyrolysis
operated to make oils**[a]

Char	56.7
Oils[b]	35.0
Gas	6.6
Aqueous liquor	1.7

[a] After McMath *et al.* [76]; expressed in weight percent of dry
feed coal.
[b] Contain 20–30% asphaltenes
and 5–20% benzene-insoluble
matter.

Typical distributions when making *oil* are summarized in Table 11.4.4
and show oil yields averaging 35 wt % of the dry feed coal. This yield corresponds to ~1.8 bbl/ton and is more than twice the tar yield determined by
Fischer assays (see Section 6.2; Note G). As in the COGAS process, the
heavy oils can be upgraded by hydrotreating.

11.5   Plasma Pyrolysis

When a gas is exposed to an electric arc or to radio-frequency radiation,
it dissociates and ionizes to form a hot plasma whose enthalpy can be used to
effect very fast high-temperature decomposition of coal. This and the fact
that above ~2450°F (1345°C) acetylene is thermodynamically more stable
than other hydrocarbons have prompted efforts to develop plasma pyrolysis
for production of acetylene and carbon blacks from coal.

The type of equipment in which a suitable plasma, usually from argon,
hydrogen, or an argon–hydrogen mixture, can be generated is illustrated in
Fig. 11.5.1. But the use of such a "gun" for coal pyrolysis is, in practice,
complicated by (a) materials problems (specifically, abrasion) associated
with injection of pulverized coal into the plasma and (b) the need to quench
reaction products quickly in order to prevent decomposition of acetylene to
carbon and hydrogen. It has therefore been found useful to resort to spinning
or magnetically rotated plasmas [78, 79]—and most important to hold the
residence time of primary decomposition products to a few milliseconds.

Subject to these precautions, acetylene yields approaching or sometimes
even exceeding the total (nominal) volatile matter contents of the test coals
have been obtained (see Table 11.5.1).

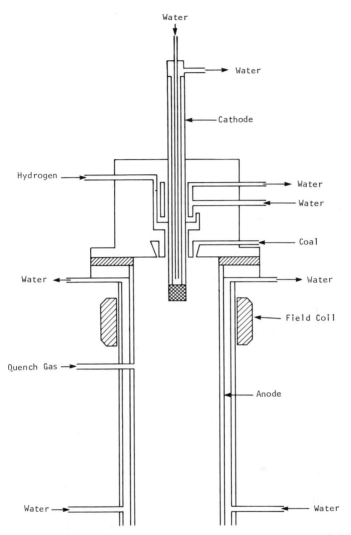

Fig. 11.5.1   Schematic of a plasma gun reactor. (After Gannon and Krukonis [79].)

However, in most cases high yields accrue only when very fine coal particles are injected into the plasma and injection rates are low. Bond *et al.* [80] thus found that formation of C_2H_2 falls drastically when particle sizes are increased from 50 to 200 μm, and the exceptionally high yields recorded by Nicholson and Littlewood [81] were contingent on feed rates as low as 0.04 gm/min. (The wide yield variations in Table 11.5.1 reflect, therefore,

Table 11.5.1

Acetylene yields from plasma pyrolysis of coal[a]

Reference	Plasma	"Best" yield
Bond *et al.* [80]	Argon;	20
	argon–10% hydrogen	40
Jimeson [82]	Argon;	6
	argon–12% hydrogen	8
Graves *et al.* [83]	Argon;	15
	argon–hydrogen	40
Kawana *et al.* [84]	Argon;	52
	hydrogen	65
Newman *et al.* [85]	Hydrogen	46
Anderson *et al.* [86]	Argon;	6
	argon–12% hydrogen	12
James [87]	Argon	18
Nicholson and Littlewood [81]	Argon–hydrogen	74
Gannon and Krukonis [79]	Argon–hydrogen	35

[a] All yields in weight percent of daf coal.

different operating conditions, as well as the influence of different coal types and the beneficial effect of hydrogen.)

Also, energy consumptions have generally been found to be high. The molar heats of conversion of benzene or naphthalene to acetylene are approximately 110 kJ per C atom; and assuming a reaction path that ultimately delivers acetylene by degradation of simple aromatics, the specific energy consumption should therefore be in the order of 2 to 3 kWh/kg C_2H_2. In fact, however, most investigators have observed that the low feed rates necessary for extensive conversion of coal material into acetylene result in far higher energy consumptions, commonly in the range of 50 to 100 kWh/kg C_2H_2.

So far, only Gannon *et al.* [79] seem to have succeeded in obtaining high C_2H_2 yields (up to 35 wt % on dry, ash-free coal substance) at reasonable feed rates (up to 600 gm/min) and acceptable specific energy consumptions ($\sim$9 kWh/kg C_2H_2; Note H). It is undoubtedly significant that these results were recorded when using magnetically rotated hydrogen plasmas and quenching the hot product gases with hydrogen.

11.6   Formed Coke

Formed coke is a carbonized *briquetted* fuel, which, even though manufactured from weakly caking or noncaking coals that are in themselves unsuitable for coke making, can replace conventional lump coke in metallurgical operations (see Section 11.2). It is therefore of particular interest where the prime coking coals that would otherwise be needed are unavailable or too costly.

Except for its greater mechanical strength (which accrues in part from higher briquetting pressures, larger additions of binder, and/or higher final carbonization temperatures), formed coke is indistinguishable from "smokeless" domestic briquettes (e.g., Phurnacite; see Section 11.1) and is, in fact, made by very similar processes. In all cases, the crushed coal is first charred at temperatures between 600 and 800°C (~1100 and 1470°F), then intimately mixed with a binder and briquetted (see Section 8.3), and finally carbonized at 900–1000°C (~1650–1830°F). The initial partial devolatilization (by charring) is designed to prevent swelling and sticking (or, alternatively, excessive shrinkage and fissuring) of the briquettes during HT treatment; and the binders needed for briquetting are usually obtained from the combined by-product tars generated during LT and HT carbonization.

But the strength of the formed coke thus made and hence its suitability for metallurgical purposes are markedly dependent on the operating variables in each of the processing steps. Tables 11.6.1–11.6.3 illustrate this

Table 11.6.1

Effects of final carbonization temperature[a]

Temperature (°C)	Yield (%)	Formed coke properfies		
		VM (%)	Crushing strength (lb)	Porosity (%)
800	86.5	4.8	660	20.5
900	84.4	3.5	820	22.5
1000	83.0	1.6	1080	24.0

[a] Adapted from Idris Jones [89].

Table 11.6.2

Effects of soaking time at final carbonization temperature[a]

Temperature (°C)	Time (hr)	VM (%)	Crushing strength (lb)	Bulk density (lb/ft³)
720	2	12.0	80	41.5
	4	6.0	470	43.0
920	4	4.4	600	43.8
	8	1.9	1150	48.3

[a] Adapted from Idris Jones [89].

Table 11.6.3

Effects of binder amounts[a,b]

Binder (%)	Apparent density[c] (gm/cm^3)	Breaking strength (lb/in.^{2})
5.0	1.261	~385
7.5	1.307	915 ± 32
10.0	1.329	1270 ± 37
12.5	1.340	1440 ± 45
15.0	N.A.[d]	1370 ± 58
20.0	N.A.[d]	1030 ± 116

[a] After Gillmore et al. [90].
[b] Briquetting pressure: 3920 lb/in.^{2}; carbonized at 10°C/min to 750°C; soaking period: 1 hr.
[c] Calculated from briquette weights and dimensions.
[d] Briquettes distended and/or extensively fissured.

Table 11.6.4

FMC formed coke properties[a]

	FMC coke	"Standard" metallurgical coke
Relative crushing strength, lb/in.^{2}	>3000	400–2000
(ASTM) apparent density, gm/cm^3	0.8–1.2	0.85–1.3
Bulk density, lb/ft^3	30–45	20–30
Hardness, moh scale	6+	6+
Surface area,[b] m^2/gm	50–200	1–25
Chemical reactivity,[c] %/hr	15–50	1–5
Volatile matter, %	<3	1–2

[a] After FMC Corporation [94].
[b] Measured by BET nitrogen sorption.
[c] CO_2 reactivity at 925°C.

Table 11.6.5

Experimental blast-furnace test data[a]

	FMC coke	"Standard" 2 × 3/4-in. metallurgical coke
Sinter/coke, lb/lb	2.96	2.82
Coke rate, lb/ton hot metal	1062	1096
Production rate, lb/hr	3601	3384
Slag volume, lb/ton hot metal	604	600
Stack dust, lb/hr	20.2	12.7

[a] After FMC Corporation [94].

with some examples of the influence of heating rates, final heat-treatment temperatures, soaking time, and binder proportions on the crushing strengths of the coked briquettes.

Among more recently proposed schemes for producing formed coke (e.g. [91, 92]), the most prominent and apparently successful is the FMC Corporation's process, which began to be researched in the mid-1950s and was later extensively tested in a 250-ton/day output demonstration plant at Kemmerer, Wyoming [93]. This installation processed a variety of subbituminous coals and lignites and yielded small ($1\frac{1}{4} \times 1\frac{1}{8} \times \frac{3}{4}$-in. or $\frac{7}{8} \times \frac{3}{4} \times \frac{1}{2}$-in.) coked briquettes that performed as well as, or better than, conventional lump coke when used in experimental blast furnaces and phosphorus furnaces [94]. A comparison of FMC formed coke and a standard metallurgical (blast-furnace) coke is shown in Table 11.6.4, and some data from tests of FMC coke in a US Steel Corporation experimental blast furnace are summarized in Table 11.6.5.

Despite this and other evidence of its practical utility, however, formed coke has not yet found significant commercial application.

Notes

A. Where, as here and later in this book, data are presented in SI *or* other conventional units, standard notations are used. Particular note should be taken that 1 MMcf = 1000 Mcf = 10^6 scf (standard cubic feet); 1 Bcf = 1000 MMcf; 1 Mcf $\simeq$ 28.32 m^3; 1 Btu/scf $\simeq$ 37.26 kJ/m^3; 1 psi $\simeq$ 6.9 kPa; 1 ton = 0.907 tonne; 1 lb $\simeq$ 0.45 kg; 1 (U.S.) gal $\simeq$ 3.785 liters.

B. In passing, it is interesting to note that in the late 1950s beehive ovens still furnished ~ 5–6% of the total annual coke production in the United States.

C. For production of pig iron in acid Bessemer processes, US specifications [30] put these maximum concentrations at S = 0.8%, Si = 1.5%, and P = 0.1%. For basic open-hearth operations, P is not specifically limited but usually held below 0.6%, while maxima for S and Si are 0.5 and 1.25%, respectively.

D. Examples of synthetic textile dyestuffs are *Perkin's mauve* (from aniline) and *alizarin* (from anthracene). Alizarin (or Turkey red) was previously obtained from the roots of the madder plant; and synthesis of this dye is estimated to have released more than 300,000 acres (122,000 hectares) in Europe and Asia Minor, on which madder had been grown, to food production. The subsequent commercial synthesis of *indigo* from coal tar chemicals (in 1897) eventually freed another 1,000,000 acres. Still greater economic importance came to attach to coal tars in the wake of Baekeland's discovery, in 1907, of phenol–formaldehyde resins (for which the tars furnished phenol). And uses for other tar chemicals developed through later synthesis of *nylon* and other artificial yarns.

E. Specifications for *diesel fuels* (b.p. < 390°C) are chiefly concerned with ensuring suitable viscosity and ignition characteristics, and are otherwise quite flexible. *Kerosene* (boiling range ~ 180–320°C), also referred to as solvent naphtha, is a cleaning fluid, heating oil, and illuminant but is also used as a jet fuel. For this latter purpose, it must be refined to consist mostly of C_{10}–C_{16} paraffins and aromatics. *Creosote* is an important wood preservative and marketed as "low-residue" or "high-residue" creosote, depending on whether it contains little or substantial amounts of material boiling at > 355°C. Therefore, as a rule, further processing of the neutral middle oils requires little more than blending of different cuts obtained by fractional distillation (which is conventionally undertaken in order to recover naphthalene).

F. In American literature, liquid hydrocarbons are frequently called *oils* even when they are heavy tarlike substances.

G. While the yield data cited here and in later sections are taken from the technical literature, a word of caution is necessary: In most cases published product yields refer to *a ton of coal actually processed and thus exclude coal or coal products used to provide process heat.* If calculated on the total weight of coal taken into the plant, yields would therefore often be significantly lower than claimed.

H. It is interesting to observe that these specific energy consumptions are very similar to specific energy consumptions entailed in plasma-conversion of *oil* to acetylene [88].

References

1. F. M. Gentry, "The Technology of Low-Temperature Carbonization." Williams and Wilkins, Baltimore, Maryland, 1928.
2. Utah Conservation and Research Foundation; "Low-Temperature Carbonization of Utah Coals." Quality Press, Salt Lake City, Utah, 1939.
3. P. J. Wilson Jr. and J. D. Clendenin, *in* "Chemistry of Coal Utilization." Suppl. Vol. Wiley, New York, 1963.
4. G. Lorenzen, *Brennst. Chem.* **32,** 324 (1951); D. T. Barritt and T. Kennaway, *J. Inst. Fuel* **27,** 229 (1954); W. Gollmer, *Glückauf* **92,** 320 (1956).
5. E. O. Rhodes, U. S. Bureau Mines Inform. Circ. No. 7490 (1949); G. L. Kennedy, *J. Inst. Fuel* **33,** 598 (1960).
6. R. J. S. Jennings, *Chem. Process Eng.* **34,** 343 (1953).
7. Anonymous, *Coke Gas* **14,** 5 (1952).
8. G. S. Pounds, *J. Inst. Fuel* **24,** 61 (1951); *Coke Gas* **21,** 395 (1959).
9. Anonymous, *Coke Gas* **16,** 383 (1954).
10. Anonymous, *Coke Gas* **8,** 77 (1946); see also Lorenzen, and Barritt and Kennaway [4], and Rhodes [5].
11. W. T. Reid, U. S. Bureau Mines Inform. Circ. No. 7430 (1948).
12. A. B. Jones, *Coke Gas* **12,** 287, 229, 331 (1950).
13. T. Nicklin and M. Redman, *Gas World* **139,** 784, 854, 924, 998 (1954).
14. C. E. Lesher, *Coke Gas* **12,** 269 (1950); C. E. Lesher and J. B. Goode, U. S. Patent 2,287,437 (1942).
15. G. V. Woody, *Ind. Eng. Chem.* **33,** 841 (1941).
16. R. L. Kimberly, Engineering Experimental Station, Univ. of Kentucky, Bull. No. 21 (1951).
17. V. I. Korobchanski, *Tr. Khim. Tekhnol., Fak. Donetsk. Ind. Inst.* 46 (1956).
18. W. M. Farafonow, U. S. Patent 2,787,584 (1957).
19. A. C. Fieldner, U. S. Bureau Mines Tech. Paper No. 396 (1926).
20. K. L. Storrs, U. S. Patent 2,239,833 (1941); 2,349,387 (1944); 2,525,051 (1950); G. W. Carter, *Chem. Eng. Progr.* **34,** 180 (1947); U. S. Patent. 2,809,154 (1957).
21. K. Baum, U. S. Patent 2,825,679 (1958).
22. F. E. Poindexter and F. W. Lowe, U. S. Patent 2,903,400 (1959).
23. V. F. Parry, U. S. Bureau Mines Rep. Invest. No. 5123 (1955); V. F. Parry, W. S. Landers, E. O. Wagner, J. B. Goodman, and G. C. Lammers, U. S. Bureau Mines Rep. Invest. No. 4954 (1953).
24. N. Schapiro, R. J. Gray, and G. R. Eusner, *Proc. Blast Furn. Raw Mater. Conf.* **20,** 89 (1961).

25. B. S. Ignasiak and N. Berkowitz, *Bull. Can. Inst. Min. Metall.*, July (1974).
26. D. Carson, B. S. Ignasiak, and N. Berkowitz, *Bull. Can. Inst. Min. Metall.*, December (1976).
27. C. D. King, *AIME Proc. Blast Furn. Coke Oven Raw Mater. Comm.* **3**, 3 (1943); **4**, 181 (1944).
28. R. L. Gray and N. Isenberg, *AIME Proc. Blast Furn. Coke Oven Raw Mater. Comm.* **13**, 107 (1954).
29. J. D. Price, *Proc. Int. Coal Prep. Conf., 2nd, Essen* Paper No. 5,111, 1 (1954).
30. D. R. Mitchell and H. B. Charmbury, *in* "Chemistry of Coal Utilization" (H. H. Lowry, ed.), Suppl. Vol., p. 317. Wiley, New York, 1963.
31. R. E. Brewer, *in* "Chemistry of Coal Utilization," Vol. 1, p. 240. Wiley, New York, 1945; C. L. Potter, *AIME Proc. Blast Furn. Coke Oven Raw Mater. Comm.* **11**, 159 (1952); D. C. Coleman and P. J. Fareley, *ibid.* **15**, 176 (1956).
32. J. G. Price, R. W. Schoenberger, and B. Perlic, *AIME Proc. Blast Furn. Coke Oven Raw Mater. Comm.* **17**, 212 (1958); H. W. Jackmann, R. J. Helfinstine, R. L. Eissler, and F. H. Reed, *ibid.* **14**, 204 (1955).
33. R. A. Mott and R. V. Wheeler, "The Quality of Coke," p. 464. Chapman & Hall, London, 1939.
34. R. V. Wheeler, *J. Chem. Soc.* **113**, 945 (1918); K. Peters and W. Picker, *Angew. Chem.* **46**, 498 (1933).
35. H. E. Blayden, W. Noble, and H. L. Riley, *J. Inst. Fuel* **7**, 139 (1934); P. J. Askey and S. M. Doble, *Fuel* **16**, 359 (1937).
36. American Iron and Steel Inst., New York (1952).
37. G. Langlois, Ch. G. Thibaut, and R. Wildenstein, *Publ. IRSid. Ser. B.* No. 35 (1957).
38. R. W. Campbell, *Blast Furn. Steel Plant* **40**, 653, 779 (1952); B. P. Mulcahy, *ibid.* **41**, 408 (1953); W. T. Rogers, *ibid.* **43**, 627 (1955).
39. ISO Tech. Comm. 27, Document 475 (1959).
40. W. L. Glowacki, *in* "Chemistry of Coal Utilization," Vol. 2, p. 1206. Wiley, New York, 1945; J. J. Lawton, U. S. Patent 2,720,486 (1955).
41. W. Urban, *Erdöl Kohle* **4**, 279 (1951); H. Nonnemacher, O. Reitz, and P. Schmidt, *ibid.* **8**, 407 (1955); F. Trefny, *ibid.* **8**, 874 (1955); Scholven Chemie A. G. British Patent 789,986 (1958).
42. W. Jäckh, *Erdöl Kohle* **11**, 625 (1958).
43. B. J. Mair, A. R. Glasgow Jr., and F. D. Rossini, *J. Res. Nat. Bur. Std.* **27**, 39 (1941).
44. G. Pierotti and C. L. Dunn, U. S. Patent 2,341,812 (1944); R. N. Shiras and A. J. Johnson, U. S. Patent 2,351,028 (1944); W. J. Sweeny, U. S. Patent 2,378,808 (1945); J. Griswold, U. S. Patent 2,537,459 (1951); W. A. Herbst, U. S. Patent 2,665,315 (1954); E. H. Lebeis and F. S. Bondor, U. S. Patent 2,768,131 (1956); Standard Oil Development Co., British Patent 571,913 (1945).
45. D. Read, *Oil Gas J.* **51** (7), 82 (1952); J. C. Reidel, *ibid.* **53** (5), 72 (1954); H. N. Grote, *Chem. Eng. Progr.* **54** (8), 43 (1958); G. C. Johnson and A. W. Francis, *Ind. Eng. Chem.* **46**, 1662 (1954).
46. R. Scott and E. H. Joscelyne, U. S. Patent 2,598,449 (1952); Z. Krzesz, U. S. Patent 2,711,432 (1955); R. S. Detrick and G. G. Laver, U. S. Patent 2,890,254 (1959).
47. A. B. Densham, G. Gough, and R. H. Griffith, British Patent 735,706 (1955).
48. J. W. Andrews, U. S. Patent 2,675,345 (1954).
49. J. Feldman and M. Orchin, U. S. Patent 2,590,096 (1952).
50. E. O. Rhodes, *Encycl. Chem. Technol.* **13**, 626 (1954).
51. R. Holroyd, U. S. Bureau Mines Inform. Circ. No. 7370 (1946).
52. R. Birthler and C. Szkibik, *Freiberger Forsch.* **A36**, 42 (1955).

53. R. D. Law, *Proc. Rocky Mountain Regional Meeting*, Tech. Publ. 57-29. Western Petroleum Refiners' Assoc., Casper, Wyoming (1957).

54. A. V. Lozovoi, D. L. Muselevich, A. I. Blonskaya, T. M. Ravikovich, and T. A. Titova, USSR Patent 111,769 (1958); A. V. Lozovoi, A. A. Kirchko, and D. P. Pchelina, USSR Patent 111,825 (1958).

55. P. LeRoux, *Goudron pour Routes* No. 7, 12 (1958).

56. H. Bahmüller, *Braunkohle* **9**, 486 (1957).

57. C. H. Chilton, *Chem. Eng.* **65** (2), 53 (1958).

58. J. Pursglove Jr., *Coal Age* **62**, 70 (1957); M. B. Dell, *Ind. Eng. Chem.* **51**, 1297 (1959).

59. R. S. Montgomery, D. L. Decker, and J. D. Mackay, *Ind. Eng. Chem.* **51**, 1293 (1959).

60. E. Terres, *Nat. Pet. News* **38** (6), R84 (1946).

61. R. W. Hiteshue, S. Friedman, and R. Madden, U. S. Bureau Mines Rep. Invest. 6376 (1964).

62. T. V. Sheehan, M. Steinberg, and Q. Lee, Brookhaven National Library, BNL 18268 R (1973).

63. R. T. Eddinger, J. F. Jones, J. F. Start, and L. Seglin, *Int. Coal Sci. Conf., 7th, Prague* (1968).

64. F. Mosely and D. Paterson, *J. Inst. Fuel* **40**, 523 (1967).

65. S. Weller, E. L. Clark, and M. G. Pelipetz, *Ind. Eng. Chem.* **42**, 334 (1950); S. Weller, M. G. Pelipetz, and S. Friedman, *ibid.* **43**, 1575 (1951).

66a. R. W. Hiteshue, S. Friedman, and R. Madden, U. S. Bureau Mines Rep. Invest. 6027 (1962).

66b. R. W. Hiteshue, S. Friedman, and R. Madden, U. S. Bureau Mines Rep. Invest. 6125 (1962).

66c. S. Friedman, R. W. Hiteshue, and M. D. Schlesinger, U. S. Bureau Mines Rep. Invest. 6470 (1964).

67. C. W. Albright and H. G. Davis, *Am. Chem. Soc. Div. Fuel Chem. Prepr.* **14**, 99 (1970).

68. A. M. Squires, U. S. Patent 3,855,070 (1975).

69. R. E. Wood and W. H. Wiser, *Ind. Eng. Chem. Process Design Dev.* **15** (1) (1976).

70. W. C. Schroeder, *Hydrocarbon Process.* **55**, 131 (1976).

71. M. Steinberg, T. V. Sheehan, and Q. Lee, *Am. Chem. Soc. Ind. Eng. Div. Ann. Meeting, New York, April* (1976).

72. B. H. Rosen, A. H. Pelofsky, and M. Greene, U. S. Patent 3,960,700 (1976).

73. J. M. Holmes, H. D. Cochran Jr., M. S. Edwards, D. S. Joy, and P. M. Lantz, U. S. Dept. Commerce, NTIS Rep. ORNL-TM-4835 (August 1975).

74. M. M. Wald, U. S. Patent 3,542,665 (1969).

75. W. A. Paige, *Synth. Fuels Coal Conf., 5th Oklahoma State Univ. May* (1975); R. T. Eddinger and M. E. Sacks, *Nat. Meeting, 79th, Am. Inst. Chem. Eng., Houston, Texas* (March 1975).

76. H. G. McMath, R. E. Lumpkin, and A. Sass, *Ann. Meeting, 66th Am. Inst. Chem. Eng., Philadelphia, Pennsylvania* (November 1973); D. E. Adam, S. Sack, and A. Sass, *ibid.*

77. A. Sass, *Ann. Meeting, 65th Am. Inst. Chem. Eng., New York* (November 1972).

78. T. Aust, W. R. Ladner, and G. I. T. McConnell, *Int. Coal Sci. Conf., 6th, Münster, Germany* (1965).

79. R. E. Gannon and V. J. Krukonis, *Am. Chem. Soc. Div. Fuel Chem. Prepr.* **14**, 15 (1970); R. E. Gannon and V. J. Krukonis, OCR Contr. No. 14-01-00001-493, Final Rep. (1972); V. J. Krukonis, R. E. Gannon, and M. Modell, *Adv. Chem. Ser.* **131**, 29 (1974).

80. R. L. Bond, I. F. Galbraith, W. R. Ladner, and G. I. T. McConnell, *Nature (London)* **200**, 1313 (1963); R. L. Bond, W. R. Ladner, and G. I. T. McConnell, *Adv. Chem. Ser.* **55**, 650 (1966); *Fuel* **45**, 381 (1966).

81. R. Nicholson and K. Littlewood, *Nature (London)* **236,** 396 (1972); K. Littlewood and I. A. McGrath, *Int. Coal Sci. Conf., 5th, Cheltenham* (1963).

82. R. M. Jimeson, *Min. Eng.* **15,** 52 (1963).

83. R. D. Graves, W. Kawa, and P. S. Lewis, *Am. Chem. Soc. Div. Fuel Chem. Prepr.* **8,** 118 (1964); R. D. Graves, W. Kawa, and R. W. Hiteshue, *Ind. Eng. Chem. Process Design Dev.* **5,** 59 (1966).

84. Y. Kawana, *Chem. Econ. Eng. Rev.* **4,** 13 (1972); Y. Kawana, M. Makino, and T. Kimura, *Int. Chem. Eng.* **7,** 359 (1967).

85. J. O. H. Newman, A. J. T. Coldrick, P. L. Evans, T. J. Kempton, D. G. O'Brien, and B. Woods, *Int. Coal Sci. Conf., 7th, Prague* (1968).

86. L. L. Anderson, G. R. Hill, E. H. McDonald, and M. J. McIntosh, *Chem. Eng. Progr. Symp. Ser.* **64,** 81 (1968).

87. A. H. James, Ph.D. Thesis, Univ. Sheffield, Dept. Fuel Technol. (1968).

88. K. Gehrmann and H. Schmidt, *Proc. World Pet. Congr., 8th* (1971).

89. W. Idris Jones, *J. Imperial College Chem. Eng. Soc.* **11,** 23 (1957); see also, D. C. Rhys Jones, *in* "Chemistry of Coal Utilization" (H. H. Lowry, ed.), Suppl. Vol., p. 718. Wiley, New York, 1963.

90. D. W. Gillmore, C. C. Wright, and C. R. Kinney, *J. Inst. Fuel* **32,** 50 (1959).

91. K. Baum, U. S. Patent 2,825,679 (1958).

92. C. H. Simpson, Australian Patent 153,801 (1953).

93. J. Work, *J. Met.* May, 635 (1966).

94. FMC Corporation, New York; FMC Coke (June 1965).

CHAPTER 12

GASIFICATION

A sharp distinction—still not always appreciated—must be made between "coal gas" and the gas mixtures accruing from coal gasification.

The former is generated by destructive distillation of coal (see Section 11.1) and is therefore a carbonization by-product even when its manufacture is the *only* purpose of carbonization. Its yield and composition always depend on coal rank and on the temperature at which the coal is carbonized. And since it is, in effect, merely a part of the volatile matter of the coal, it never represents more than a relatively small fraction ($< 10\%$) of the original coal substance. In contrast, gasification converts *all* organic material of coal into gaseous form; coal rank and temperature only affect the rate of gasification; and where desired, it can be made to yield a gas that, except for impurities such as hydrogen sulfide, consists entirely of CO, CO_2, and hydrogen.

The reason for this radical difference lies in the fact that gasification entails far-reaching and controllable secondary interactions of volatile matter and residual char (or coke) with oxygen.

250

12.1 Gasification Reactions

The chemical processes that, in various combinations and to different extents, gasify coal[1] usually proceed at significant rates only at temperatures above 1500°F (815°C) and can be formally represented by

$$C + O_2 \rightleftharpoons CO_2 \qquad (+170.0 \times 10^3 \text{ Btu}; 395.4 \text{ MJ}) \qquad \text{combustion} \qquad (1)$$

$$C + CO_2 \rightleftharpoons 2CO \qquad (-72.19 \times 10^3 \text{ Btu}; 167.9 \text{ MJ}) \qquad \text{Boudouard reaction} \qquad (2)$$

$$C + H_2O \rightleftharpoons CO + H_2 \qquad (-58.35 \times 10^3 \text{ Btu}; 135.7 \text{ MJ}) \qquad \text{carbon–steam reaction} \qquad (3)$$

$$CO + H_2O \rightleftharpoons CO_2 + H_2 \qquad (+13.83 \times 10^3 \text{ Btu}; 32.18 \text{ MJ}) \qquad \text{shift reaction} \qquad (4)$$

The parenthetical terms show the accompanying heat changes per pound and kilogram-mole of carbon gasified (with + and − indicating release or absorption of heat, respectively) and refer to the standard form of solid carbon (β-graphite) reacting at 1000°C ($\sim$1800°F).

At <2000°F (1150°C) and elevated pressures, hydrogen will also interact with carbon to form methane via

$$C + 2H_2 \rightleftharpoons CH_4 \qquad (+39.38 \times 10^3 \text{ Btu}; 91.6 \text{ MJ}) \qquad \text{carbon hydrogenation} \qquad (5)$$

and whenever the carbon source generates volatile matter, further quantities of CH_4 will form by thermal cracking, which is qualitatively expressed by

$$C_m H_n \rightleftharpoons \frac{n}{4} CH_4 + \frac{m-n}{4} C \qquad (6)$$

The reaction equations do not, of course, reveal any of the inherent mechanistic complexities of the processes they represent. For example, reaction (1), although very fast, actually involves [1] initial chemisorption of oxygen at active carbon sites and subsequent detachment of CO, which then interacts with gaseous oxygen; i.e.,

$$\begin{cases} C_{act} + O_2 \rightarrow C(O) + O \\ C_{act} + O \rightarrow C(O) \\ \quad C(O) \rightarrow CO + C_{act} \quad (\text{or} + C_{inact}) \end{cases}$$

and

$$\begin{cases} CO + O_2 \rightarrow CO_2 + O \\ CO + O \rightarrow CO_2 \end{cases}$$

---

[1] In principle, these processes will, of course, also gasify any other source of organic carbon, such as coke or wood.

The carbon–steam reaction (3) has been postulated [2] to follow an analogous course; i.e.,

$$C_{act} + H_2O \rightarrow C(O) + H_2$$

and

$$C(O) \rightarrow CO + C_{act} \quad (or + C_{inact})$$

with

$$C(O) + CO \rightarrow CO_2 + C_{act}$$

in effect contributing to the shift reaction (4). And carbon hydrogenation (5) is believed [3] to proceed via successive hydrogen additions at the edges of carbon crystallites, e.g., as in Fig. 12.1.1.

These reaction mechanisms and the thermodynamic properties of the various processes (Note A) impose important constraints on how fast or how efficiently a coal or other carbon source can be gasified and also determine the composition of the product gas obtained in any particular instance. Even though much has been learned from studies of individual gasification reactions and from modeling of gasification systems [4, 5], it is therefore in practice generally still necessary to establish optimum processing parameters by tests under actual operating conditions.

But Eqs. (1)–(5) do make it possible to identify the several options that gasification offers.

The simplest case is represented by reactions (1) and (2), which together, i.e., as $2C + O_2 = 2CO$, constitute incomplete combustion due to an insufficiency of oxygen. If $C \rightarrow CO_2$ is sustained with air, appropriate control over airflow in a deep fuel bed allows close approach to the theoretical

$$C + \tfrac{1}{2}(O_2 + 4N_2) = CO + 2N_2$$

and the off-gas, termed *producer gas*, will contain $\sim 33\%$ carbon monoxide, equivalent to a heat content of ~ 150 Btu/scf.

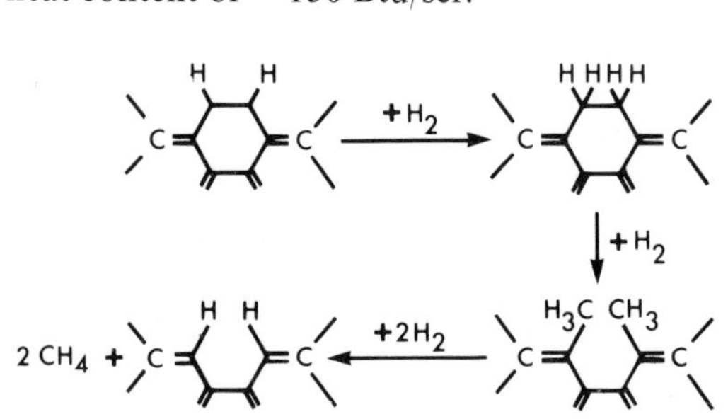

Fig. 12.1.1   Carbon hydrogenation by successive hydrogen additions.

Table 12.1.1

Effects of steam injection on producer gas compositions

Steam, lb/lb of coal	0.45	0.80	1.55
Gas composition, %			
Carbon dioxide	5.25	9.15	13.25
Carbon monoxide	27.30	21.70	16.05
Hydrogen	16.60	19.65	22.65
Methane[a]	3.35	3.40	3.50
Nitrogen	47.50	46.10	44.55
Gross heating value, Btu/ft^3	185	177	169
Gas yield, Mcf/ton of coal	138	141	147

[a] The near-constancy of the methane concentration suggests that CH_4 forms from volatile matter rather than by carbon hydrogenation (N.B.).

If steam as well as air is admitted to the fuel bed, the carbon–steam reaction (3) will contribute CO and H_2 to the product gas and yield a so-called *water gas* with correspondingly greater heat content. But since this process is highly endothermic and tends to lower the fuel bed temperature, it will also promote the shift reaction (4) and, unless steam rates are carefully controlled, thereby raise the CO_2 content of the gas at the expense of its CO. Table 12.1.1, adapted from Bone and Wheeler [7], illustrates this effect. The principal benefit of using air–steam mixtures lies thus in the greater gas yield per unit weight of gasified carbon.

However, if the same reactions are carried out with *oxygen* and steam rather than with air and steam, another, much more important benefit accrues. The product gas will then be a nitrogen-free raw *synthesis gas*,[2] which consists almost entirely of CO_2, CO, and H_2 and, after removal of CO_2, offers an essential feedstock for a wide variety of chemical syntheses (see Section 12.7).

12.2   Early and Current Commercial Gasifiers

Contrary to impressions that might be gained from the intense interest now expressed in it, coal gasification is not a new or even a particularly novel concept. The technical principles underlying it were well understood before the middle of the nineteenth century, and some industrial gasification processes were, in fact, operated as early as 1860.

Most of the first plants, built in England, were designed by K. W. (later

[2] Now generally contracted to "syngas," this appellation reflects the unique versatility of $CO + H_2$ as a chemical building block.

Sir William) Siemens to manufacture *producer gas* and consisted of a brick chamber with a steeply sloping side that incorporated an ash grate (see Fig. 12.2.1). Coal was continuously gravity-fed from a hopper to a deep fuel bed resting against the slope; process heat was supplied from a furnace connected to the bottom of the chamber; and gas was taken off through a chimney that also served to induce sufficient draught in the "producer."

However, by the 1870s, rapidly growing demand for gas also led to construction of facilities for producing *water gas* by a process developed by Gaillard (1848) and du Motay and Lowe (1873). In these plants, an ignited coke bed was first "blown" with air until the required temperature was reached, then "steamed" to make gas, and returned to an air-blast when falling temperatures resulted in excessive generation of CO_2 (by the shift reaction). Since all gas formed during air-blowing was vented as waste, the water gas produced in this manner had a calorific value of ~ 300 Btu/scf.

Modern gas producers are mostly direct descendants of these early versions but embody refinements that simplify gasification and make for greater efficiencies.

The immediate antecedents of two contemporary gasifiers can be seen in pressure producers, which were introduced in the 1920s. These were usually

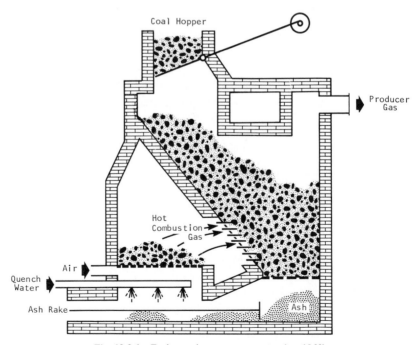

Fig. 12.2.1   Early producer gas generator (ca. 1860).

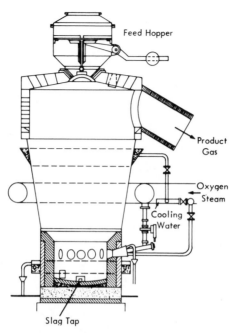

Fig. 12.2.2   Schematic of an early pressure gasifier. The type shown here is the Leuna slagging gasifier that was designed to discharge ash in its molten state.

15- to 20-ft-high cylindrical steel vessels fitted with grates through which air and steam could be injected, and ash was discharged as molten slag or into revolving bowls so arranged that they sheared clinker from the bottom of the ash column (see Fig. 12.2.2). Important features were their high throughput capacity, which accrued from continuous rather than cyclical "steaming" of the fuel bed, and their ability to meet all their own steam requirements from their water jackets. Some pressure producers of this type were still in operation in the 1960s and, on average, yielded 122 Mcf of 170 Btu gas as well as 20–22 gal of by-product tars per ton of (dry) coal.

One obvious elaboration of these pressure producers is the modern fixed-bed *Lurgi* gasifier [8], which was introduced in 1936 and is now mainly used for production of syngas. This unit (see Fig. 12.2.3) consists of a vertical water-cooled pressure shell to which coal is intermittently charged from above through a lock-hopper. Steam and oxygen (or air) are injected through a rotary bottom grate through which crushed clinker is also continuously withdrawn into a lock-hopper under the reactor, and gasification is normally effected under ~3 to 3.5 MPa at temperatures between 1700 and 1900°F (~925 and 1035°C). Since these conditions favor carbon hydrogenation, the cleaned CO_2-free gas produced from an oxygen-and-steam-blown reactor

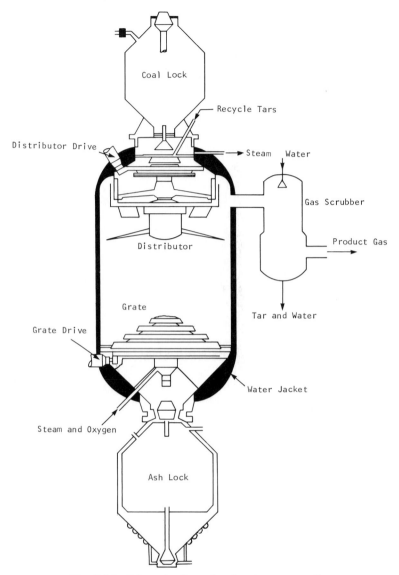

Fig. 12.2.3   Schematic of a modern Lurgi pressure gasifier.

typically contains $\sim 50\%$ hydrogen, 35% carbon monoxide, and 15% methane.

Lurgi gasifiers have been progressively improved over the years by in-corporation of a rotary coal distributor, a steam-cooled grate, and an optional mechanical stirrer which enables processing of weakly caking as well as of noncaking coals; and their principal limitation is their inability to accept fine (-28-mesh) coal.

Current standard (12-ft-diam) Lurgi units can gasify up to 600 tons of coal per day, and a 16-ft-diam version now being designed is expected to handle up to 1000 tons.

The other contemporary commercial gasifier closely related to the pressure producers of the 1920s, though operated at atmospheric pressure, is the *Wellman* reactor [9]. This is a 6- to 12-ft-diam, double-walled water-cooled steel vessel to which coal is continuously charged from a hopper through one or more pipes and rotary drum feeders (see Fig. 12.2.4). Steam and air are admitted through tuyeres in a revolving basal grate through which ash is simultaneously withdrawn into a bottom dump.

Wellman gasifiers have so far only been used to make producer gas but are now available in two series, one comprising variously sized single-stage (Wellman–Galusha) units and the other offering two-stage reactors that

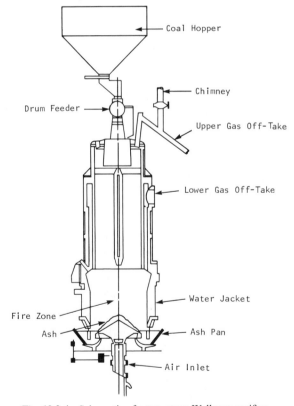

Fig. 12.2.4   Schematic of a two-stage Wellman gasifier.

promote the formation of tar in the first stage. The heat values of the product gas depend therefore on how the gas is made and on whether or not tars are separated from it. Typical values range from 170 Btu/scf for cold clean gas (from which tar has been extracted) to 185 Btu/scf for hot detarred gas and 200 Btu/scf for hot raw gas (in which tar vapors remain).

In the more than 20 Wellman plants built in South Africa in the past two decades to provide on-site fuel for pottery, glass, steel, and cement manufacture, each gasifier can be automatically adjusted to operate anywhere between full and one-third of rated capacity without loss of efficiency, and the largest two-stage units can each process up to 72 tons of coal per day with production of 8.4×10^6 scf of cold clean gas (equivalent to 1.4×10^9 Btu). Like Lurgi reactors, however, Wellman gasifiers are unable to accept $-\frac{1}{4}$-in. (6.3-mm) coal and also require coals with ash fusion temperatures above 2200°F (1200°C). Where such fuels are not available, small coke or coke breeze can be used.

Since the mid 1930s, two other gasifier systems, both radically different from Lurgi and Wellman reactors, have also firmly established themselves in industrial practice. One of these is the *Koppers–Totzek* generator [10], which produces syngas from oxygen-entrained coal, and the other is the *Winkler* generator, which gasifies a fluidized coal bed.

The Koppers–Totzek generator, which has been commercially used since 1938, is a squat, cylindrical, refractory-lined vessel into which a suspension of pulverized coal in oxygen is injected through two or four tangentially disposed "burner heads." Steam is introduced around these heads in order to shroud the reaction zone and protect the vessel walls from excessive heat; product gas is taken off through a collector pipe at the top (Note B); and ash leaves as a molten slag through the bottom (Fig. 12.2.5). The two-burner-head unit has a maximum internal diameter of 10 to 12 ft, tapering to 6–8 ft at either end; an overall length of ~ 25 ft; and a volume of approximately 1000 ft^3. The four-burner-head model is similarly shaped, but has a volume of ~ 2000 ft^3 (~ 56 m^3).

Unlike the Lurgi gasifier, the Koppers–Totzek generator operates at atmospheric pressure, generates no tars, and yields a substantially methane-free synthesis gas that typically contains 30–32% hydrogen, 55% carbon monoxide, and 12% carbon dioxide.

Among its technical advantages are its ability to accept all types of coal without any prior treatment other than pulverization and some flexibility in ash removal: Instead of being tapped off as a slag, ash can also, where so desired, be removed as fly ash with the raw gas and then separated from it in an external cyclone. The high operating temperature, usually 3300–3500°F (1815–1925°C), also allows very fast gasification, so that units with four burner heads can process up to 850 tons of coal per day. In view of this

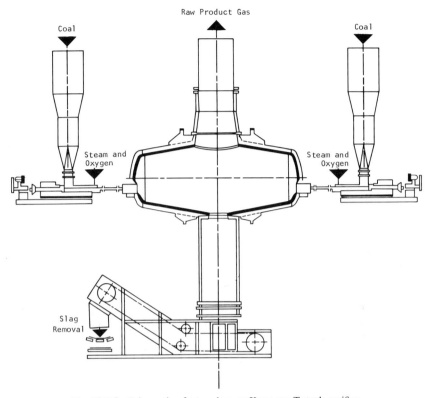

Fig. 12.2.5   Schematic of a two-burner Koppers–Totzek gasifier.

exceptionally high capacity and the modest cost of the Koppers–Totzek generator, its relatively low carbon inventory and consequent above-average oxygen consumption (which are often cited against it) are generally not serious disadvantages. But some disincentives arise from the loss of sensible heat when the raw gas, which exits at 2300–2700°F (1260–1480°C), is cooled to ambient temperatures for cleaning and from the need to compress the clean gas for downstream chemical syntheses (see Section 12.7).

Modern versions of the Winkler gasifier [11], which also operates at atmospheric pressure, are 75-ft (~22.9-m) -high, 18-ft (~5.5-m) -diam, refractory-lined shaft reactors, which can process up to 1100 tons of coal per day. −3/8-in. (−9.5-mm) coal is continuously fed by a screw from a pressurized hopper and fluidized by a primary blast immediately above the bottom grate; and a secondary blast just above the fluidized bed serves to gasify any unreacted char in the product gas stream. Ash not leaving with this stream (and removed in an external cyclone) is withdrawn through the base of the reactor by means of a rotating scraper (see Fig. 12.2.6).

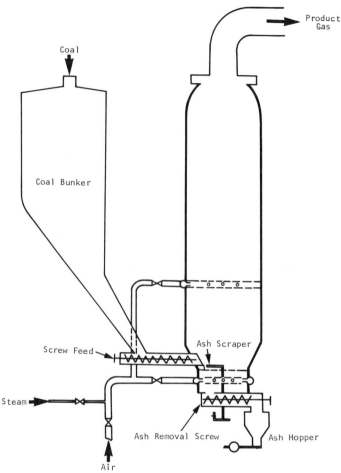

Fig. 12.2.6   Simplified schematic of a modern Winkler gasifier.

Depending on whether a producer gas or syngas is to be made, either an air–steam or an oxygen–steam blast is used, and outputs can be varied over a relatively wide range without appreciable loss of efficiency. A typical syngas will contain 36–37% hydrogen, 45% carbon monoxide, and 16% carbon dioxide. However, owing to the construction of the reactor, maximum temperatures must not exceed ~1800°F (~980°C), and this restricts fuels for Winkler gasification to lignites and subbituminous coals. Higher-rank coals are, as a rule, insufficiently reactive at such comparatively low temperatures.

12.3   Second-Generation Gasifiers

Although existing industrially proven gasifiers are perfectly capable of furnishing syngas suitable for conversion to substitute natural gas (SNG) and other petrochemicals, massive efforts are currently being devoted, particularly in the United States, to the development of so-called *second-generation* reactors. This work is mainly directed toward creation of improved technology for very large-scale SNG production for which available methods are held to be inadequate,[3] and centers therefore on gasifiers that could be

(a)   sized for daily throughputs as high as 5000–10,000 tons of coal, and
(b)   operated at or near conventional pipeline pressure (~ 1000 psig).

The greater throughputs, which would far exceed the capacity of existing equipment, would simplify plant operation by eliminating the need for extensive gasifier *batteries*;[4] and high-pressure gasification would enhance formation of methane (and thereby correspondingly reduce the extent of downstream "shifting" of syngas to the $1:3$ $CO:H_2$ ratio required for $CO + 3H_2 \rightarrow CH_4 + H_2O$; see Section 12.5), as well as obviate final compression of the SNG prior to its injection into a pipeline distribution system.

But pursuit of these objectives, which, if attained, would result in significantly lower SNG costs and make for greater technical reliability of large-scale gasification, has also led to the emergence of some important novel design and processing concepts.

Most of the reactor configurations and gasification procedures tested in the United States (Note C) fall into one or other of three groups. Group I comprises techniques that gasify fluidized or gas-entrained coal and generate a syngas that, while containing a substantial proportion of methane, still requires fairly massive "shifting" and methanation before it can be used as a "high-Btu pipeline gas."[5] The second group is made up of hydrogasification methods, i.e., processes that gasify fluidized or entrained coal with *hydrogen* and steam and thereby yield a methane-rich gas that needs little, if any, further "shifting." And Group-III techniques produce much the same type of syngas as those in Group I but gasify the coal in a liquid medium that may also catalyze the gasification reaction (see Section 12.4).

---

[3] In view of South Africa's coal gasification practices and experiences (to which brief reference is made in Section 12.7), this is perhaps a somewhat debatable contention.

[4] For example, a facility producing 250 million scf/day SNG would, if based on 12-ft-diam Lurgi reactors, require continuous operation of 30 such units, with another 5 on standby.

[5] In American terminology this is for all practical purposes synonymous with SNG and means a gas with > 940 Btu/scf.

The processes summarized below illustrate the different approaches taken in each of these categories (Note D).

Of the Group-I gasifier systems, technically the simplest is Bituminous Coal Research Inc.'s *Bi-Gas* process [13], which employs a two-stage reactor operated at ∼1100 psi (see Fig. 12.3.1) and has been undergoing testing in

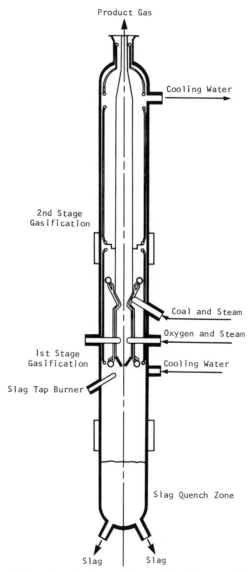

Fig. 12.3.1 Schematic of the Bi-Gas gasifier. (Courtesy NTIS.)

a 120-ton/day pilot plant at Homer City, Pennsylvania, since 1976. Pulverized coal and steam (or a coal slurry) is injected into the upper (dense-phase, second-stage) part of the reactor and there contacted with hot syngas moving up from the lower section to form additional CO, H_2, and CH_4 at $\sim 1700°F$ ($\sim 925°C$). Residual unreacted char is then carried out with product gas through the top, separated in a cyclone, and reintroduced near the bottom (first stage) of the reactor where it is almost completely gasified in a more dilute phase with oxygen and steam at $\sim 2700°F$ ($\sim 1480°C$). Ash is withdrawn as a molten slag into a water reservoir below the reactor, and residual clinker particles are periodically withdrawn through a lock-hopper device. Crude product gas leaves with $\sim 19–20\%$ CH_4.

A similar approach is taken in the US Bureau of Mines' *Synthane* process [14], which is schematically illustrated in Fig. 12.3.2 and is presently being developed in a 75-ton/day pilot plant at Bruceton, Pennsylvania. In this scheme crushed (-20-mesh) dry coal is first fed from a pressurized lock-hopper into a pretreatment unit where it is fluidized with oxygen and steam at $\sim 750°F/1000$ psi ($\sim 400°C/7$ MPa) in order to devolatilize it and destroy any caking properties. The resultant char, together with free volatile matter and unreacted steam, then moves to the generator where it is partially gasified in a dense fluidized state at $1100–1450°F/1000$ psi ($\sim 590–790°C/7$ MPa) and finally reacted with additional oxygen and steam in a more dilute phase (in the constricted lower section of the generator) at $1750–1850°F/1000$ psi ($\sim 950–1000°C/7$ MPa). Unconsumed char and some ash are withdrawn from the bottom of the generator, while raw syngas (with $22–23\%$ CH_4) leaves at the top for cyclone separators and a water scrubber, which free the gas of entrained dust and tarry matter.

In the Consolidation Coal Company's CO_2 *Acceptor* process [15], an important novel feature is the use of coal and calcined dolomite ($MgO \cdot CaO$) as a gasification feedstock, with the dolomite removing CO_2 by carbonate formation and thereby also generating some heat that is utilized to sustain the carbon–steam reaction. In the first step of this process, for which a simplified flow sheet is shown in Fig. 12.3.3, crushed coal and calcined dolomite (from the *regenerator*) are fed to the *devolatilizer* and fluidized with steam (and gas from the gasifier; see Fig. 12.3.4) at $\sim 1500°F/300–400$ psi ($\sim 815°C/2–3$ MPa). The net reactions that take place here are

$$2C + H_2 + H_2O \rightarrow CH_4 + CO$$
$$CO + H_2O \rightarrow CO_2 + H_2$$
$$MgO \cdot CaO + CO_2 \rightarrow MgO \cdot CaCO_3$$

and as these reactions proceed, much of the spent dolomite, being heavier than the partly gasified char, moves down toward the bottom of the fluidized

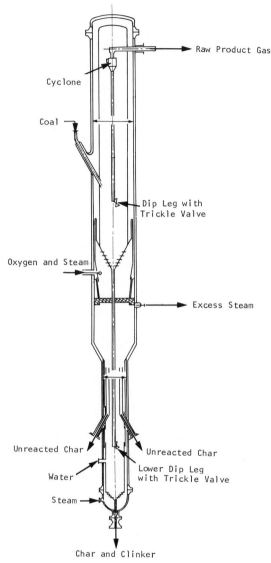

Raw Product Gas

Cyclone

Coal

Dip Leg with
Trickle Valve

Oxygen and Steam

Excess Steam

Unreacted Char

Unreacted Char

Water

Lower Dip Leg
with Trickle Valve

Steam

Char and Clinker

Fig. 12.3.2 Schematic of the Synthane gasifier. This unit has aa overall height of ~ 100 ft (30.5 m) and an internal diameter of ~ 5 ft (1.5 m). (Courtesy NTIS.)

bed. From the upper part of the bed, unconsumed char is then sent to the gasifier where it is contacted with steam and more calcined dolomite to allow

$$C + H_2O \rightarrow CO + H_2$$
$$CO + H_2O \rightarrow CO_2 + H_2$$
$$MgO \cdot CaO + CO_2 \rightarrow MgO \cdot CaCO_3$$

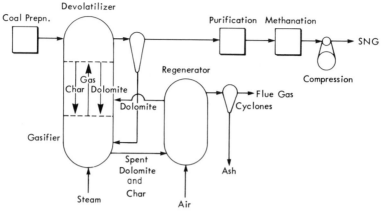

Fig. 12.3.3   Simplified flow sheet of the CO_2 Acceptor process.

and the resultant product gases are returned to the devolatilizer from which they finally exit as raw syngas.

Unreacted char from the gasifier, together with spent dolomite from the devolatilizer and the gasifier, is sent to the regenerator where the char is burned in air at ~1950°F (~1065°C) and the heat is used to calcine the dolomite before recycling it.

The fact that the CO_2 Acceptor process, which began undergoing pilot-plant tests in a 40-ton/day facility at Rapid City, North Dakota, in 1972, requires no off-site oxygen or hydrogen and yields a substantially CO_2-free syngas endows it with some advantages. However, the use of dolomite in the devolatilizer and gasifier makes it necessary to operate these units at temperatures below 1600°F (870°C); and since steam-gasification of bituminous coals at such temperatures proceeds rather slowly, the process may, like Winkler gasification, be restricted to lignites and subbituminous coals. Also, some difficulties may, in practice, arise from the need to balance material flows between three independent fluidized beds.

A final example of Group-I systems is Union Carbide's *UCC* process [16], which uses hot agglomerated ash (from combustion of residual char in a separate reactor) as a heat source in the gasifier in much the same way that the CO_2-Acceptor process uses calcined dolomite. In 1974 this scheme was undergoing tests in a process development unit that produced some 800 Mcf/day of syngas at 100 psi and 1880°F (~0.7 MPa/1025°C).

Of the Group-II processes, the most actively pursued at present is the *HYGAS* process [17], which the Illinois Institute of Gas Technology (IGT) began to develop in the late 1940s and which is now being tested in an 80-ton/day pilot plant in Chicago. Through direct hydrogenation of coal with steam and hydrogen under pressures of 1100 to 1500 psi (~7.8 to 10.5 MPa),

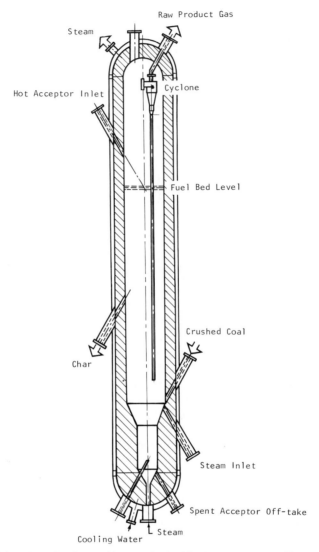

Fig. 12.3.4   Schematic of the gasifier used in the CO$_2$ Acceptor process. (Courtesy NTIS.)

Fig. 12.3.5   (a) Simplified flow sheet of the HYGAS process. Electrothermal generation of process hydrogen, shown here, is one of three alternative generation methods. The other two, now preferred because of lower cost, are illustrated in Figs. 12.3.5b and 12.3.5c: (b) flow sheet of the HYGAS steam–iron process for hydrogen generation: (c) flow sheet of the HYGAS steam–oxygen process for hydrogen generation. (Figure 12.3.5 on facing page.)

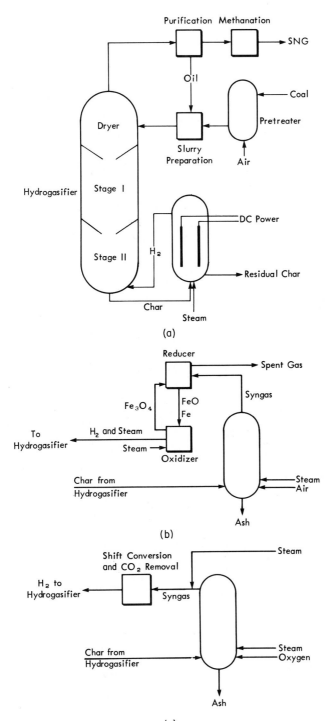

(a)

(b)

(c)

this yields a raw gas that contains approximately 37.5% CH_4, 45% H_2, and 13% CO and can therefore be sent to purification and final methanation without prior shifting.

HYGAS gasification (see Fig. 12.3.5a) begins with finely crushed dry coal, which, where necessary, is pretreated with air at $\sim 800°F$ (425°C) to destroy caking properties and thereafter slurried with oil. The slurry is then pumped (at 1100–1500 psi) into the dryer section of the gasifier (see Fig. 12.3.6) from where, as the oil vaporizes and is carried out with product gas, the coal

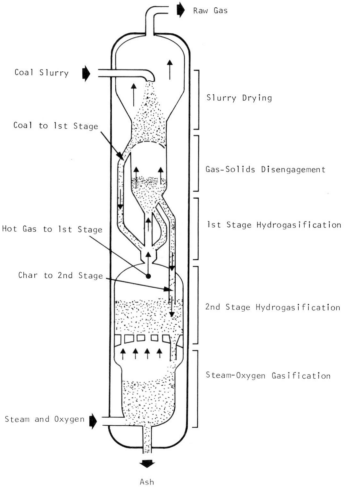

Fig. 12.3.6 Schematic of the HYGAS gasifier. This unit has an overall height of ~ 130 ft (39.5 m) and a maximum internal diameter of ~ 5.5 ft (1.7 m).

moves downward through a rising stream of hydrogen and steam, reacting in the first stage at 1200°F (650°C) and in the second at 1700–1800°F (925–980°C). Unconsumed char, representing $\sim 50\%$ of the original fuel charge, is withdrawn from the bottom of the gasifier and passed to a hydrogen generator where it is fluidized with steam to produce $CO + H_2$ that is returned to the gasifier.

As initially conceived, the hydrogen generator employed an electrothermal process, with heat for $C + H_2O \rightarrow CO + H_2$ supplied by passing an electric current through the fluidized char. However, since this was estimated to require some 32 kWh per Mcf of product SNG, later work centered on producing $CO + H_2$ by the steam–iron process or by conventional steam–oxygen gasification of the char.

In the steam–iron process (see Fig. 12.3.5b), which is now well-established technology, hot char from the gasifier is transferred to a gas producer and fluidized with air and steam at 3000°F/1000–1500 psi. The resultant gas is then taken through a "reducer" in which it reacts with magnetite (Fe_3O_4) at ~ 1500°F/1500 psi; spent gas is vented; and the reduced $FeO + Fe$ are sent to an oxidizer where they are contacted with steam at ~ 1500°F/1500 psi. Fe_3O_4 regenerated thereby is returned to the reducer, while steam + hydrogen are sent to the gasifier.

In the alternative steam–oxygen process (see Fig. 12.3.5c), unconsumed char from the gasifier is fluidized with steam and oxygen in a separate reactor to produce CO, H_2, and CO_2. The crude syngas is then "shifted" in order to enrich it in hydrogen and, after removal of CO_2 (see Section 12.6), returned to the second stage of the main gasifier.

Aside from the favorable composition of the gas it makes, the HYGAS process possesses significant advantages through not requiring gas compression at the line injection point and making efficient use of sensible heats in and between the various process units. Also to be noted is that it has no need of an oxygen plant unless hydrogen is to be produced by the steam–oxygen method.

On the basis of studies in a 10-lb/hr bench unit, US Bureau of Mines workers have outlined a two-stage hydrogasification scheme that resembles the HYGAS process but can accept caking as well as noncaking coals without prior pretreatment [18]. The heart of this scheme, known as the *Hydrane* process, is a hydrogasifier in which a dilute-phase suspension of coal in a methane- and hydrogen-rich hot gas first devolatilizes at 1650°F/1000 psi and then flows concurrently with the gas into stage 2 where it is partially gasified by hydrogen. Satisfactory temperature control in this stage, which is important because of the heat release associated with accompanying methanation via $CO + 3H_2 \rightarrow CH_4 + H_2O$, is achieved by water coils in which the excess heat load serves to generate superheated steam.

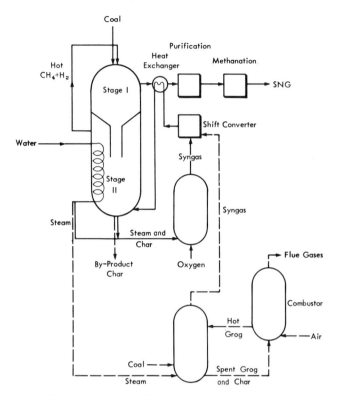

Fig. 12.3.7 Simplified flow sheet of the Hydrane process. The broken lines show an alternative syngas generation method that employs a solid heat carrier (grog).

Hydrogen-rich gas required for the lower second stage of the hydrogasifier derives from one of two alternative sources. In the first, the so-called oxygen process, char and superheated steam from stage 2 are transferred with oxygen to a conventional fluid-bed syngas producer, and the resultant product gas is moved through shift converters and a CO_2 absorption train before being returned to the hydrogasifier. The alternative method, which would be employed when other uses for the surplus char are projected, produces hydrogen from raw coal and steam, with process heat supplied by a solid carrier or "grog." The grog itself comes from a separate combustor in which it is prepared by adding ~ 5% of carbon to it and burning it in air. A Hydrane flow diagram, including the two options for making hydrogen, is shown in Fig. 12.3.7.

Group-III processes are exemplified by the M. W. Kellogg Company's *Molten Salt* process (see Fig. 12.3.8). Utilizing the fact that sodium carbonate tends to catalyze the carbon–steam reaction, this process [19] gasifies

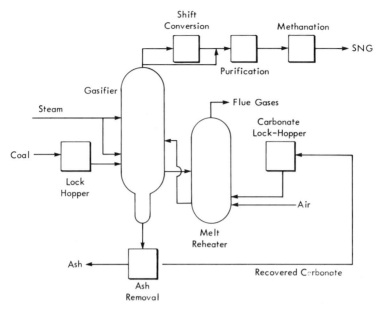

Fig. 12.3.8   Simplified flow sheet of the Kellogg Molten Salt process.

coal with steam in a recirculating Na_2CO_3 melt at 1835°F (1000°C) and
400–1200 psi ($\sim$3–8.5 MPa),[6] with raw syngas being removed through the
top of the reactor and unconsumed char as well as all ash carried by the
melt into a reheat section. Through interaction of Na_2CO_3 with H_2S and
other sulfur-bearing compounds, all sulfur is likewise carried into this
section. Here, air is injected and the residual fuel burned at $\sim$2200°F
(1200°C) to reheat the melt before moving it back to the gasifier. At the same
time, a slipstream with $\sim$8% ash is continuously withdrawn from the returning melt into an ash removal unit where it is quenched to 444°F (228°C)
with 100°F (38°C) aqueous Na_2CO_3. The resultant slurry of ash and solidified melt particles is then flashed to atmospheric pressure, held until all
Na_2CO_3 has dissolved, and finally filtered to separate ash and residual
carbon. The clear solution is moved to a carbonation tower where it reacts
with CO_2 and precipitates solid bicarbonate, which is recycled to the gasifier
while the solution is sent back to the ash removal unit.

   However, although otherwise technically attractive, development of this
process has been slowed by problems arising from the highly corrosive nature

---

[6] Since first reported in 1968, the conceptual flow diagram and operating parameters for the
process have been repeatedly modified. High-pressure operation is proposed in the more recent
versions.

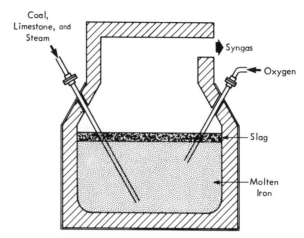

Fig. 12.3.9   Schematic of the ATGas generator.

of the melt and from attendant containment difficulties. These do not yet appear to have been entirely resolved.

Another Group-III scheme, also still in an early stage of development, is the *ATGas* process, which has been proposed by the Applied Technology Corporation [20]. This envisages gasifying coal by injecting it together with steam, limestone, and oxygen or air into a bath of molten iron (see Fig. 12.3.9). Syngas (or, if air is used, producer gas) is then obtained by carbon–steam interaction and by thermal cracking of volatile matter. Sulfur in the coal is captured by iron, transferred to a lime slag, and recovered as a by-product, while carbon dissolved in the iron is oxidized to carbon monoxide and contributes to the product gas.

Most of the technology needed for the ATGas scheme exists as unit processes in the iron and steel industry, but it has not yet been demonstrated that these can be practically combined into a single operation.

Which, if any, of these second-generation concepts, or others not referred to here [12], will eventually become *commercial* technologies will remain uncertain for several years and will undoubtedly be influenced by the outcome of parallel efforts to widen the utility of existing industrially proven reactors. Particularly noteworthy, in this connection, are major R & D projects directed toward (a) modifying the Koppers–Totzek gasifier for operation at elevated pressures,[7] (b) increasing the capacity and flexibility of the Lurgi reactor by adapting it for operation under slagging conditions

[7] No details of this project, which has been initiated by a Shell–Krupp consortium, have so far been released, but a 150-ton/day prototype pressure Koppers–Totzek gasifier was scheduled to begin undergoing testing in the summer of 1978.

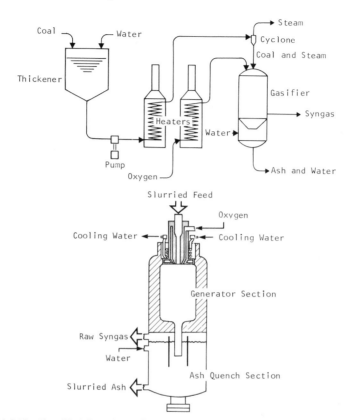

Fig. 12.3.10   Simplified flow sheet of the Texaco gasification process (top) and schematic of the Texaco gasifier (bottom).

(Note E), and (c) adapting the *Texaco Partial Oxidation* process, which has been widely used to gasify *oils*, for coal gasification. In the version now undergoing tests in a 15-ton/day pilot plant at Montebello, California, this latter process [6, 12, 22] involves injecting preheated oxygen and an aqueous coal slurry (with at least 70% of the coal sized to −200 mesh) at ~1000°F/225 psi (~540°C/1.5 MPa) into a vertical shell-type gasifier, which, near its bottom, contains a water-filled quench section. Gasification in the upper part proceeds at 2000–2500°F/300–1200 psi (~1100–1370°C/2–8.5 MPa); product gas is taken off through the side of the reactor; and ash, which forms a molten slag under these conditions, flows down along the reactor walls into the quench section from where it is withdrawn as a granular suspension in water. Figure 12.3.10 shows a flow sheet of this operation. A 15-ft (~4.5-m) -high, 9-ft (~2.7-m) -diam Texaco gasifier is expected to be capable of gasifying up to 1900 tons of coal per day with production of 100 MMscf of syngas.

12.4 Catalyzed Gasification

Many compounds catalyze steam decomposition [23] and/or the carbon–steam reaction [24] and thereby increase the specific gasification rate, i.e., the rate at which carbon is "lost" to gaseous reaction products. Especially effective are alkali metal compounds, notably chlorides and carbonates of sodium or potassium, which can raise the gasification rate by as much as 35–60%; but a wide range of metal oxides, e.g., oxides of calcium, iron, magnesium, or zinc, are also fairly potent, accelerating gasification by 20–30%.

Potassium carbonate, potassium bicarbonate, and zinc chloride have similarly been found [25] to increase the rate of hydrogasification.

These observations have prompted attempts to improve the performance of second-generation gasification processes by incorporating cheap catalysts into the coal charged to the reactors.

Aside from Kellogg's Molten Salt process (see Section 12.3), in which sodium carbonate is primarily used as a heat-transfer medium, the Exxon Corporation has thus reported that impregnation of 10 to 20 wt % of potassium carbonate will lower the optimum temperatures and pressure for steam gasification of bituminous coals from 1800–1900°F (~980–1040°C) and 1000 psi (7 MPa) to 1200–1400°F (~650–760°C) and 500 psi (3.5 MPa). And the Battelle Memorial Institute has shown that addition of 5 to 10 wt % of calcium hydroxide reduces optimum temperatures for hydrogasification by as much as 300–500°F (~150–260°C).

However, current schemes for catalyzed gasification demand fairly elaborate preparation of the coal; and considerable difficulties attach also to the recovery and regeneration of the catalysts that the massive amounts in which they must be deployed makes obligatory on economic grounds. Whether catalyzed gasification will offer any real advantages over noncatalyzed operations remains therefore to be demonstrated.

12.5 Gas Shifting

Where a syngas is to be converted to SNG or to ammonia, methanol, or liquid hydrocarbons (see Section 12.7), its $CO:H_2$ ratio must first be adjusted to the required stoichiometric value. Thus, for synthesis of ammonia via

$$N_2 + 3H_2 \rightleftharpoons 2NH_3$$

for which only the hydrogen in the syngas is needed, all CO must be abstracted. For synthesis of methane (SNG) or methanol via

$$CO + 3H_2 \rightleftharpoons CH_4 + H_2O$$

or

$$CO + 2H_2 \rightleftharpoons CH_3OH$$

the required $CO:H_2$ ratios are $1:3$ and $1:2$, respectively. And for synthesis of liquid hydrocarbons, which is a less specific process and always yields a fairly complex product mixture, experience indicates optimum $CO:H_2$ ratios that, depending on the particular process, lie between $1:1\frac{1}{2}$ and $1:4$.

These changes in gas composition are brought about by making use of the "shift" reaction

$$CO + H_2O \rightleftharpoons CO_2 + H_2$$

i.e., by treating the required fraction of the raw syngas stream with steam and then removing CO_2 from the product (see Section 12.6).

Although there is considerable freedom in conducting this operation, extensive shifting is usually effected in two stages, with reaction at 700–800°F (370–425°C) over a chromium-promoted iron catalyst followed by reaction at 375–450°F (190–230°C) over a reduced copper–zinc catalyst. The bulk of the shift is accomplished in the first stage.

As a rule, uncooled reactors with fixed catalyst beds are used; and in such equipment, the heat released by the shift reaction, $\sim 14 \times 10^3$ Btu/lb-mole of CO converted to CO_2, can serve to enhance catalyst activity and increase the reaction rate. However, since higher temperatures will also reduce the equilibrium conversion of CO, it is generally preferable to speed the reaction by raising the system pressure and secure maximum oxidation of CO by appropriate adjustment of the steam:feed-gas ratio. (It is, in this connection, worth noting that an improved cobalt–molybdenum shift catalyst has been claimed to be active at substantially lower temperatures than conventional formulations and therefore to effect more complete CO conversion [26].)

12.6   Gas Purification

If a "synthetic" gas is destined for use as a fuel, it is usually only necessary to free it of suspended particulate matter (i.e., unreacted carbon and fly ash) and, where environmental ordinances require it, to lower its content of sulfur-bearing compounds (mainly H_2S) to acceptable levels. However, where a syngas is to be used as a petrochemical feedstock, much more rigorous purification is mandatory. In such cases it is also necessary to remove CO_2 and residual hydrocarbons and to eliminate almost all H_2S, which would otherwise quickly poison the downstream catalysts.

These needs can be met by a variety of industrially proven techniques that are widely used in natural gas processing and related petrochemical operations.

Suspended solids, which are always extracted from the gas as it leaves the gasifier, are conveniently and efficiently removed with cyclones, electrostatic precipitators, or filter bags (see Section 15.2). The bulk of entrained water and tar can be separated by cooling the gas to near-ambient temperatures, and water itself can also be removed by glycol dehydration [29]. And residual light hydrocarbons can be stripped out by scrubbing the gas with a wash oil and then passing it through metal-oxide-impregnated activated carbon, which will also remove all remaining traces of carbon disulfide, thiophenes, and mercaptans.

Ammonia can, where necessary or desirable, be eliminated by interposing a wash tower in which the gas is scrubbed with fresh or acidified water.

For removal of acid gases, principally CO_2 and H_2S, from syngas several (mostly proprietary) processes that extract the unwanted components by weak chemical interaction with a basic compound or by absorption in a solvent are available. The most important of these are listed in Table 12.6.1. (The *Benfield* and *Catacarb* processes, which are also widely used but not identified in this table, are modifications of the hot carbonate system that improve the performance of K_2CO_3 by means of catalysts.)

In all cases, the acid components are removed by countercurrent scrubbing of the incoming "sour" gas with the solution or solvent, and the spent absorbent is continuously regenerated by slightly raising its temperature and/or flashing it to a lower pressure. Figure 12.6.1, which shows a simplified flow sheet of the hot carbonate process, illustrates this mode of operation. The terms *rich*, *lean*, and *very lean* here refer to the more or less CO_2- and H_2S-saturated solution, and to the partly and fully regenerated streams that are recycled to the absorber. Special mention must, however, be made of Lurgi's *Rectisol* process, which operates at temperatures between $0°$ and $-80°F$ ($\sim -18°$ and $-62°C$) and provides for separate recovery of CO_2 and H_2S.

Disposal of the CO_2- and H_2S-rich waste streams generated by purification of "sour" gas depends on their volumes and composition. Small volumes or lean gases with relatively low concentrations of H_2S are usually incinerated, and the resultant SO_2 is either vented to the atmosphere or, where SO_2 emission standards do not allow this, fixed by adsorption on activated carbon from which it can subsequently be recovered as elemental sulfur.

Table 12.6.1

Acid gas removal processes[a]

Process	Scrubbing solution or solvent	Attainable purity (ppm)	
		CO$_2$	H$_2$S
Chemical absorption			
Alkazid	25% aqueous solutions of "M"[b] or "DIK"[c]	<0.2%	15
Amine	15% aqueous monoethanolamine or	20	1
	25% aqueous diethanolamine	200	1
Economine	50–70% aqueous diglycolamine	100	4
Hot carbonate	20–30% aqueous K$_2$CO$_3$	0.2%	15
Physical absorption			
Purisol	N-methyl-2-pyrrolidone	10	2
Rectisol	Methanol	10	1
Selexc!	Dimethyl ether of polyethylene glycol	0.5%	<4
Sulfinol	Sulfolene plus diisopropanolamine	200	1

[a] For detailed descriptions of these and other processes, see [27] and [28].

[b] Potassium salt of methyl-amino-propionic acid.

[c] Potassium salt of dimethyl-amino-acetic acid; this is a more selective absorbent than "M" for H$_2$S in the presence of CO$_2$.

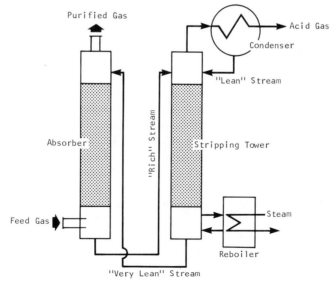

Fig. 12.6.1 Flow sheet of the hot carbonate process for gas purification.

However, if the waste gas holds more than $\sim 10\%$ H_2S, it is more conveniently and economically processed in a Claus plant, where H_2S is converted to high-purity elemental sulfur by partial combustion according to

$$H_2S + \tfrac{1}{2}O_2 = S + H_2O$$

The first stage of this process proceeds in a furnace section where the gas is burned in limited air to yield sulfur (in amounts corresponding to about two-thirds conversion) plus a mixture of SO_2 and H_2S; and this mixture is then passed over a bauxite or alumina catalyst at $\sim 420°F$ ($\sim 215°C$) in order to generate more free sulfur via

$$2H_2S + SO_2 = 3S + 2H_2O$$

With two or three catalytic converters in the second stage and gas reheating between them, up to 96% of the H_2S contained in the raw waste gas stream can thus be drawn off as liquid elemental sulfur.

Where only H_2S needs to be removed from the gas,[8] purification can also be accomplished by means of either of two older schemes, the *Giammarco-Vetrocoke* or the *Stretford* process [27]. Both are capable of reducing H_2S concentrations in the gas to 1 ppm, and both furnish elemental sulfur directly, i.e., without subsequent Claus treatment. But the sulfur tends to be less pure; and for this, as well as other (mainly economic) reasons, they would now only be preferable over alternative ($CO_2 + H_2S$) absorption methods where recovery of salable sulfur is of little importance to the operation.

In the Giammarco-Vetrocoke system, H_2S in the incoming gas contacts an aqueous solution of potassium arsenite (K_3AsO_3) and forms the thioarsenite (K_3AsS_3) which is subsequently reacted with potassium arsenate to yield a monothioarsenate; and when this is air-blown, it is reconverted to the arsenite and elemental sulfur.

The Stretford process employs a mixed solution of Na_2CO_3, Na_2VO_3, and an organic oxidation catalyst and makes use of the fact that H_2S will interact with this solution to form a sodium acid sulfide that is immediately converted to elemental sulfur by reduction of V^{5+} to V^{4+}. Air-blowing then reoxidizes the vanadium to the higher valency state.

12.7   Carbon Monoxide–Hydrogen Reactions

Much of the interest now shown in coal gasification reflects recognition that (low-Btu) producer gas and cleaned but otherwise not further processed

---

[8] A case in point might be a syngas that is to be used as a fuel gas rather than as a petrochemical feedstock.

(medium-Btu) syngas from coal are useful industrial heating and steam-raising fuels, and that coal gasification could therefore meet essential demands for gaseous fuels wherever natural gas or light hydrocarbons (such as propane or butane) are unavailable or too costly.

The practicality of deploying "synthetic" gases from coal as fuels where economic circumstances make it appropriate is well established by earlier European practices (see Section 12.2) and confirmed by more sophisticated recent commercial operations. Thus, in South Africa, a 175–200 Btu/scf producer gas generated in Wellman reactors is now routinely used to fuel plants manufacturing steel, pottery, glass, and other products. In Scotland, 180–250 Btu gas made in air- and steam-blown Lurgi gasifiers at the British Gas Corporation's Westfield plant [30] was used between 1960 and 1974 to augment supplies of domestic gas (Note F). And at Lünen, West Germany, a 170-MW combined-cycle generating facility is fueled by 140-Btu gas manufactured in Lurgi reactors. A flow sheet of this installation [31], which began operating in 1971 and is seen as a prototype for a 400-MW plant scheduled to come on stream in 1982, is shown in Fig. 12.7.1.

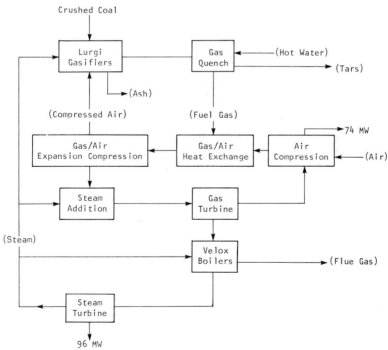

Fig. 12.7.1 Flow sheet of the STEAG combined-cycle generating plant at Lünen, West Germany. (After ERCB [32].)

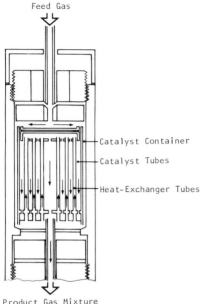

Feed Gas

Catalyst Container

Catalyst Tubes

Heat-Exchanger Tubes

Product Gas Mixture
(to Refrigeration)

Fig. 12.7.2   Schematic of an ammonia synthesis unit.

Equally important, however, and from some points of view perhaps even more significant, is the fact that a suitably shifted coal-derived syngas can stand in lieu of "reformed" natural gas or syngas obtained by partial oxidation of light distillate oils (such as naphtha), and that this interchangeability also allows coal to be used as a petrochemical feedstock.

Commercially proven examples of such use are provided by synthesis of ammonia, methanol, liquid hydrocarbons (by Fischer–Tropsch processes), and "substitute natural gas" (by methanation).

Direct synthesis of *ammonia* is accomplished by passing a 3:1 mixture of purified hydrogen and nitrogen through a converter (see Fig. 12.7.2) in which $N_2 + 3H_2 \rightarrow 2NH_3$ proceeds at 500–600°C/20–100 MPa over an iron oxide catalyst promoted by an acid and an alkaline oxide (e.g., Al_2O_3 + K_2O). The exiting gases, which, depending on reaction conditions, contain 8–80% NH_3, are then refrigerated in order to extract anhydrous ammonia[9] and recycled. Nitrogen for this process is conventionally obtained from an air separation plant; and where natural gas or a light distillate fraction is the source of hydrogen, it is first reformed by reaction with steam over a nickel catalyst to yield H_2, CO, and CO_2 via

$$CH_4 + H_2O \rightarrow CO + 3H_2$$
$$CH_4 + 2H_2O \rightarrow CO_2 + 4H_2$$

[9] For production of *aqueous* ammonia, refrigeration is replaced by cooling to near-atmospheric temperature and absorbing NH_3 in water. This practice, however, is now rare.

CO_2 is then absorbed into hot carbonate or methanol (see Section 12.6), and CO is removed by compressing the CO_2-free gas at 20 MPa and scrubbing it with an ammoniacal solution of a cuprous salt. Alternatively, where the hydrogen comes from coal-derived syngas, all CO is abstracted by shifting, and the CO_2 thereby produced is removed by compression or Rectisol purification.

Commercial coal-based ammonia plants built between 1950 and 1975 are identified in Table 12.7.1. Most employ variously sized Koppers–Totzek reactors (which generate little, if any, methane and consequently simplify purification of the syngas for downstream processing), and all gasify the coal by utilizing oxygen from the air separation plant. A flow sheet for a modern 1000-ton/day NH_3 installation, together with some data on material requirements and gas compositions at various process stages, is presented in Fig. 12.7.3.

Table 12.7.1

Coal-based ammonia plants, 1950–1975

Company and location	Start-up	Number and type of gasifiers	Raw gas capacity (10^6 scf/day)	NH_3 output (tons/day)
Typpi Oy, Oulu, Finland	1952	5 Koppers–Totzek	10.0	120
Azot Gorazde, Yugoslavia	1953	1 Winkler	4.5	50
Nippon Suiso, Onahama, Japan	1956	3 Koppers–Totzek	7.8	100
Empreso Nacional Calvo Sotelo Puentes, Spain	1956	3 Koppers–Totzek	9.0	100
Daud Khel, Pakistan	1958	2 Lurgi	N. A.	60
Azot Sanyyll Tas, Kutahya, Turkey	1959	2 Winkler	2.2	
	1968	2 Koppers–Totzek	28.8	370
Nitrogenous Fertilizer Company, Ptolemais, Greece	1961	6 Koppers–Totzek	38.6	525
Neyveli, South Arcot, India	1962	3 Winkler	11.2	300
Naju Fertilizer Company, Seoul, Korea	1964	3 Lurgi	N. A.	150
Chemical Fertilizer Company, Mae Moh, Lampang, Thailand	1965	1 Koppers–Totzek	8.0	100
Industrial Development Corporation, Kafue, Lusaka, Zambia	1969	2 Koppers–Totzek	16.0	200
Fertilizer Corporation of India, Ramagundam, India	1971	3 Koppers–Totzek	84.0	900
Fertilizer Corporation of India, Talher, Orissa, India	1972	3 Koppers–Totzek	84.0	900
Fertilizer Corporation of India, Korba, Madhya Pradesh, India	1974	3 Koppers–Totzek	84.0	900
AD & CI Ltd., Modderfontein, South Africa	1974	6 Koppers–Totzek	90.0	1000

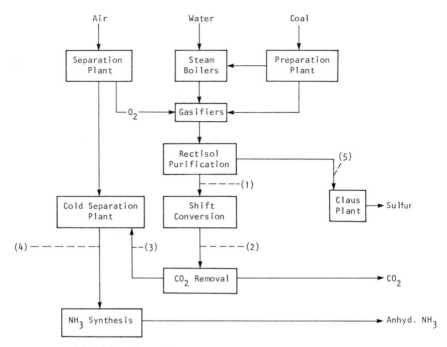

Material requirements
Coal: for gasification 1800 tons/day (at ~8500 Btu/lb)
          for steam-raising 830 tons/day
Water: 4.8 × 10⁶ gal/day

Gas composition, %	Gas stream				
	1	2	3	4	5
CO_2	12.0	41.3	—	—	75.0
CO	55.0	3.0	—	4.7	—
H_2	32.0	55.0	94.1	75.0	—
N_2	1.0	0.7	1.2	20.3	—
H_2S					22.0
COS					3.0
Gas flow rate, 10⁶ scf/day	93.3	97.3	80.2	99.7	21.0

Fig. 12.7.3   Flow sheet and related data for a 1000-ton/day ammonia plant (with Koppers–Totzek gasifiers). (After ERCB [32].)

Methanol synthesis from carbon monoxide and hydrogen goes back to 1913, when the Badische Anilin & Soda Fabrik (BASF) was granted a German patent for it, and was for many years thereafter accomplished by high-pressure (~100–150 MPa) reaction over cerium, cobalt, chromium, molybdenum, or manganese catalysts. In modern plants, which mostly use

Imperial Chemical Industries' (ICI) medium-pressure process, it is achieved by catalyzed reaction at 10–60 MPa and 250–400°C. If natural gas is used as feedstock, some CO_2 must be added during steam-reforming in order to generate the stoichiometrically required $1:2$ $CO:H_2$ ratio via

$$CH_4 + H_2O \rightarrow CO + 3H_2$$
$$CH_4 + 2H_2O \rightarrow CO_2 + 4H_2$$
$$CO_2 + H_2 \rightarrow CO + H_2O$$

or, if a coal-derived syngas is used, that ratio is adjusted by an appropriate extent of shifting.

In either case, all plant operations other than the final synthesis step are thus very similar to those involved in synthesis of ammonia.

Because of the availability of relatively cheap natural gas and/or light petroleum distillates after World War II, little methanol is currently manufactured from coal, but growing demand for methanol and more favorable pricing of coal vis-à-vis oil and gas are making this alternative increasingly attractive (Note G).

Methanation, i.e., conversion of carbon monoxide and hydrogen to (high-Btu) substitute natural gas (SNG) predominantly composed of methane, occurs over nickel or iron at 500–700°F (260–370°C) and is, in effect, the reverse of steam-reforming (see above). For practical purposes, freshly precipitated Raney nickel or commercially available Ni–kieselguhr or Ni–alumina preparations are used; and since the performance of these catalysts improves with increasing pressure, the reaction is always conducted at 200–1000 psi (~ 1.5–7 MPa). However, as $CO + 3H_2 \rightarrow CH_4 + H_2O$ is strongly exothermic (and will release some 65 Btu/scf of feed gas converted to CH_4), paramount importance attaches to close temperature control in order to avoid catalyst deterioration through sintering and/or deposition of carbon on it; and recent technical studies of methanation have therefore given much attention to the design of efficient reactors. These have included fixed-catalyst-bed systems cooled by heat-exchange surfaces [33] or by gas recycling [34–36]; fluidized catalyst beds indirectly cooled by heat-exchange surfaces [37, 38]; tubular reactors in which the catalyst is supported on and cooled by heat-exchanger tubes[10] [39, 40]; and a liquid-phase methanation reactor in which the catalyst is fluidized by an inert liquid that also serves as a heat sink [41]. Among the advantages claimed for this latter method are nearly isothermal operation (with a maximum temperature spread of $< 80°F$) and almost constant catalyst activity at optimum level.

In efficiently cooled methanators, catalysts have been found to remain

[10] An adaptation of such a reactor has been used in the Synthane process (see Section 12.3). The heat-exchanger tubes in this unit were closed-end steel tubes onto which a Raney nickel catalyst had been sprayed, and the coolant was "Dowtherm."

active for up to five years and capable of accomplishing almost 100%
conversion.

Although no commercial plants have so far been built, the technical
feasibility of large-scale methanation has been demonstrated,[11] and several
coal-to-methane plants, mostly for production of 250×10^6 scf SNG per
day (Note H), are in various stages of design in the United States. These
"first-generation" facilities will use Lurgi reactors (see Section 12.2),
Rectisol gas purification, and either Conoco's or Lurgi's proprietary
methanation technology to manufacture ~ 950 Btu gas. "Second-generation"
systems (see Section 12.3) that will simplify the gasifier batteries and greatly
reduce downstream shifting are not expected to be commercially available
before the mid-1980s but will from then on also make increasingly important
contributions to high-Btu gas supplies.

Conversion of carbon monoxide and hydrogen to *liquid hydrocarbons* and
related oxygenated compounds over variously promoted Group-VIII
catalysts has its origins in the now classic researches of Fischer and Tropsch
in the 1920s; and by the late 1930s, commercial technology for such hydrocarbon synthesis was sufficiently well developed to be used at nine German
plants (with a combined output of 675,000 tons/yr) and in several similar
installations elsewhere, notably in Britain and Japan. Some of these facilities
were destroyed during World War II, and those that were not were later
gradually phased out as low-cost petroleum and natural gas became available. However, oil price movements since 1973 and efforts toward a greater
degree of energy self-sufficiency in oil-importing industrialized countries
with access to large coal reserves are again making Fischer–Tropsch (FT)
syntheses viable options.

The principal products of such operations, of which the synthesis of
MeOH and CH_4 are special (and particularly *stereo-specific*) instances, are
as a rule straight-chain alkanes, alkenes, ketones, acids, and smaller amounts
of alcohols that are generated by processes formally represented as

$$(2n + 1)H_2 + nCO \rightarrow C_nH_{2n+2} + nH_2O$$

$$(n + 1)H_2 + 2nCO \rightarrow C_nH_{2n+2} + nCO_2$$

$$2nH_2 + nCO \begin{cases} \nearrow C_nH_{2n} + nH_2O \\ \searrow C_nH_{2n+1}\cdot OH + (n - 1)H_2O \end{cases}$$

$$(n + 1)H_2 + (2n - 1)CO \rightarrow C_nH_{2n+1}\cdot OH + (n - 1)CO_2$$

$$nH_2 + 2nCO \rightarrow (C_nH_{2n+1} -)_x + nCO_2$$

[11] In 1973–1974, Conoco Methanation Company, a Continental Oil Company subsidiary
with which 12 other US energy companies were associated, demonstrated sustained conversion
of shifted 370-Btu Lurgi syngas to methane at the British Gas Board's Westfield plant by producing 2.6 MMcf/day of pipeline gas with heat contents in excess of 950 Btu/scf. A similarly
successful program was concurrently carried out by Lurgi GmbH of Frankfurt, West Germany,
at South Africa's SASOL complex (to which reference is made later in this section). However,
neither company has so far released details of these development programs.

Additionally, since all these processes are accompanied by a concurrent shift reaction and a reverse Boudouard reaction, i.e., $2CO \rightarrow C + CO_2$,[12] reaction conditions that promote *high* CO conversion will also promote some hydrogenation of carbon dioxide by reverse shifting via

$$(3n + 1)H_2 + nCO_2 \rightarrow C_nH_{2n+2} + 2nH_2O$$

And under appropriate conditions, straight-chain hydrocarbons and their oxygenated homologs will, to some extent, also undergo secondary isomerization and/or cyclization to branched-chain and/or aromatic compounds.

But while FT syntheses can thus, in principle, furnish almost all hydrocarbons conventionally obtained from petroleum, the actual product mixes depend on temperature–pressure–catalyst combinations and on reactor design (Note I); and this dependence makes it necessary to distinguish among five quite different CO hydrogenation processes.

Medium-pressure synthesis, conducted at 220–340°C and 0.5–5 MPa over iron catalysts, yields mainly gasoline, diesel oils, and heavier paraffins (Note J). Either fused or nitrided fused iron, in each case optimally promoted by 0.5% K_2O, can be used, but nitrided fused iron generally produces more low-molecular-weight hydrocarbon material and is preferred in fluid-bed reactors because of its greater resistance to attrition. As a rule, the proportion of gasoline in the product mix increases and gasoline quality improves as the H_2:CO ratio in the feed gas is raised; and overall mix compositions are also markedly affected by whether synthesis is carried out over fixed catalyst beds or in a fluid-bed reactor. Table 12.7.2 illustrates this. The qualitative effects of other operating variables are summarized in Table 12.7.3.

High-Pressure synthesis, conducted at 100–150°C and 5–100 MPa over ruthenium catalysts [60], furnishes mainly straight-chain paraffin waxes with molecular weights up to 105,000 and melting ranges up to 132–134°C. However, formation of lower-molecular-weight hydrocarbons can be encouraged by increasing the proportion of hydrogen in the feed gas, raising the reaction temperature, and lowering pressures. The observation [61] that reaction of a 4:1 H_2:CO syngas at 225°C over 0.5% Ru on alumina yields ~96% methane with ~82% CO conversion indicates how far such product modification could be driven.

The third FT process, *iso-synthesis,* is a reaction that converts CO and H_2 to branched-chain hydrocarbons. Developed by Pichler and Ziesecke [62], this synthesis is usually conducted at 400–500°C and 10–100 MPa over thoria or K_2CO_3-promoted thoria–alumina catalysts and yields predomi-

---

[12] Unless inhibited by proper choice of operating conditions and catalysts that selectively promote formation of higher alkanes, this reaction can have serious detrimental effects. It will not only deactivate the catalyst by depositing carbon on its surface, but also tend to plug the catalyst bed.

Table 12.7.2

**Distribution of hydrocarbons in product mix of medium-pressure
FT synthesis**[a]

	ARGE reactor[b]		SYNTHOL reactor[c]	
	Wt % of total	Wt % of olefins	Wt % of total	Wt % of olefins
C_1	7.8	—	13.1	—
C_2	3.2	23	10.2	43
C_3	6.1	64	16.2	79
C_4	4.9	51	13.2	76
C_4–C_{11}	24.8	50	33.4	70
C_{11}–C_{20}	14.7	40	5.1	60
C_{20+}	36.2	15	—	—
Alcohols ⎫ Ketones ⎭	2.3		7.8	
Acids	—		1.0	

[a] After Frohning and Cornils [59].
[b] Fixed catalyst beds: H_2:CO = 1.7:1; 220–240°C; gasoline yield ~32%.
[c] Fluidized catalyst beds: H_2:CO = 3.5:1; 320–340°C; gasoline yield ~70%.

Table 12.7.3

**Influence of operating parameters on composition of product mix
of medium-pressure FT synthesis**[a]

	Mean molecular weight of products	Yield	
		Oxygenated products	Olefins
Increasing			
Pressure	+	+	−
Temperature	−	−	+
CO conversion	+	−	−
Flow rate	−	+	+
Gas recycle ratio	−	+	+
H_2:CO ratio	−	−	−

[a] +, increasing; −, decreasing.

nantly low-molecular-weight (C_4 and C_5) isoparaffins. Under suitable conditions, 1000 m³ of feed gas can produce up to 85 kg of isobutane and 25 kg of higher branched-chain hydrocarbons. Iso-synthesis at temperatures much above 400°C promotes coproduction of aromatics, while temperatures below 400°C tend to encourage formation of *oxygenated* compounds.

For production of oxygenated compounds it is, however, generally more convenient to make use of *synthol synthesis* [63], in which the feed gas is reacted at 400–450°C and ~ 14 MPa over alkalized iron turnings, or to employ either of two modifications of this procedure. One of these is *synol synthesis* [64], which involves interaction of CO and H_2 at 180–200°C and 0.5–5 MPs in the presence of highly reduced ammonia catalysts. If a recycling method is used and conversion per pass is deliberately kept low, a product mix with 40–50% oxygenated compounds, mostly straight-chain alcohols, can be obtained. The other modification is the so-called *oxyl synthesis* [65], which is carried out at 180–200°C and 2–5 MPa over a precipitated-iron catalyst. In a two-stage process, this results in up to 95% CO conversion and yields a product mix with over 30% oxygenated compounds, likewise mostly straight-chain alcohols.

Finally, mention must be made of *oxo-synthesis*, sometimes also termed oxonation or, more correctly, hydroformylation. This reaction between an olefin and syngas takes place at 100–200°C and 10–50 MPa over cobalt carbonyl catalysts and adds CO and H_2 across the olefinic bond to form an aldehyde via

$$R\!-\!CH\!=\!\!CH_2 + CO + H_2 \rightarrow R\!-\!\underset{\underset{H}{|}}{C}H\!-\!\underset{\underset{CHO}{|}}{C}H_2$$

The syngas used for this purpose is an approximately equal molar mixture of CO and H_2, and the ratio of syngas to olefin is $\sim 2:1$.

Since 1950, oxo-synthesis has become the by far most common industrial method for producing C_3–C_{16} aldehydes (Note K). (The aldehydes are then reduced to the corresponding alcohols and, for the greater part, used in the manufacture of detergents or as esters (e.g., phthalates) for plasticizing vinyl and other resins. Substantial volumes are also marketed as solvents.)

Because of the progressive phasing out of European and Japanese coal-based Fischer–Tropsch plants in the decade following the end of World War II, the only commercial facility presently operating (and illustrating the versatility of FT synthesis) is the South African Coal, Oil & Gas Corporation's (SASOL) chemical complex near Johannesburg. This installation, which has been repeatedly expanded since first coming on stream in 1955, now converts some 2 million tons of subbituminous coal per year into motor fuels, fuel oils, industrial solvents, and a range of by-products that include a hydrocarbon-rich fuel gas (with a heating value of ~ 650 Btu/scf) which is distributed to residential consumers. Figure 12.7.4 shows a simplified flow diagram for this complex. The gasifier section consists of 13 continuously operated Lurgi generators that deliver $\sim 300 \times 10^6$ scf (10.5×10^6 m^3) of raw gas per day; and hydrocarbon synthesis is conducted in ARGE (fixed-catalyst-bed) reactors over alkaline precipitated iron, as well as in SYNTHOL

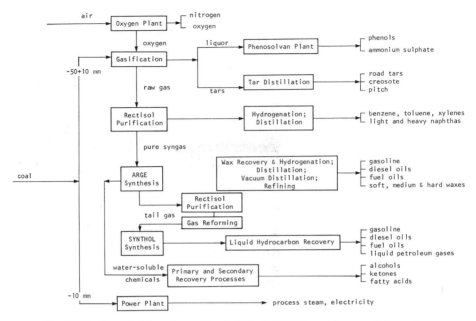

Fig. 12.7.4   Simplified flow sheet of South Africa's SASOL complex, near Johannesburg.

(entrained catalyst) reactors over alkaline highly reduced magnetite. Tables 12.7.4 and 12.7.5 summarize the processing conditions in these systems. (For compositions of the respective product streams, see Table 12.7.2.)

In the economic and political circumstances in which South Africa finds itself, the SASOL complex has proved so attractive that the government in

Table 12.7.4

Typical operating conditions for FT synthesis in SASOL ARGE reactors

Catalyst charge	32–36 tons					
Temperature	220–225°C ($\sim$430–440°F)					
Pressure	2.5 MPa ($\sim$350 psi)					
Fresh feed gas rate	20,000–22,000 m^3/hr ($\sim$700–770 Mcf/hr)					
Recycle ratio (tail gas:fresh gas)	2.2–2.5					
CO + H$_2$ conversion, %	60–66					
Gas composition, %	H$_2$	CO	CO$_2$	CH$_4$	C$_m$H$_n$	N$_2$
Feed gas	53.7	31.6	1.0	13.2	—	0.5
Tail gas	40.5	23.5	7.1	27.0	0.9	1.0

Table 12.7.5

Typical operating conditions for FT synthesis in SASOL SYNTHOL reactors

Catalyst charge	100–140 tons (circulating at an hourly rate of 6000 to 8000 tons)					
Temperature	320–330°C ($\sim$610–625°F)					
Pressure	2.3 MPa ($\sim$325 psi)					
Fresh gas feed rate	80,000–110,000 m³/hr ($\sim$2.8–3.85 MMcf/hr)					
Recycle ratio (tail gas:fresh gas)	2					
CO + H₂ conversion, %	70–75					
Gas composition, %	H_2	CO	CO_2	CH_4	C_mH_n	N_2
Feed gas	61.6	22.2	7.2	5.0	—	4.0
Tail gas	29.7	1.9	17.5	26.2	11.0	13.7

1975 decided to proceed to construction of a similar second facility, dubbed SASOL II. Estimated to cost some \$1.8 billion and scheduled to commence full operations in 1981, this will be capable of gasifying over 6 million tons of coal and thus, except for relatively minor engineering changes, be a $\times 3$ replica of the existing installation.

12.8 Hydrocarbon Synthesis from Carbon Monoxide and Steam

A significant and potentially important variant of CO hydrogenation, but fundamentally different from Fischer–Tropsch processing since it does not require hydrogen per se, is the synthesis of hydrocarbons from carbon monoxide and *steam* [66]. This reaction can be formally represented by

$$3CO + H_2O \rightarrow (-CH_2-) + 2CO_2$$

and proceeds at 190–250°C ($\sim$375–480°F) over cobalt, nickel, or ruthenium or at 250–300°C ($\sim$480–570°F) over iron. Optimum pressures lie between 0.8 and 1.0 MPa ($\sim$120 and 150 psi) for pure CO but increase with falling CO concentrations in an impure feed gas. The main product fraction, representing some 45–60% of the total, is comprised of C_8–C_{10} hydrocarbons, and up to 75% of the total product mix is olefinic (see Tables 12.8.1 and 12.8.2).

Except for the related shift reaction (see Section 12.5), CO–steam reactions have not so far been used in commercial operations. But they are now attracting increasing attention from researchers and are certainly worthy of closer scruntiny. They could, for example, offer means for producing hydrocarbons from *waste carbon dioxide* (which could be reduced to CO

Table 12.8.1

**Proportions of unsaturated hydrocarbons in products
of the CO–steam reaction**[a]

Boiling range (°C)	Yield (wt %)	Specific gravity (gm/cc)	Unsaturates (wt %)
70–200	33.7–34.6	0.760–0.771	36.8–50.4
200–250	20.8–22.3	0.784–0.795	35.0–56.0
250–300	15.2–29.0	0.800–0.820	30.0–49.0

[a] After Kravtsov *et al.* [68].

Table 12.8.2

Composition of C_5–C_8 hydrocarbons from the CO–steam reaction[a]

	Yield, wt %, over		
	Fe–Cu–kieselguhr	Fused iron	Fe–Cu–kaolin
Paraffins	25.6	25.6	50.7
Olefins	74.3	74.2	49.3

[a] After Kravtsov *et al.* [68].

via the Boudouard reaction) or from *producer gas*, from which CO could be recovered by the *COsorb process* recently developed by Tenneco Chemicals Co. [67]. In this, an aromatic solvent "activated" by $CuAlCl_4$ selectively absorbs CO and releases it when the solvent is heated in a stripper unit.

12.9   Underground (In-Situ) Gasification

The possibility that coal might be gasified underground was first envisaged in 1868 by Sir William Siemens [69] and independently suggested by Mendeleev [70], who discussed conceptual gasification methods in a series of papers published in 1888 in St. Petersburg (now Leningrad). But even though a patent covering injection of air and steam into an ignited coal seam through boreholes was issued in Britain in 1909 [71], almost nothing was done [13] until 1931, when the Soviet government, influenced by Lenin's enthusiasm for the idea [72], established several field stations for the purpose of developing a

[13] In 1913, Sir William Ramsay caused preparatory work for an underground gasification test to be started in Durham (Britain), but this was abandoned upon outbreak of World War I and never resumed—perhaps because Ramsay's death in 1916 removed a prime proponent of underground gasification from the scene.

workable underground gasification technology. Since then, the USSR has conducted numerous extended test programs and successfully operated several relatively large industrial installations that have supplied a low-Btu fuel gas to power stations and other industrial consumers [73].

Outside the USSR, underground coal gasification experiments were undertaken between 1945 and 1955 in the United States (at Gorgas, Alabama [74]), Britain (at Newman Spinney, Derbyshire [75]), Belgium (at Bois-la-Dame [76]), Italy (at Banco-Casino [77]), and under French auspices in Morocco (at Djerada [78]). But these were discontinued when availability of low-cost natural gas removed incentives for further, more systematic development work; and active programs, prompted by rapid escalation of natural gas prices since the early 1970s, have so far only been resumed in the United States [79] and Canada [80].

The principles of underground gasification are very similar to those governing the operation of gasifiers (see Sections 12.1 and 12.2) and are illustrated in Fig. 12.9.1 where two boreholes (I, II) are assumed to be sufficiently well "linked" (see below) to permit essentially unhindered passage of gas from I to II. If combustion is initiated at the bottom of I and sustained by continuous downhole injection of air, the ensuing reactions in such a system will quickly establish three distinctive zones. In the *combustion zone*, carbon dioxide will be generated by $C + O_2 \rightarrow CO_2$; and as this is pushed toward II by the incoming air (which will also, though more slowly, drive combustion ahead), it reacts with partially devolatilized coal in the *reduction zone* to form carbon monoxide via $C + CO_2 \rightarrow 2CO$. If the temperature is sufficiently high, and this depends largely on the air injection rate, any moisture held in the coal or supplied by injecting steam with the

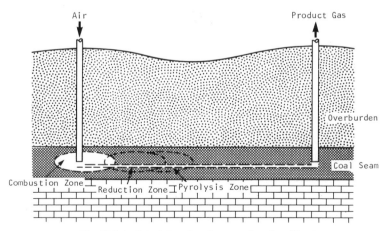

Fig. 12.9.1   Principles of underground coal gasification.

combustion air will also react with the coal via $C + H_2O \rightarrow CO + H_2$. And at the same time, coal heated ahead of the reduction zone will decompose and release volatile matter which, depending on whether or not it can move to II as fast as the hot CO and hydrogen, will either escape as tar or thermally crack and then contribute additional CO and H_2, as well as light hydrocarbons, to the product gas mixture. In either event, these processes will continue until all coal physically within their reach has been consumed; and what eventually surfaces through II is a fuel gas that is mainly composed of nitrogen, CO_2, CO, and H_2 and that may have heat values ranging up to 150 Btu/scf (~ 5.6 MJ/m^3).

Apparently proceeding from Siemens's notion of gasifying broken coal and coal pillars left behind in old mine workings rather than from Mendeleev's concepts or the Betts patent, initial efforts to apply these principles centered on attempts to gasify coal rubble in walled-off chambers. But aside from requiring extensive underground preparation, this method was found to promote combustion with little or only intermittent production of com-

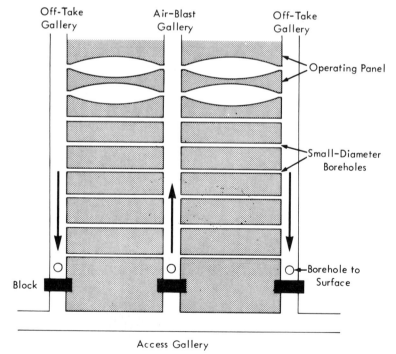

Fig. 12.9.2 Early "borehole producer" method for underground gasification of flat-lying coal seams. This method was used in the Soviet Union in the 1930s but was later abandoned in favor of "shaftless" techniques. (By permission of J. Wiley and Sons Inc.)

bustible gas; and development work since 1935 has therefore been concentrated first on *borehole producers* and *stream* systems and later on so-called *shaftless* techniques.

Borehole producers (see Fig. 12.9.2) were designed to gasify substantially flat-lying seams and were usually constructed by driving three parallel galleries ($\sim$150 m apart) into the coal from an access road and then connecting them by $\sim$10-cm-diam drill holes at 4- to 5-m intervals. Incoming combustion air was directed to operating panels by control valves placed at the drill hole inlets.

For gasification by the stream method (see Fig. 12.9.3), which was used in inclined seams, parallel galleries were driven 50–60 m down-dip into the seam at 80- to 100-m intervals and joined at the bottom by a "fire drift" in which blocks were placed to isolate coal panels. With combustion initiated in the fire drift, the flame front then moved up-dip in the coal, while ash and collapsed roof material collected in the burned-out zones.

Modern shaftless underground gasifiers, which began supplanting borehole producers and early types of stream systems as improved drilling tech-

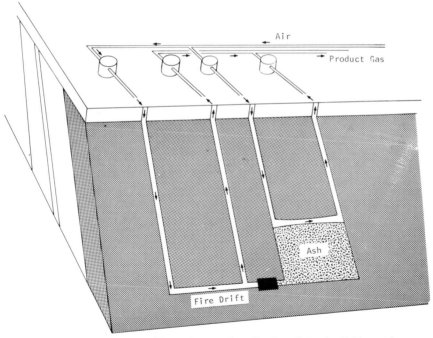

Fig. 12.9.3 "Stream" method for underground gasification of steeply pitching coal seams. In more modern versions (see Fig. 10.9.5), the injection holes, production holes, and fire drifts are directionally drilled. (By permission of J. Wiley and Sons Inc.)

niques and more sophisticated remote sensing devices became available, operate by identical modes, but simplify matters by allowing all necessary preparatory work to be done through boreholes from the surface.

An essential prerequisite for efficient coal gasification in *horizontal* shaftless systems is enhancement of the natural permeability of the seam between injection holes and production holes; and while this can, in principle, be achieved by recourse to conventional formation fracturing (e.g., by "hydrofracking" or by electrolinking [81], it is generally better accomplished by drilling a narrow connecting channel between the two boreholes or by establishing such a channel by combustion (Note L). In more recent Soviet practice, much use is made of directional drilling for this purpose [82, 83]; i.e., a drill hole is started at an angle to the surface and then *curved* into the coal. Guidance for the drill bit is provided by a downhole compass-and-pendulum device that transmits instructions to the drive mechanism at the surface. Up to 100-m-long holes with curvature radii as small as 600 m have reportedly been completed in this manner. The alternative method involves reverse combustion, i.e., igniting the coal at one borehole and then drawing oxygen to the fire by injecting air through the other. This technique advances the channel much more slowly than directional drilling (at ~6 cm/hr as against 50 cm/hr or better [83]), but requires no special equipment and has been found to be quite satisfactory as long as the boreholes are not spaced further apart than 25–30 m (Note M).

In order to maximize the amount of coal that can be efficiently gasified between any two holes, the initial channel is normally located just above the floor of the seam. This allows coal to cave into the otherwise progressively widening channel as gasification proceeds, and consequently extends the period in which a useful combustible gas can be produced. It also helps to offset a natural tendency of the flame front to move toward and along the seam roof whenever liquids, i.e., infiltrating water and/or by-product tars, transiently migrate *with* rather than ahead of the product gas.[14]

The manner in which an industrial-scale shaftless gas generator in horizontal coal would be operated is illustrated by a system worked for some years near Moscow (see Fig. 12.9.4).

Shaftless operation in a *pitching* seam differs from the earlier stream method primarily in substituting a curved directional hole for the "fire drift," and has been used in an installation at Lisichansk ([83]; see Fig. 12.9.5).

Mainly as a result of experiences in the USSR, which have been extensively reviewed in a number of Western publications [83, 84] and were recently

---

[14] This phenomenon, sometimes called *gravity override*, is due to the fact that the denser liquids tend to constrict the gas-flow path and force the gas up. Where it occurs, in-place coal will be effectively bypassed and gas quality quickly decays.

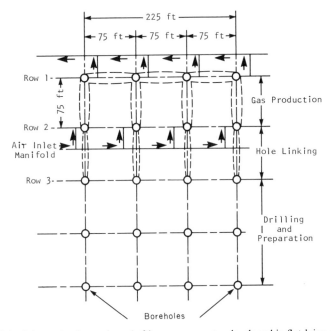

Fig. 12.9.4   Schematic of a modern shaftless gas generator developed in flat-lying coal. This generator was operated near Moscow, USSR, as a fuel source for a thermal generating station. (By permission of J. Wiley and Sons Inc.)

reexamined by Lawrence Livermore Laboratories [85], the major operating variables that affect gas yields, gas compositions, and coal conversion efficiencies are now fairly clearly defined and, in part, reasonably well collated. Soviet authorities do, in fact, seem to regard their underground gasification technology so well developed as to require no further R & D.[15] But some of their procedures, aside from being designed for thicker seams

---

[15] This has been repeatedly emphasized by Soviet experts (inter alios, to the author) and is reflected in current efforts to license Soviet technology in the United States. It may, however, also reflect waning interest in underground coal gasification in the USSR in the wake of the discovery of the huge Siberian natural gas reservoirs. It is, in this connection, undoubtedly significant that several previously well-publicized gasification installations (e.g., those at Lisichansk and Tula) have been shut down, and that a number of projected developments have been shelved. At present, only two such installations are, in fact, still active. One, an Angren (near Tashkent), was commissioned in 1961 and currently produces some 600×10^6 m³ (~ 21 Bcf) per year of 90 to 100 Btu/scf gas which fuels 50-MW capacity in a nearby 600-MW coal-fired power station; and the other, at Yuzhno-Abinsk (near Novosibirsk), generates an average 400×10^6 m³ (~ 14 Bcf) per year of similar quality gas for industrial heating and steam-raising purposes.

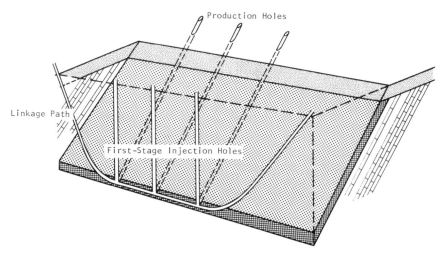

Fig. 12.9.5 Schematic of a shaftless gas generator designed for underground gasification of pitching coal seams (see also Fig. 12.9.3). This system, in which all linkages are established by directional drilling, was first used at Lisichansk, USSR, in 1955.

than commonly occur in other countries, are costly and/or may be objectionable on environmental or resource conservation grounds.

The basic problems arise from the fact that the design parameters of an underground gas generator tend to be site-specific, and that the properties of the immediate geological environment invariably require compromises with thermodynamic and kinetic optima for coal gasification reactions. For example: The need to prevent undue influx of formation waters (which, if they reached the gasification zone, would lower temperatures and cause deterioration of gas quality by promotion of the shift reaction) may call for injection pressures that could cause fracturing of the overlying strata and create serious gas containment difficulties. And efficient but not appropriately controlled gasification in a relatively thick seam may trigger subsidence, which not only allows gas leakages to the surface, but could also block communication between an injection- and production- hole.

These matters and the important economic advantages that would accrue from wider spacing of boreholes than is now feasible support the belief that further systematic field studies and mathematical modeling of underground coal gasification schemes [86] still have much scope for improving existing technology. Having regard to anticipated North American demands for gas, considerable interest would also attach to determining whether it might now, in some circumstances, be practical to inject oxygen rather than air and thereby produce a richer fuel gas (theoretically, with heat contents up to

360–370 Btu/scf) that could be economically transported over limited distances and also meet feedstock requirements of petrochemical industries (see Section 12.7). While the Soviets have occasionally experimented with oxygen-enriched air, the improvements in gas quality reported from these tests were generally only marginal.

Notes

A. An excellent review of the thermodynamics and kinetics of gasification reactions has been published by von Fredersdorff and Elliott [6].

B. In order to improve overall energy efficiencies, modern Koppers–Totzek installations take the hot product gas off through a waste heat boiler mounted directly above the burners.

C. In addition to projects sponsored by the US Energy Research and Development Administration (ERDA), now merged into the US Department of Energy, and by the American Gas Association (AGA), about which information is made public, many gasification R & D projects are being conducted by private companies as in-house activities, and about these little is known beyond what can be gleaned from outlines in patent disclosures.

D. A recent review [12] lists over 20 different gasifier systems currently under development. However, if previously pilot-tested schemes are included, nearly 60 could be identified.

E. Some experimental work with small slagging pressure gasifiers (from which ash is withdrawn as a melt) was reported in the late 1950s [21] and indicated that while such reactors required more oxygen than dry-bottom systems, they permitted substantially greater throughputs per unit reactor space and also made for simpler gas cleanup (see Section 12.6). Renewed commercial interest in coal gasification has therefore prompted an extensive US-financed program in Britain and a similar one by Lurgi GmbH in West Germany to develop industrial-size fixed-bed slagging gasifiers. Indications are that the prototypes now undergoing trials can efficiently gasify all types of coal and, unlike the dry-bottom Lurgi, accept fine as well as coarse coal. Judged by present rates of progress, slagging Lurgi gasifiers could be available for commercial use by the early 1980s.

F. This operation was shut down when North Sea natural gas became available. But since 1974, the Westfield gasifiers have been used to test the performance of a wide variety of US coals in Lurgi units and to develop commercial-scale methanation of synthesis gas.

G. What emphasizes this point is the growing interest in fuel-grade methanol as an *automotive* fuel, either as such or in admixture with gasoline. Also noteworthy in this connection is Mobil Oil Company's development of a process for *converting methanol to gasoline* over a zeolite catalyst by a reaction that can be formally written as

$$CH_3OH \rightarrow (CH_2)_x + xH_2O$$

A useful account of this process has been given by Meisel *et al.* [42].

H. Notwithstanding the fact that almost all recently published design and economic studies relating to methanation plants assume an installation capable of producing 250 MMcf SNG/day (8.8×10^6 m^3), a plant of this size does not necessarily represent an optimum economic unit. As little as 80 MMcf/day ($\sim 2.8 \times 10^6$ m^3/day) could be manufactured without significant loss of economy of scale.

I. In the 1930s and 1940s, some plants made synthetic gasolines and heating oils at atmospheric pressure over a standard cobalt catalyst (Co:ThO$_2$:MgO:kieselguhr = 100:5:8:200). Synthesis under these conditions is, however, relatively inefficient and no longer regarded as potentially viable technology.

J. Continuing interest in this and other FT syntheses and recognition of their potential industrial importance have created an extensive technical literature to which reference should be made for details. Several excellent general reviews have been published (e.g. [43–48]); and numerous papers (e.g. [49–58]) discuss reaction mechanisms and kinetics.

K. By the mid-1970s, installed and planned annual oxo-capacity, mostly in the United States and Western Europe, totaled over 2 million tons. Syngas for operating plants was in all cases produced by reforming natural gas or gasifying oils.

L. The major disadvantage of "hydrofracking" and electrolinking methods is that they usually make for somewhat uncertain and rather tortuous communication paths.

M. To facilitate subsequent gasification and promote rapid movement of product gas to the production hole, it is often also helpful to align injection and production holes with the major *cleat* in the coal. Cleat is a natural, highly regular fracture system—believed to have been caused by tectonic forces—in which smaller fissures, known as butt cleat, lie at very nearly 90° to the larger (major) fractures.

References

1. J. R. Arthur, D. H. Bangham, and M. W. Thring, *J. Soc. Chem. Ind.* **68,** 1 (1949).
2. R. F. Strickland-Constable, *J. Chim. Phys.* **47,** 356 (1950).
3. C. H. Zielke and E. Gorin, *Ind. Eng. Chem.* **47,** 820 (1955).
4. C. Y. Wen and J. Huebler, *Ind. Eng. Chem. Process Des. Dev.* **4,** 142 (1965); F. Moseley and D. Paterson, *J. Inst. Fuel* **38,** 378 (1965), **40,** 523 (1967).
5. R. L. Zahradnik and R. A. Glenn, *Fuel* **50,** 77 (1971); K. H. van Heek, H. Jüntgen, and H. Peters, *J. Inst. Fuel* **46,** 249 (1973); J. L. Johnson, *Adv. Chem. Ser.* **131,** 145 (1974).
6. C. G. von Fredersdorff and M. A. Elliott, *in* "Chemistry of Coal Utilization" (H. H. Lowry, ed.), Suppl. Vol. Wiley, New York, 1963.
7. W. A. Bone and R. V. Wheeler, *Engineering* **86,** 837 (1908).
8. J. Cooperman, D. Davis, W. Seymour, and W. L. Ruckes, U. S. Bureau Mines Bull. 498 (1951); T. S. Ricketts, *J. Inst. Fuel* **34,** 177 (1961); P. E. H. Rudolph, *Proc. Synth. Pipeline Gas Symp., 4th, Chicago, Illinois, October* (1972); D. C. Elgin and H. R. Perks, *Proc. Synth. Pipeline Gas Symp., 6th, Chicago, Illinois, October* (1974); Lurgi Express Information Brochure No. 1018/10.75 (1975).
9. McDowell-Wellman Eng. Co., Brochure No. 576; G. M. Hamilton, *Cost Eng.*, July, 4 (1963); *Am. Inst. Min. Metall. Pet. Eng., Ann. Meeting, St. Louis, Missouri* (1961).
10. R. G. Dressler, H. R. Batchelder, R. F. Tenney, L. P. Wenzell, and L. L. Hirst, U. S. Bureau Mines, Rep. Invest. 5038 (1954); J. F. Farnsworth, H. F. Leonard, D. M. Mitsak, and R. Wintrell, Koppers Co. Publ. (August 1973); R. Wintrell, *Am. Inst. Chem. Eng., Nat. Meeting, Salt Lake City, Utah, August* (1974).
11. W. Odell, U. S. Bureau Mines Information Circ. 7415 (1947); I. N. Banchik, *Symp. Coal Gasif. and Liquefact., Pittsburgh, Pennsylvania* (1974); Davy Powergas Inc. Publ. (1974).
12. Electric Power Research Institute, Evaluation of Coal Conversion Processes to Provide Clean Fuels, Part II, NTIS PB-234 203 (1974).
13. R. A. Glenn, *Proc. Synth. Pipeline Gas Symp., 1st, Pittsburgh, Pennsylvania* (1966); R. J. Grace, *Clean Fuels Coal Symp.* Institute Gas Technology, Chicago, Illinois (September 1973); R. A. Glenn and R. J. Grace, *Proc. Synth. Pipeline Gas Symp., 2nd, Pittsburgh, Pennsylvania* (1968).
14. A. J. Forney and J. P. McGee, *Proc. Synth. Pipeline Gas Symp., 4th, Chicago, Illinois* (1972); S. J. Gasior, A. J. Forney, W. P. Haynes, and R. F. Kenny, *AIChE Ann. Meeting, 78th,*

Salt Lake City, Utah, August (1974); S. E. Carson, *Proc. Synth. Pipeline Gas Symp., 7th, Chicago, Illinois, October* (1975).

15. G. P. Curran, *Chem. Eng.* **62** (2), 80 (1966); C. E. Fink and M. H. Vardaman, *Proc. Lignite Symp., Grand Forks, North Dakota, May* (1973); C. E. Fink, G. P. Curran, and J. D. Sudbury, *Proc. Synth. Pipeline Gas Symp., 7th, Chicago, Illinois, October* (1975).

16. U. S. Patent No. 3,171,369 (1965); W. M. Goldberger, *Proc. Synth. Pipeline Gas Symp., 4th, Chicago, Illinois, October* (1972); W. C. Corder, H. R. Batchelder, and W. M. Goldberger, *Proc. Synth. Pipeline Gas Symp., 5th, Chicago, Illinois, October* (1973).

17. F. C. Schora, B. S. Lee, and J. Huebler, *Proc. World Gas Conf., 12th, Nice, June* (1973); B. S. Lee, *Proc. Synth. Pipeline Gas Symp., 5th, Chicago, Illinois, October* (1973); *Proc. Synth. Pipeline Gas Symp., 7th, Chicago, Illinois, October* (1975).

18. H. F. Feldmann, *AIChE Nat. Meeting, 71st, Dallas, Texas, February* (1972); H. F. Feldmann and P. M. Yavorsky, *Proc. Synth. Pipeline Gas Symp., 5th, Chicago, Illinois, October* (1973).

19. A. E. Cover, W. C. Schreiner, and G. T. Skapendas, *Chem. Eng. Progr.* **69** (3), (1973); *Proc. Clean Fuels Coal Symp.* Institute Gas Technology, Chicago, Illinois (September 1973).

20. J. A. Karnavos, P. J. LaRosa, and E. A. Pelczarski, *Chem. Eng. Progr.* **69** (3) (1973); P. J. LaRosa and R. J. McGarvey, *Proc. Clean Fuels Coal Symp.* Institute Gas Technology, Chicago, Illinois (September 1973).

21. D. Hebden and R. F. Edge, Gas Council, Great Britain, Research Comm. GC 50 (1958); Anonymous, BCURA Quarterly Gazette No. 37/38, p. 18 (1959).

22. D. Eastman, *AIME Ann. Meeting, New York, February* (1952); W. A. MacFarlane, *J. Inst. Fuel* **28**, 109 (1955); E. T. Child, *Symp. Coal Gasif. Liquefact.*, Univ. of Pittsburgh School of Eng., Pittsburgh, Pennsylvania (August 1974).

23. C. Kröger and G. Melhorn, *Brennst. Chem.* **19**, 257 (1938).

24. C. Kröger and E. Fingas, *Z. Angew. Allgem. Chem.* **197,** 321 (1931); W. P. Haynes, S. J. Gasior, and A. J. Forney, *Adv. Chem. Ser.* **131,** 179 (1974); W. G. Willson, L. J. Sealock Jr., F. C. Hoodmaker, R. W. Hoffman, and D. L. Stinson, *Adv. Chem. Ser.* **131,** 203 (1974).

25. N. Gardner, E. Samuels, and K. Wilks, *Adv. Chem. Ser.* **131,** 217 (1974).

26. W. Auer, *AIChE Nat. Meeting, 68th, Houston, Texas, February* (1971).

27. F. C. Riesenfeld and A. L. Kohl, "Gas Purification," 2nd ed. Gulf Publ. Co., Houston, Texas, 1974.

28. U. S. Department of Commerce, NTIS Publ. No. FE-1772-11 (February 1976).

29. U. S. Department of Commerce, NTIS Publ. No. PB-234 203 (1974).

30. D. C. Elgin and H. R. Perks, *Proc. Synth. Pipeline Gas Symp., 6th, Chicago, Illinois, October* (1974).

31. H. Puhr-Westerheide and H. W. S. Marshall, *Fall Meeting, Can. Elec. Assoc., Montreal, September* (1974).

32. Energy Resource Conservation Board, Province of Alberta; ERCB Rep. 75-M (1975).

33. H. A. Dirksen and H. R. Linden, *Illinois Inst. Gas Technol. Res. Bull.* 31 (1963).

34. F. J. Dent, A. H. Moignard, A. H. Eastwood, W. H. Blackburn, and D. Hebden, *Gas Res. Bd.*, Great Britain, Comm. No. 21/10 (1945).

35. H. W. Wainwright, G. C. Egleson, and C. M. Brock, U. S. Bureau Mines Rep. Invest. 5046 (1954).

36. A. J. Forney, R. J. Denski, D. Bienstock, and J. H. Field, U. S. Bureau Mines Rep. Invest. 6609 (1965).

37. H. A. Dirksen and H. R. Linden, *Ind. Eng. Chem.* **52,** 584 (1960).

38. M. D. Schlesinger, J. J. Demester, and M. Greyson, *Ind. Eng. Chem.* **48,** 68 (1956).

39. M. M. Gilkeson, R. R. White, and C. M. Sliepcevich, *Ind. Eng. Chem.* **45,** 460 (1953).

40. J. H. Field and A. J. Forney, *AGA Synth. Pipeline Gas Symp.*, *Pittsburgh, Pennsylvania, November* (1966).

41. R. L. Espino and M. B. Sherwin, *AIChE Ann. Meeting, 65th, New York, November* (1972).

42. S. L. Meisel, J. P. McCullough, C. H. Lechthaler, and P. B. Weisz, *Chem. Tech.* February 86 (1976).

43. H. H. Storch, R. B. Anderson, L. J. E. Hofer, C. O. Hawk, H. C. Anderson, and N. Golumbic, U. S. Bureau Mines Tech. Paper 709 (1948).

44. H. H. Storch, N. Golumbic, and R. B. Anderson, "Fischer-Tropsch and Related Syntheses." Wiley, New York, 1951.

45. R. B. Anderson, *Adv. Catal.* **5,** 355 (1953).

46. R. B. Anderson, *in* "Catalysis" (P. H. Emmett, ed.), Vol. 4, Chapters 1–3. Van Nostrand-Reinhold, Princeton, New Jersey, 1956.

47. J. F. Shultz, L. J. E. Hofer, E. M. Cohn, K. C. Stein, and R. B. Anderson, U. S. Bureau Mines Bull. 578 (1959).

48. H. Pichler and A. Hector, *Ency. Chem. Technol.* **4,** 446 (1964).

49. J. T. Kummer, H. H. Podgurski, W. B. Spencer, and P. H. Emmett, *J. Am. Chem. Soc.* **73,** 564 (1951).

50. J. T. Kummer and P. H. Emmett, *J. Am. Chem. Soc.* **75,** 5177 (1953).

51. W. K. Hall, R. J. Kokes, and P. H. Emmett, *J. Am. Chem. Soc.* **79,** 2983, 2989 (1957).

52. G. Blyholder and P. H. Emmett, *J. Phys. Chem.* **63,** 962 (1959); **64,** 470 (1960).

53. W. K. Hall, R. J. Kokes, and P. H. Emmett, *J. Am. Chem. Soc.* **82,** 1027 (1960).

54. E. J. Gibson, *Chem. Ind.* 649 (1957); E. J. Gibson and R. W. Clark, *J. Appl. Chem.* **11,** 293 (1961).

55. V. M. Vlasenko and G. E. Yuzefovich, *Russ. Chem. Rev.* **38,** 728 (1969).

56. H. Pichler, *Erdöl Kohle* **26,** 625 (1973).

57. M. V. C. Sastri, R. B. Gupta, and B. Viswanathan, *J. Indian Chem. Soc.* **51,** 140 (1974).

58. M. A. Vannice, *J. Catal.* **37,** 449, 462 (1975).

59. C. D. Frohning and B. Cornils, *Hydrocarbon Proc.* November (1974).

60. H. Pichler, *Brennst. Chem.* **19,** 226 (1938); H. Pichler and B. Firnhaber, *ibid.* **44,** 33 (1963).

61. J. F. Shultz, F. S. Karn, and R. B. Anderson, U. S. Bureau Mines Rep. No. 6974 (1967).

62. H. Pichler and K. H. Ziesecke, *Brennst. Chem.* **30,** 13, 60, 81 (1949).

63. F. Fischer and H. Tropsch, *Chem. Ber.* **56,** 2428 (1923).

64. W. Wensel, *Angew. Chem.* **B21,** 225 (1948).

65. O. Roelen, H. Heckel, and F. Martin, German Patent 902,851 (1943).

66. H. Kölbel and F. Engelhardt. *Brennst. Chem.* **32,** 150 (1951); **33,** 13 (1952); H. Kölbel and E. Vorwerk, *ibid.* **38,** 2 (1957); H. Kölbel and K. Bhattacharyya, *Liebigs Ann. Chem.* **618,** 67 (1958); V. M. Mironov, A. V. Kravtsov, S. I. Smol'yaninov, and I. V. Goncharov, *Izv. Tomsk Politekh. Inst.* **257,** 200 (1973); G. T. Grozev and D. G. Ivanov, *Dokl. Bolk Akad. Nauk* **27,** 349 (1974); Y. Maekawa, S. K. Chakrabartty, and N. Berkowitz, *Can. Symp. Catal., 5th, Calgary, Alberta, October* (1977).

67. D. G. Walker, *Chem. Tech.* May, 308 (1975).

68. A. V. Kravtsov, I. V. Goncharov, S. I. Smol'yaninov, I. F. Bogdanov, and N. V. Lavrov, *Khim. Tverd. Top.* **8,** 106 (1974).

69. W. Siemens, *Trans. Chem. Soc.* **21,** 279 (1868).

70. D. I. Mendeleev, *Severny Vestn. St. Petersburg* No. 8, Sect. 2, 27; No. 9, Sect. 2, 1; No. 10, Sect. 2, 1; No. 11, Sect. 2, 1; No. 12, Sect. 2, 1 (1888).

71. A. G. Betts, British Patent 21,674 (1909).

72. V. I. Lenin, Pravda, No. 19 (April 21, 1913).

73. J. L. Elder, *in* "Chemistry of Coal Utilization" (H. H. Lowry, ed.), Suppl. Vol. p. 1023 et

seq. Wiley, New York, 1963; Organization for European Economic Cooperation, The Gasification of Coal, TAR/99(52)1, Paris (1953).

74. J. J. Dowd, J. L. Elder, J. P. Capp, and P. Cohen, U. S. Bureau Mines, Rep. Invest. 4164 (1947); J. L. Elder, M. H. Fies, H. G. Graham, R. C. Montgomery, L. D. Schmidt, and E. T. Wilkins, U. S. Bureau Mines Rep. Invest. 4808 (1951).

75. Ministry of Fuel and Power, "British Trials in Underground Gasification." H.M. Stationary Office, London, 1956.

76. Institut National de l'Industrie Charbonniere, "Underground Gasification: Belgian Experiments at Bois-la-Dame." Liège, Belgium, 1952.

77. R. Loison and J. Venter, "Underground Gasification: Russian, Italian, Belgian and French Experiments." CERCHAR, Paris and INICHAR, Liège, 1952.

78. R. Loison, *Proc. Int. Congr. Underground Gasif., 1st, Birmingham, Alabama* CERCHAR, Paris, 1952.

79. L. A. Schrider, J. W. Jennings, C. F. Brandenburg, and D. D. Fischer, *AIME Soc. Pet. Eng., Fall Meeting, Houston, Texas, October*, Paper No. SPE 4993 (1974). C. F. Brandenburg, D. D. Fischer, G. G. Campbell, R. M. Boyd, and J. K. Eastlack, *Am. Chem. Soc.., Div. Fuel Chem., Nat. Spring Meeting, Philadelphia, Pennsylvania, April* (1975). C. F. Brandenburg, R. P. Reed, R. M. Boyd, D. A. Northrop, and J. W. Jennings, *AIME Soc. Pet. Eng., Ann. Fall Meeting, 50th, Dallas, Texas, September* (1975). D. R. Stephens, F. O. Beane, and R. W. Hill, *Ann. Underground Coal Gasif. Symp., 2nd, Morgantown, West Virginia, August* (1976).

80. N. Berkowitz and R. A. S. Brown, *Ann. Gen. Meeting, 79th*, Canadian Institute Mining and Metallurgy, Ottawa, April 1977; *Bull. Can. Inst. Min. Metall.*, **70**, 92 (1977).

81. J. D. Forrester and E. Sarapuu, Univ. Missouri School of Mines and Metallurgy Bull. No. 78 (1952); J. L. Elder, M. H. Fies, H. G. Graham, J. P. Capp, and E. Sarapuu, U. S. Bureau Mines Rep. Invest. No. 5367 (1957); S. T. Bondarenko, B. Kh. Brodskaya, S. N. Lyandres, E. A. Meerovich, V. I. Pan'kovskii, and A. D. Reznikov, G.M. Krzhizhanovskii Power Inst., Akad. Nauk SSR, 1959, Lawrence Livermore Laboratory, Livermore, California, UCRL Trans-11050 (1976).

82. D. M. Arinenkov and L. M. Markman, Knizhnoe Izdatel'stvo Stalino-Donbas, 1960. Lawrence Livermore Laboratory, UCRL-Trans. 11007 (1976).

83. O. de Crombrugghe, *Ann͏ᵉ Mines Belges* No. 5, 478 (1959).

84. See Elder [73] and Loison and Venter [77]; D. W. Gregg and D. U. Olness, Lawrence Livermore Laboratories, UCRL-52107 (1976); R. M. Nadkami, C. Bliss, and W. I. Watson, *Chem. Technol.* **4** (4), 230 (1974); C. Wang, Dissertation, Columbia Univ., New York (1969); J. P. Capp, W. L. Rober, and D. W. Simon, U. S. Bureau Mines Information Circ. No. 8193 (1963).

85. E. Kreinin and M. Revva, Lawrence Livermore Laboratory Rept. UCRL-Trans-1080 (1966).

86. C. G. Thorsness and R. B. Rosza, *AIME Soc. Petrol. Eng.*, New Orleans, Paper No. SPE 6182 (1976); R. D. Gunn and D. L. Whitman, Laramie Energy Research Center, Wyoming, Rep. LERC/RI-76/2 (1976); C. F. Magnani and S. M. Farouq Ali, *J. Soc. Pet. Eng.* October 425 (1975); I. M. Stewart and T. F. Wall, *Proc. Int. Symp. Combust., 16th*, Boston, Massachusetts (1976).

CHAPTER 13

LIQUEFACTION

From a chemical viewpoint, the principal differences between coal and petroleum are ultimately all due to the much lower H/C ratio of coal (~ 0.7 as against > 1.2 for petroleum); and it is therefore possible to transform coal into liquid hydrocarbons by *direct* hydrogen addition as well as by indirect hydrogenation (via gasification and subsequent CO hydrogenation; see Chapter 12).

The first practical methods that accomplished such conversion were largely based on the work of Bergius and his collaborators [1][1] and involved reacting pulverized coal or coal–oil slurries with gaseous hydrogen at high pressures and temperatures. These processes were used in Europe (notably in Germany until the end of World War II and in the USSR and Czechoslo-

---

[1] Much of this work, for which F. Bergius was awarded the 1931 Nobel Prize in Chemistry, is reported in numerous papers in the German literature between 1918 and 1925 (e.g., in *Angew. Chem., Liebig's Ann.*, etc.). For a summary and account of its practical applications, see, e.g., [1].

vakia for some years afterward) for producing synthetic gasolines, aviation fuels, and heating oils. However, since the early 1960s, major advances have been made in understanding and controlling coal liquefaction by hydrogen transfer from donor liquids (see Section 5.5), and recent work indicates that this technique offers several important advantages over conventional Bergius hydrogenation.

13.1 Bergius Hydrogenation

In order to facilitate reaction and avoid severe erosion of high-pressure equipment by dry coal, the feedstock for large-scale Bergius hydrogenation was a coal–oil slurry, and the process was operated in two stages—with *liquid-phase* hydrogenation first transforming the coal into "middle oils" (boiling between ~180 and 325°C) and subsequent *vapor-phase* hydrogenation then converting these oils to gasoline, diesel fuel, and other relatively light hydrocarbons.

Liquid-phase hydrogenation of *lignites* and other similar low-rank coals was usually accomplished at 475–485°C (890–905°F) under pressures of 25 to 30 MPa (3500 to 4300 psi) by cycling feedstock and products through a series of four 18-m-tall, 1-m-diam reactors and removing gases and light liquid hydrocarbons while the mixture passed from one reactor to the next. The hydrogenation catalyst was iron oxide,[2] which was mixed into the coal slurry in amounts of 2 to 5% on the coal; but in some operations the coal was also impregnated with a 1.3% aqueous solution of ferrous sulfate.

Under these conditions, 50–55% of the (moisture- and ash-free) coal material could be converted to middle oils with average hydrogen expenditures of ~9 Mcf/ft^3 of reactor volume hr^{-1} (~270–280 kg/m^3 hr^{-1}). In addition, heavy oils representing some 40–42% of the coal were obtained, and these were mostly recycled and used for slurrying fresh coal.

For liquid-phase hydrogenation of *bituminous* coals, very similar procedures were employed, but pressures were in the range of 35 to 70 MPa (5000 to 10000 psi); and with hydrogen expenditures of ~11 Mcf (product) bbl (~195 m^3/100 liters), some 60–62% of the (moisture- and ash-free) coal substance was converted into middle oils. Depending on whether hydrogenation was conducted at 30–40 MPa or at higher pressures, either NH$_4$Cl-promoted tin oxalate or iron oxide was used as catalyst.

Table 13.1.1 shows typical yields of liquid products manufactured in this manner.

---

[2] Commonly termed red mud, *luxmasse*, or *Bayermasse*, this was a residual by-product of bauxite processing and sufficiently cheap to be massively used as a throwaway "once-through" catalyst.

Table 13.1.1

Liquid-phase hydrogenation of coal[a]

	Lignite[b]	Bituminous coal[c]
Feedstock		
Raw coal	100.0	100.0
Slurrying oil	51.4	132.5
Products		
Middle oils (b.p. <325°C)	21.5	52.4
Heavy distillate oils	33.6	66.7
Heavy centrifuge oils	42.9	96.6
Heavy coker oils	10.9	12.8
Hydrocarbon gases	10.5	20.5

[a] After Krönig [2]; all quantities in tons.
[b] With 52% moisture; hydrogenated at 30 MPa.
[c] With 8.5% moisture; hydrogenated ay 70 MPa.

Vapor-phase hydrogenation [2, 3] was generally conducted in downflow reactors over fixed catalyst beds at 30 MPa. Average residence times in the reactor were 2–4 min. The products were then cooled to ~30°C by heat exchange against fresh feedstock, freed from unreacted hydrogen, depressurized in stages in order to obtain separate streams of "rich" and "lean" by-product gas, and finally fractionated to gasolines, etc. Distillate middle oils were recycled. On tungsten sulfide catalysts prepared from ammonium sulfo-tungstate or over WS_2 supported on HF-activated Fuller's earth, better than 50% of the middle oils could be converted to (70–75 octane) gasoline, and only 8% appeared as gaseous hydrocarbons (Note A). Hourly space–time yields of gasoline were typically 30 lb/ft³ of catalyst (~480 kg/m³).

Although now only of historical interest, it is worth noting that a first commercial plant for such two-stage coal hydrogenation was built as early as 1927 at the Leuna (Central Germany) works of I. G. Farbenindustrie AG. This was initially designed to produce 100,000 metric tons/yr (~2500 bbl/day) of gasoline by hydrogenating lignites (and some lignite tars) at 20 MPa, but proved so attractive that its capacity was later expanded to 650,000 tons/yr (~16,250 bbl/day) and then adapted for manufacture of aviation fuels as well as gasolines.

Evidently spurred by Leuna's success, several essentially similar installations (each incorporating improvements that reflected the experiences of its predecessors and the results of an intensive continuing research program conducted by I. G. Farben and its associates) were subsequently commissioned elsewhere. In 1935, Imperial Chemical Industries (ICI) thus built a 150,000-ton/yr plant at Billingham (County Durham) that processed bituminous coal (but, in response to demands created by the outbreak of World War II, later shifted to creosote feedstocks). And from the mid-1930s on, three more coal hydrogenation plants (with a combined capacity of 930,000 tons/yr) were constructed in Germany at Scholven (1936), Gelsenberg (1939), and Wesseling (1941). In the same period, five large installations—at Magdeburg and Böhlen (1936), Zeitz (1939), Most (1941), and Blechhammer (1943)—also began using much the same technology for manufacturing gasolines, aviation fuels, and heating oils from coal *tars*.

Since 1945, the coal hydrogenation plants have been phased out or converted to upgrading of heavy crude oils. But until the mid-1950s, further laboratory and pilot plant work on direct coal hydrogenation was carried out, particularly in the United States, in efforts to improve the efficiency and economics of this technology. The extensive US Bureau of Mines program of those years, which included operation of a 200-bbl/day synthetic liquid fuels demonstration plant at Louisiana, Missouri, has been described in a series of Bureau reports and information circulars [5]; and results obtained in Union Carbide's test facility at Institute, West Virginia, which was designed to convert 300 tons of coal/day into chemicals rather than into liquid fuels [6], have been reported by, inter alios, Woodcock and Tenney [7]. The information gained from this work and from hydrogenation studies directed toward elucidation of coal structure [8, 9] has proved of considerable importance in the subsequent development of H-transfer hydrogenation methods.

13.2 "Second-Generation" Liquefaction Processes

Liquefaction techniques now under development are direct descendents of classic Bergius hydrogenation technology and often bear close resemblance to it as well as to each other (Note B). Like their predecessors, most involve more or less actively catalyzed interaction between molecular hydrogen and coal–oil slurries at elevated temperatures and pressures. But because they place greater reliance on H transfer from the slurry oil to the coal, they can be operated under less severe conditions;[3] and by more careful attention to liquids recycling and to heat-coupling of their component steps, they also promise to achieve significant economies and greater overall efficiencies. So far, however, few of the second-generation coal liquefaction processes described in recent technical and patent literature [10] have pro-

---

[3] Although the slurrying oils used in Bergius hydrogenation were also potential H-donors, the higher temperatures employed in the German processes tended to operate against efficient hydrogen transfer, and the classic plants depended almost entirely on reaction between the coal and molecular hydrogen.

gressed beyond testing in relatively small "process development units," and judgment on their ultimate performance in full-size industrial installations must therefore still be withheld.

Farthest advanced of the new liquefaction techniques—and from some points of view one of the most interesting—is the *H-Coal* process [11, 12], which is an outgrowth of Hydrocarbon Research Inc.'s earlier development of a method for hydrogenating petroleum feedstocks.[4]

The core of the process is a unique "ebullated bed" reactor in which a coal–oil slurry reacts with hydrogen-rich gas while passing through an expanded, i.e., partially fluidized or "ebullated," bed of cobalt molybdate on alumina. This configuration is achieved by continuously admitting fresh slurry at the bottom of the reactor and choosing sizes of coal (-100 mesh) and catalyst pellets ($1\frac{1}{2}$–6 mm) which ensure that the product mixture, i.e., unreacted coal, ash, liquids, and gases, can leave through the top of the reactor without carrying catalyst with it. The height of the catalyst bed is controlled by regulating the quantity of catalyst and the concentration of coal in the slurry; and constant catalyst activity is secured by continuously withdrawing spent catalyst and replacing it with fresh material, usually at a rate of 1 lb/ton of coal passed through the reactor. Uniform reactor temperatures are maintained by the internal circulation of the catalyst and by returning most of the heavier, nonvolatile oils to a chamber below the distributor plate, where they are used to thin the incoming slurry to the desired consistency.

Hydrogenation is normally conducted at $\sim 850°F$ (455°C) and 2500 psi (17.5 MPa) with dry coal feed rates of up to 100 lb/ft^3 of expanded catalyst per hour, and liquid yields are reportedly 3–4 bbl/ton of (moisture- and ash-free) coal material. Overall conversion of coal to oils and C_1–C_3 hydrocarbon gases is better than 90%.

For downstream processing of the crude oils, conventional refining technology is used, but adapted to the particular raw material needs of the process. In the currently favored scheme, oil *vapors* leaving near the top of the reactor are first cooled and depressurized in order to recover unreacted hydrogen and hydrocarbon gases, and then sent to atmospheric pressure distillation from which a synthetic light crude and a heavy fuel oil fraction are obtained. *Liquids*, separately withdrawn and cooled, are dispatched to a hydrocyclone from which the low-solids-content "overflow" stream is recycled to slurry preparation, while the "underflow" is taken to vacuum-distillation. The heavy distillate oils produced in this unit are combined with the heavy fraction from atmospheric pressure distillation, and the vacuum

---

[4] This is currently used in two units with a combined capacity of 80,000 bbl/day at Cities Services' Lake Charles, Louisiana, refinery.

bottoms are gasified with oxygen to produce the syngas that supplies hydrogen to the reactor.

A simplified flow sheet for the complete operation is shown in Fig. 13.2.1.

Since 1968, the H-coal process has undergone intensive testing in a 250-lb/hr ($\sim$110-kg/hr), $8\frac{1}{2}$-in. (21.5-cm) -i.d. development unit in which runs with bituminous as well as with subbituminous coal have been successfully completed. A 600-ton/day pilot plant with a 5-ft (1.5-m) -i.d. reactor is now under construction at Catlettsburg, Kentucky, and expected to provide engineering data for later scale-up to commercial sizes.

In parallel work, Exxon Corporation is developing its EDS (*Exxon Donor Solvent*) process [13], which liquefies coal by noncatalytic H transfer from recycle oil and regenerates the donor by hydrotreating and returning a portion of the coal liquids.

The first step of this process takes place in a plug flow reactor, or so-called dissolver, where gaseous hydrogen and hydrogenated distillate oils, usually boiling between 400° and 850°F ($\sim$205 and 455°C) interact with pulverized coal at 800–870°F ($\sim$425–465°C) and 1500–2000 psi (10.5–14 MPa). The hydrogen atmosphere promotes dissolution of the coal substance by participating in the stabilization of free coal radicals (see Section 5.5), and also tends to prolong the effective "donor life" of the oils by partly rehydrogenating hydrogen-depleted liquids. The product mixture is then taken through heat exchangers, flashed to atmospheric pressure in order to extract hydrogen

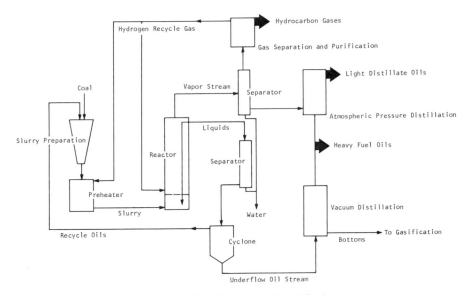

Fig. 13.2.1 Simplified flow sheet of the H-Coal process.

and C_1–C_3 gases, and, after conventional hydrotreating, fractionated to yield naphtha and heavier distillate oils of which a portion is returned to the dissolver for reaction with fresh coal. The vacuum bottoms from distillation are subjected to Exxon's proprietary "flexi-coking" [14], which delivers additional heavy oil; and the residue cokes are gasified with oxygen or air, depending on whether hydrogen or a fuel gas for in-plant use is required.

Overall, this procedure yields ~ 3 bbl of distillate oils per ton of (dry) coal with boiling points below 1000°F (~ 537°C), and because coal dissolution and upgrading can be conducted as entirely independent operations, the final products are virtually sulfur-free oils (see Table 13.2.1).

The EDS process is currently being further developed in a 1-ton/day test facility, and a 250-ton/day demonstration plant is under construction at Baytown, Texas. Exxon anticipates that all necessary engineering and economic data for a commercial plant will be available in the early 1980s.

Gulf Oil is meanwhile pursuing a similar conversion route with its *SRC-II* process [15], which grew out of SRC (*solvent refined coal*) technology (see Section 14.2) but is designed to produce synthetic oils rather than substantially sulfur- and mineral matter-free solid fuels.

SRC-II starts by pressurizing a slurry of dry, pulverized coal in recycle oil with hydrogen at 1500–2500 psi (10.5–17.5 MPa) and briefly preheating the mixture in a digestor to 700–750°F (~ 370–400°C) in order to initiate thermal decomposition and exothermic dissolution of the coal material. The slurry is then transferred to a "dissolver" where heat released by liquefaction is allowed to raise the temperature to 820–870°F (~ 440–465°C) and excess heat is dissipated by injection of cold hydrogen. Overhead vapors generated in the dissolver are withdrawn through a series of separators and finally fractionated at atmospheric pressure into light and heavy oils boiling,

Table 13.2.1

Products from EDS liquefaction of Illinois bituminous coal

	Naphtha (b.p. < 200°C)		Heavy oils (b.p. > 200°C)	
	Crude	Hydrotreated	Crude	Hydrotreated
Composition, wt %				
Carbon	85.6	86.8	89.4	90.8
Hydrogen	10.9	12.9	7.7	8.6
Nitrogen	0.21	0.06	0.66	0.24
Oxygen	2.82	0.23	1.83	0.32
Sulfur	0.47	0.005	0.41	0.04
Density, gm/cm³	0.87	0.80	1.08	1.01
Boiling range, °C (°F)	70–200 (~ 160–390)		200–540 (~ 390–1000)	

respectively, below and above 400°F (~ 205°C), while liquids (with entrained solid matter) from the dissolver are taken to vacuum-distillation. As in other liquefaction processes, a portion of the distillates recovered from this unit is returned to the digestor for slurrying fresh coal, and vacuum bottoms (typically with $\sim 30\%$ mineral matter) are gasified with oxygen to produce hydrogen for the process. Tables 13.2.2 and 13.2.3 list some of the properties of oils obtained in this manner in laboratory rigs.

So far, SRC-II has only been investigated in relatively small process development units, but larger-scale development has commenced in a 50-ton/day pilot plant which came on stream in Tacoma, Washington, in late

Table 13.2.2

SRC-II oil yields from bituminous coal[a]

C_1–C_4 hydrocarbon gases	16.6	
Naphtha, b.p. <380°F (193°C)		11.4
Middle distillates, b.p. 380–480°F		9.5
Heavy distillates, b.p. 480–850°F		22.8
Total liquids to 850°F (450°C)	43.7	
Residue oils, b.p. >850°F	20.2	
Carbonaceous solids	3.7	
Water	7.2	
Ash	9.9	
CO, CO_2, and H_2S	3.4	
	104.7	

[a] After Schmid and Jackson [15]; expressed as wt % of dry coal.

Table 13.2.3

SRC-II oil properties[a]

	Light distillates	Fuel oils
Boiling range, °F (°C)	100–400 (~ 38–200)	400–900 (~ 200–480)
API gravity	39	5
Flash point		168°F (75.5°C)
Composition, wt %		
Carbon	84.0	87.0
Hydrogen	11.5	7.9
Nitrogen	0.4	0.9
Sulfur	0.2	0.3
Oxygen	3.9	3.9

[a] After Schmid and Jackson [15]; oils from bituminous coal.

1977. An important observation in laboratory studies was that coal mineral matter in the digestor and dissolver displays significant catalytic activity for dissolution [16].

Other liquefaction procedures identified in recent technical literature incorporate one or more of the features illustrated by the H-coal, EDS, and/or SRC-II process and differ only in the details of their operating sequence or in the particular manner in which they handle and upgrade the crude liquefied product mixture. However, some of these processes are also specifically designed to produce environmentally acceptable *heavy fuel oils* rather than feedstocks from which synthetic light distillate oils and gasolines can be obtained.

One example is the CCL (*Catalytic Coal Liquids*) process [17], which Gulf Oil is seeking to develop concurrently with its SRC-II technology in order to create capability for converting eastern US high-sulfur bituminous coals into clean heating oils. Although Gulf has so far released few details, this process is known to center on reacting a coal–oil slurry with gaseous hydrogen over specially designed fixed beds of a proprietary catalyst that is claimed to have an exceptionally long active life and to be substantially unaffected by coal ash constituents that would poison other catalysts. A 1-ton/day test facility has been more or less continuously operated at Harmarville, Pennsylvania, since 1975 and yielded ~3 bbl of specification fuel oil per ton of (dry) coal. A larger pilot plant is being planned.

In the Lummus Corporation's CFFC (*Clean Fuels from Coal*) process [18], which is also primarily intended to produce heavy fuel oils with $<0.5\%$ S and $<0.1\%$ ash from high-sulfur coals, the unique feature is a proprietary technique for removing mineral matter from the crude coal liquids. After liquefaction in an upflow reactor over an ebullated catalyst bed, the product mixture is stripped of middle oils (b.p. $<600°$F) and mixed with a 425–500°F kerosene cut that promotes rapid settling of dispersed solid matter. This mixture is then separated at 300–600°F and 100 psi in an inclined gravity settler from which an ash-free overflow and a high-ash underflow are obtained. The overflow is subsequently fractionated at atmospheric pressure into the kerosene cut and a heavy oil (b.p. $>600°$F), while the underflow is vacuum-distilled in order to recover additional oil and vacuum bottoms that are gasified for production of process hydrogen. In small test units, up to 95% of bituminous coal material could in this manner be converted into low-ash, quinoline-soluble heavy liquids, and up to 75% of the sulfur in the feed coal could be eliminated as H_2S.

Consolidation Coal Company's CSF (*Consolidation Synthetic Fuels*) process [19] resembles Exxon's EDS technology in that it decouples liquefaction and downstream hydrotreating, but limits the first step to *solvent extraction* of the coal with recycle light distillate oils at 750–800°F/150–400 psi (~ 400–$425°C/1$–3 MPa). The extract solution is then taken through a pressure filter at $650°F/150$ psi in order to remove residual solid matter and dispatched to catalytic hydrogenation at 800–$850°F/3500$–4500 psi (~ 425–$455°C/25$–32 MPa). Components of this sequence have been successfully demonstrated in bench-scale facilities, and the entire process is now to be further developed in a redesigned 20-ton/day pilot plant at Cresap, West Virginia, where Consol studied hydrogenation of coal tars in the 1950s [20].

Although further (pilot plant) development of all liquefaction processes now thought to be candidates for eventual commercialization is extensively funded by US government agencies (and successful technologies needed in the future for large-scale conversion of coal to liquid hydrocarbons are therefore likely to be readily accessible), the Pittsburgh Energy Research Center[5] is also actively developing two liquefaction methods.

One of these is the *Synthoil* process [21], which primarily differs from other liquefaction techniques in its operating conditions. Like SRC-II, Synthoil initiates liquefaction by preheating a 1:2 coal:oil slurry in a hydrogen-rich atmosphere, but seeks to accelerate reaction by continuously recirculating hydrogen through the digestor and agitating the slurry thereby. Further conversion is then accomplished in a packed-bed upflow reactor over $\frac{1}{8}$ in. (3 mm) pellets of cobalt molybdate on SiO_2/Al_2O_3 at $\sim 870°/4000$ psi ($\sim 465°C/28$ MPa); and here, too, fast recycling of hydrogen ensures rapid reaction by maintaining turbulent flow conditions. (The high turbulence also serves to clean catalyst surfaces by attrition.) As a result, liquefaction, and some concurrent hydrocracking of heavy liquids, is essentially complete within 2 to 5 min.

In a 400-lb/day reactor system, this process has yielded ~ 3.3 bbl of relatively low-boiling oils (with $\sim 0.3\%$ sulfur and $\sim 0.2\%$ ash) per ton of dry coal. But in extended test runs, progressive abrasion of the catalyst has tended to gradually raise the ash content of the product oil to as much as 2%, and it may therefore be necessary to insert an additional ash removal step. An 8-ton/day pilot plant in which Synthoil is to be perfected is under construction at Bruceton, Pennsylvania.

The other technique under active investigation at the Pittsburgh Center is the *COSTEAM* process [22], which is intended to produce low-sulfur fuel oils from lignites and subbituminous coals and for this purpose reacts coal–oil slurries with crude syngas (50–60% $CO + 30$–50% H_2). Conversion is assisted by the natural moisture content of the coal, which increases the hydrogen partial pressure in the reactor by the shift reaction, and by mild catalytic activity of iron-bearing compounds in mineral matter.

---

[5] Formerly a part of the US Bureau of Mines and now administered by the US Department of Energy.

In tests to date, pulverized coal (with at least 70% -200 mesh) was slurried with 30–45 wt % recycle oil, pressurized at 4000 psi (~ 28 MPa) with syngas at rates of 25 to 75 Mcf/ton of coal, and then liquefied by heating for 1–2 hr at 850°F. The product mixture was then moved through a separator from which unconsumed recycle gas was recovered and flashed to atmospheric pressure in order to extract light condensates; and the heavy liquid fraction was freed of entrained solids in a centrifuge before it was fractionated. So far, however, COSTEAM has only been studied in 10-lb/hr bench units, and several components of the process are still conceptual.

Notes

A. Donath has traced the development of catalysts for coal and coal tar hydrogenation in two reviews [3, 4] and there also described in some detail how different catalysts affected the yield and quality (i.e., the octane numbers) of synthetic gasolines.

B. Like coal gasification (see Chapter 12), coal liquefaction has attracted the interest of numerous energy companies who are seeking to develop *proprietary* technology; and most of the procedures disclosed in patents [10] differ from each other merely in matters of detail, e.g., handling of the crude "primary" coal liquids. The methods outlined in this section illustrate the principal directions followed in current R & D.

References

1. E. E. Donath, *in* "Chemistry of Coal Utilization" (H. H. Lowry, ed.), Vol. 2. Wiley, New York, 1945.
2. W. Krönig, "Katalytische Druckhydrierung." Springer, New York, 1950.
3. E. E. Donath, *Adv. Catal.* **8,** 239 (1956).
4. E. E. Donath, *in* "Chemistry of Coal Utilization" (H. H. Lowry, ed.), Suppl. Vol., p. 1041. Wiley, New York, 1963, *Fuel Processing Technol.* **1,** 3 (1977).
5. A. C. Fieldner, H. H. Storch, and L. L. Hirst, Bureau Mines Information Circular No. 7322 (1945), 7352 (1946), 7417 (1947), 7446 (1948), 7518 (1949), 7565 (1950), 7618 (1951), 7647 (1952), 7663 (1953), 7699 (1953), 7756 (1954), 7794 (1955), 7904 (1959); U. S. Bureau Mines Rep. Invest. No. 4456 (1948), 4651 (1949), 4770 (1950), 4865 (1951), 4942 (1952), 5043 (1953), 5118 (1954), 5236 (1955).
6. J. R. Callaham, *Chem. Eng.* **59** (6), 152 (1952); Anonymous, *ibid.* **60** (3), 195; **60** (9), 122; **60** (12), 180 (1953).
7. W. A. Woodcock and A. H. Tenney, *Jt. Symp. Future Arom. Hydrocarbons, April* American Chemical Society (1955).
8. R. A. Glenn, *Fuel* **33,** 419 (1954); **34,** 201 (1955).
9. J. P. Schumacher, H. A. van Vucht, M. P. Groenewege, and L. Blom, *Brennst. Chem.* **36,** 215 (1955).
10. G. K. Goldman, Chemical Process Rev. No. 57. Noyes Data Corp., Park Ridge, New Jersey, 1972.
11. U. S. Dept. of Commerce, NTIS PB-234 203, pp. 254 et seq. (1974).
12. A. L. Coun and J. B. Corns, Evaluation of Project H-Coal; U. S. Dept. of the Interior, Office of Coal Research, Washington, PB-177 068 (1967).
13. L. E. Furlong, E. Effron, L. W. Vernon, and E. L. Wilson, *Chem. Eng. Progr.* **72** (8), 69 (1976).

14. D. E. Blaser and A. M. Edelman, Preprint No. 27–28, 43rd Mid-Year Meeting, API, Toronto, Ontario, 1978.
15. B. K. Schmid and D. M. Jackson, *Ann. Conf. Coal Gasif. Liquef.*, *3rd* Univ. of Pittsburgh, Pittsburgh, Pennsylvania, August 1976.
16. C. H. Wright and D. E. Severson, *Am. Chem. Soc. Div. Fuel Chem.* **16** (2), 68 (1972).
17. S. Chung, *Conf. Mater. Probl. Res. Opportunities Coal Conversion, Ohio State Univ.* p. 263 (1974); cf. also South Africa Patents No. 05793X/04 and No. 2A7502-628.
18. M. C. Sze and G. J. Snell, U. S. Patent No. 3,932,266 (1976); 3,856,675 (1974); 3,852,182 (1974); *Proc. Am. Power Conf.* **37,** 315 (1975).
19. E. Gorin, H. E. Lebowitz, C. H. Rice, and R. J. Struck, *Proc. World Pet. Congr., 8th, Moscow* (1971); cf. also U. S. Patent No. 3,143,489.
20. J. A. Phinney, *AIChE Meeting, Pittsburgh, Pennsylvania, June* (1974).
21. P. M. Yavorsky, S. Akhtar, and S. Friedman, *Chem. Eng. Progr.* **69** (3), 31 (1973); S. Akhtar, S. Friedman, and P. M. Yavorsky, *AIChE Symp. Ser.* **70** (137), 101; P. M. Yavorsky, S. Akhtar, J. Lacey, M. Weintraub, and A. Reznick, *Chem. Eng. Progr.* **71** (4), 79 (1975).
22. H. Appell, I. Wender, and R. Miller, *Chem. Ind.* (*London*) No. 47 (1969); H. Appell, E. Moroni, and E. Miller, Preprints, *Am. Chem. Soc. Fuel Chem. Div.* **20** (1), 58 (1975).

CHAPTER 14

SOLVENT EXTRACTION

Since solvent extraction of coal at temperatures below the onset of active thermal decomposition delivers no commercially valuable materials other than montan waxes and resins (see Section 7.3), which can be more conveniently and economically obtained from other sources, *partial* extraction of coal now commands little interest. But very considerable practical importance attaches to the possibility of almost completely dissolving, or colloidally dispersing, thermally decomposing coal in H-donors under less severe conditions than liquefaction requires.

In principle, a distinction between near-total coal dissolution and liquefaction is, of course, arbitrary. Whether the products of interaction between coal and a suitable solvent at 350–450°C are solids or liquids depends ultimately, as noted earlier, merely on how much hydrogen is transferred to the coal; and some liquefaction techniques do, in fact, prefer to produce the liquids by downstream hydrotreating of a first-stage coal *extract* rather than by more drastic hydrogenation of the raw feed coal. However, because energy markets demand solid as well as liquid (and gaseous) fuels, and a coal extract solution can be readily filtered, dissolution of coal is becoming increasingly attractive in its own right as a means for converting environmentally unacceptable high-ash and/or high-sulfur coals into clean, virtually ash-free, solid fuels at minimum processing costs.

14.1   Pott–Broche and Related Extraction Processes

The range of conditions under which coal will almost totally dissolve, but not liquefy, is illustrated by two classic German processes that served as points of departure for the subsequent development of "solvent refining" techniques in the United States (see Section 14.2).

The first of these was the Pott–Broche process [1, 2], which was commercially exploited in a 125-ton/day plant at Welheim between 1938 and 1944. In that installation, finely pulverized dry coal was slurried with 2–3 times its own weight of a 4:1 tetralin:cresol mixture or, later, a middle oil from coal and coal tar hydrogenation (see Section 13.1), and then transferred to a digester in which the slurry was pressurized at 10–15 MPa and heated for ~ 1 hr at 415–430°C. Thereafter, the reaction mixture was cooled to 150°C and filtered under 300–400 kPa pressure at ~ 8.5 l/min m^2 of filter surface in order to remove residual solids, and the filtrate was vacuum-distilled to recover "spent" solvent that was subsequently rehydrogenated before being recycled. The yield of ash-free extract obtained in this manner depended to some extent on the type and petrographic composition of the coal, but generally ranged between 78 and 84 wt % of (dry, ash-free) coal material charged to the process.

In the Welheim plant, the regenerated extract was initially seen as a more convenient hydrogenation feedstock than raw coal and was, in fact, for some time processed into gasoline and diesel fuels by high-pressure (70 MPa) Bergius technology (see Section 13.1). In later years, however, it was more profitably used as raw material in the manufacture of carbon electrodes.

The other scheme was developed by Ude and Pfirrmann [3, 4] and differed from the Pott–Broche process primarily in digesting the coal-solvent mixture under gaseous hydrogen and somewhat higher pressures (~ 30 MPa instead of 10–15 MPa). This, as noted in Section 13.2 in reference to Exxon's EDS liquefaction method, had the effect of extending the useful life of the H-donor solvent by slowing its net rate of hydrogen depletion, and also shortened the time required for complete dissolution of the coal from ~ 60 to ~ 30 min. As a result, the regenerated extracts contained more hydrogen, as well as less oxygen, nitrogen, and sulfur than the Pott–Broche products; and this composition endowed them with melting ranges and combustion characteristics that even made it possible to use them as fuels in stationary diesel engines.

14.2   Solvent Refining of Coal

The feasibility of using "total" extraction as means for producing clean fuels from eastern US coals was briefly investigated in the early 1960s by Spencer Chemical Company under a contract with the then US Office of Coal Research [5]; and a Spencer affiliate, the Pittsburgh & Midway Coal Mining Company (Pamco), at that time modified the Ude–Pfirrmann process so that it could be effectively operated at much lower pressures (~ 10 instead of ~ 30 MPa). Pamco also improved some design details of the filtration step. But further efforts to develop economically viable large-scale extraction processes, now dubbed "solvent refining" of coal, began only some 10 years later, when it became evident that environmental concerns would militate against substitution of coal for increasingly costly oil and natural gas by disallowing the use of many of the coals otherwise most readily at hand.[1]

Because it is, technically, merely a prematurely arrested liquefaction process, solvent refining is a uniquely flexible conversion scheme whose product slate can, within wide limits, be varied at will through regulating hydrogen consumption. Figures 14.2.1 and 14.2.2 illustrate this with reference to typical behavior of high-volatile bituminous (hvb) coals. In such cases, $\sim 2\%$ hydrogen additions suffice to allow dissolution of 90% of the organic coal substance, and $\sim 80\%$ of this material appears in the solvent-refined coal (SRC) product. More extensive hydrogen transfer to the coal marginally improves overall efficiencies (to as high as 95%), but greater proportions of the feedstock then end up in liquids and hydrocarbon gases, and the melting range of the regenerated solid product is progressively lowered. Lesser hydrogen additions, while increasing the SRC fraction of the product stream, leave increasingly large unconverted coal residues, as well as, in most cases, unacceptably high concentrations of organic sulfur in the solvent-refined material.

The choice of specific operating conditions, i.e., pressure, temperature, coal-to-solvent ratio, and residence time in the reactor, is therefore largely determined by what product distribution is desired; and where SRC yields are to be maximized in the interests of greatest hydrogen economy, that

---

[1] The major obstacle is commonly the high organic sulfur content, which, when the coal is burned, would lead to excessive SO_2 emissions (see Section 15.1) and which cannot be reduced by physical coal-cleaning methods (see Section 8.1).

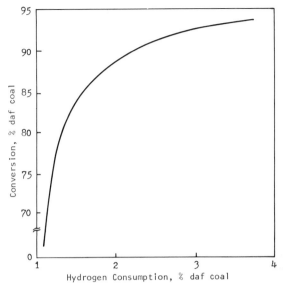

Fig. 14.2.1 Typical variation of coal conversion efficiency with hydrogen consumption. High-volatile bituminous coal, treated at 820–840°F/1000–2300 psi (~440–450°C/7–16.5 MPa); solvent:coal ratio = 2:1–3:1; residence time 40–45 min.

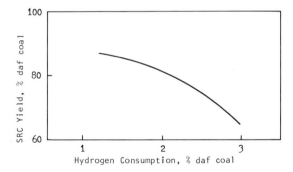

Fig. 14.2.2 Typical variation of SRC yield with hydrogen consumption. High-volatile bituminous coal, treated at 830°F/1700 psi (~445°C/12 MPa); solvent:coal ratio = 2:1–3:1; residence time 40 min.

choice also requires a compromise between conversion efficiency and hydrogen consumption.

Two major test facilities are currently seeking to develop solvent refining for commercial use in the United States. One is a 50-ton/day installation at North Fort Lewis, Washington, which Pamco (now a subsidiary of Gulf Oil Corporation) is operating under ERDA sponsorship [6], and the other is a 6-ton/day pilot plant run by Catalytic Inc. at Wilsonville, Alabama [7], on behalf of the Electric Power Research Institute (EPRI). Both have been on stream since 1974, use virtually identical process technology, and meet their solvent requirements by recycling a portion of the coal liquids, but they complement each other in exploring alternative ancillary techniques (e.g., filtration and SRC regeneration methods) that may improve equipment performance or reduce operating costs.

In what is currently the preferred mode of operation (see Fig. 14.2.3), solvent refining begins with pulverized (90% -200-mesh) coal, which is dried to $\sim 3\%$ moisture, slurried with 3–4 times its own weight of recycle solvent, pressurized at 1700 psi (~ 12 MPa) with gaseous hydrogen, and briefly heated at 825–850°F (440–455°C). This mixture is then sent through a dissolver, which provides an average retention time of ~ 40 min; cooled to ~ 600°F (315°C); taken through a gas separator in order to remove unreacted hydrogen, hydrocarbon gases, H_2S, and water vapor; and depressurized to ~ 115 psi (~ 0.8 MPa) before being passed through a precoated rotary drum or leaf filter. (The filter precoat normally consists of diatomaceous earth.) Separation of solvent from dissolved coal material is accomplished by vacuum-distillation to 600°F at 2 psi (~ 14 kPa); and the SRC product (which, depending on its hydrogen content, melts somewhere between 300° and 400°F and therefore leaves the distillation column as liquid vacuum bottoms) is then solidified by cooling.

Some typical operating conditions (at the Wilsonville plant [8]) are shown

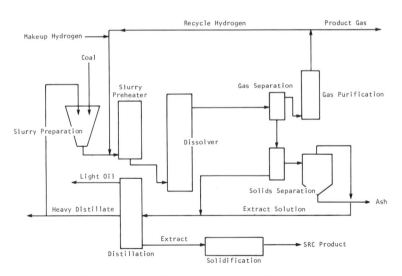

Fig. 14.2.3 Simplified flow sheet of the SRC process.

in Table 14.2.1, and typical results reported from that plant with hvb coals are summarized in Table 14.2.2.

The fact that the Fort Lewis facility attained a better than 85% on-stream factor within 1 year of start-up, and that the Wilsonville plant could conduct continuous test runs for periods up to 75 days, points to the generally high technical reliability of the SRC process and the equipment used for it. It must, however, be noted that some subbituminous coals have created difficulties by tending to polymerize with the recycle solvent during slurry preparation and, more important, that overall coal conversion levels have been found to correlate *inversely* with performance of the downstream filter units [7]. When more severe reaction conditions were necessary in order to enhance H transfer to the coal and thereby ensure adequate removal of sulfur (as H_2S), conversion efficiencies increased (see Fig. 14.2.1), but the rates at which the reaction product mixture could be satisfactorily filtered

Table 14.2.1

SRC processing conditions, Wilsonville pilot plant[a]

Coal feed rate, lb/hr	500
Coal concentration in slurry, %	25–33
Slurry feed rate, gal/min	3.6
Dissolver temperature, in/out, °F (°C)	815/825 ($\sim$435/440)
Dissolver pressure, psi (MPa)	1700 ($\sim$11.5)
Filtration rate, gal/hr/ft^2	11
Hydrogen consumption, % of daf coal	2–2.2

[a] After Deppe [8].

Table 14.2.2

Typical operating results, Wilsonville pilot plant[a]

	Illinois hvb	Kentucky hvb	Pittsburgh hvb
SRC product	63	71	63
Unreacted coal residue	7	7	7
C_1–C_4 gases	6	7	7
C_5–350°F distillates	6	4	4
350–750°F distillates	10	6	12
Hydrogen sulfide	2	2	2
CO + CO_2	6	3	2
Water	6	3	3
Hydrogen consumption	2.4	2.0–2.2	2.5
Conversion efficiency	93	92	92

[a] After Deppe [8]; all data expressed as percent of dry, ash-free coal material into process.

typically declined. Current work is therefore giving special attention to possible improvement of filter press operations and to potential alternative methods for removing solid matter from SRC solutions.

The most promising of these alternatives appears to be a solvent precipitation technique, similar to that used by the Lummus Corporation in its CFFC (liquefaction) process (see Section 13.2). This technique is based on observations [9] that dispersed solid matter in coal liquids tends to agglomerate and settle out quickly when the heaviest liquefaction products are precipitated by cooling or, better, by addition of a predominantly aliphatic "antisolvent"; and tests in a 30-lb/hr development unit [10] have confirmed that SRC solutions can be clarified to $<0.1\%$ ash by adding as little as 10% antisolvent at $400°$ to $650°F$ (~ 200 to $340°C$) and passing the mixture through a cyclone or an inclined gravity settler. As a rule, the efficiency of solids removal improves with increasing aliphatic character of the antisolvent, and the most useful are therefore solvents with a Watson characterization number of 10.5 or greater.[2]

Notwithstanding the fact that SRC coprecipitated with particulate matter (and amounting to 5–15% of the total SRC product) can*not* be recovered, solvent precipitation has been claimed [11] to be economically more attractive than conventional pressure-filtration.

Some interest has also been shown in the possibility of stripping SRC solutions of particulates by high-gradient magnetic separation [12]. Originally developed for upgrading kaolins, this method can remove 70–90% of FeS_x-bearing constituents, but its practical value is rendered doubtful by its inability to extract other ash material.

Meanwhile, Japan is joining the United States in efforts to further refine SRC technology and improve process economics. Mitsui & Co. Ltd. has announced [13] that it is building a 5-ton/day pilot plant, scheduled to begin operations in mid-1978, at its Omuta coke works. Plant design will utilize laboratory studies on some 50 different coals carried out since 1972 by a Mitsui SRC research consortium, and agreement has been reached with Gulf Oil Corporation, the principal SRC developer in the United States, for an exchange of technical information relating to solvent refining. Mitsui expects that SRC material will be used for production of high-quality coke and as a binder in the manufacture of formed coke (see Section 11.6), as well as for "clean" thermal generation of electric energy.

[2] This number, sometimes also referred to as the UOP characterization factor K, is defined as

$$K = (\text{molal average boiling point})^{1/3}/\text{specific gravity}$$

with the b.p. expressed in degrees Rankin ($= °F + 460$), and the specific gravity measured at $60°F$.

14.3 Supercritical Gas Extraction

For almost 100 years, it has been known that the vapor pressure of a solid or liquid can be greatly increased by contacting it with a compressed gas [14] and that this effect is the more pronounced the greater the gas density (which is always highest at the critical temperature[3]). Under appropriate conditions and with a suitable gas, it is therefore possible to transfer into a vapor phase substances that are otherwise substantially nonvolatile.

This technique, termed *supercritical gas extraction* and first used in the 1950s for deasphalting petroleum fractions, has been investigated at the British National Coal Board's Research Establishment as a means for extracting thermally generated nondistillable coal liquids at temperatures at which they suffer little secondary degradation [15]. With toluene or other light coal-derived liquids at $\sim 750°F$ ($\sim 400°C$) and a pressure of 10 MPa, up to 35 wt % of bituminous coal could be extracted into the vapor phase, and the extract could be recovered by merely transferring the mixed vapors to another vessel and there reducing the pressure. (This lowers the density and, hence, the "solvent power" of the extractor gas and causes the extract to precipitate.)

As a general rule, extract yields varied with the (nominal) volatile matter contents of the feed coals; and the extracts proved to be virtually ash-free low-melting solids with a ring-and-ball index of $\sim 70°C$ and significantly greater hydrogen content than the feed coal. Table 14.3.1, which compares a typical hvb coal extract and residue with the coal from which they derive, illustrates this.

Unlike SRC processes, supercritical gas extraction does not transfer external hydrogen to the coal or alter its chemical composition in any other way, and the somewhat lower oxygen, nitrogen, and sulfur contents of the extract are entirely due to pyrolytic reactions in the coal while it is being raised to extraction temperature. For this reason, gas extraction can*not* be used to prepare low-sulfur products from coals that contain relatively large amounts of organic sulfur. And since the extract represents at best only 35% of the coal material fed to the process, any future commercial application of the technique would also require profitable disposal of the residual char. But in this connection it is important to observe that the char typically still retains two-thirds of the (nominal) volatile matter content of the feed coal (see Table 14.3.1), and that it should therefore be at least as useful as the raw coal for generation of electric energy or as a gasification feedstock.

---

[3] The critical temperature (T_c) is the temperature above which the gas cannot exist as a liquid and the critical pressure is the pressure required to cause liquefaction at T_c.

Table 14.3.1

**Compositions of a supercritical gas extract
and residue from hvb coal**[a]

	Feed coal	Extract	Residue
Carbon, %	82.7	84.0	84.6
Hydrogen, %	5.0	6.9	4.4
Nitrogen, %	1.85	1.25	1.90
Sulfur, %	1.55	0.95	1.45
Oxygen, %	9.0	6.8	7.8
Atomic H/C ratio	0.72	0.98	0.63
Volatile matter, %	37.4	—	25.0

[a] After Whitehead and Williams [15]; all data expressed on dry,
ash-free coal.

On the assumption that the extracted residue can command the same per million Btu price as the raw coal, an economic analysis of a conceptual commercial plant has indicated that supercritical gas extraction could compare favorably with an SRC process; and the National Coal Board is therefore proceeding to develop the technique in a pilot plant under the auspices of the European Coal and Steel Community.

References

1. A. Pott and H. Broche, *Glückauf* **69**, 903 (1933).
2. H. Broche and W. Reinmerth, Kohlenextraktion, *in* "Ullmann's Enzykl. d. Techn. Chemie," Vol. 10, 3rd ed., p. 570. Urban & Schwarzenberg, Munich, 1958. C. Kröger, *Erdöl Kohle* **9**, 441, 516, 620, 839 (1956).
3. F. Ude, French Patent No. 800,920 (1936).
4. T. W. Pfirrmann, U. S. Patent 2,167,250 (1939).
5. Spencer Chemical Co., R & D Rep. No. 9, Office of Coal Research, Dept. of the Interior, Washington, D.C. (1965).
6. G. R. Pastor, D. J. Keetley, and J. D. Naylor, *Chem. Eng. Progr.* **72** (8), 27 (1976).
7. R. Wolk, N. Stewart, and S. Albert, *EPRI J.* May (1976).
8. W. L. Deppe, "Electrical World." McGraw-Hill, New York, 1975.
9. H. J. Rose, U. S. Patent 1,875,502 (1932); M. Pier *et al.*, U. S. Patent 1,993,226 (1935); K. Schönemann, U. S. Patent 2,060,447 (1936).
10. E. Gorin, C. J. Kulik, and H. E. Lebowitz, *Nat. Meeting, 169th*, American Chemical Society, Fuel Chemistry Div., Philadelphia, Pennsylvania, April 1975.
11. J. D. Batchelor and C. Shih, *Nat. Meeting, 68th*, American Institute Chemical Engineers, Los Angeles, California, November 1975.
12. R. R. Oder, *IEEE Trans. Magn.* September (1976); E. Maxwell *et al.*, *ibid.*; C. J. Lin *et al.*, *ibid.*
13. Mitsui Trade News, Vol. 14, No. 3 (1977).
14. P. F. M. Paul and W. S. Wise, "The Principles of Gas Extraction," M & B Monograph CE/5. Mills & Boon, London, 1971.
15. J. C. Whitehead and D. F. Williams, *J. Inst. Fuel* **49**, 182 (1975).

CHAPTER 15

ENVIRONMENTAL ASPECTS
OF COAL UTILIZATION

Regardless of how it is used, coal, like other hydrocarbon fuels, always generates some process wastes that, if freely discharged or carelessly disposed of, would cause serious air pollution and attendant ecological damage and, in some instances, even pose major health hazards. As industrial uses of coal expanded and coal plants grew in size and complexity, as well as in number, much attention was therefore given to operating procedures that mitigated these ill effects; and increasingly concerted efforts to improve these procedures (which now include chemical capture of harmful components in waste streams, as well as physical containment of effluent materials) have provided means for keeping coal operations as environmentally "clean" as contemporary pollutant emission standards require.

15.1   The Nature of Coal Pollutants

Because of their very different environmental impact, it is important not only to distinguish between solid, liquid, and gaseous pollutant wastes that can form from mined coal (Note A), but also to discriminate among pollutants in each category.

(a)   Of the two kinds of particulate pollutants on which attention has been focused, *coal dusts* (raised during transport of coal in open trucks or rail cars [see Section 9.1], from coal storage piles [see Section 9.3], or during coal preparation [notably by grinding and thermal drying; see Section 8.2]) are especially objectionable on aesthetic grounds, but as a rule only (or mainly) of concern because of their adverse effects on ambient air quality (and the intolerable nuisance that their fallout in settled areas would create). There is no substantive evidence that such dusts are otherwise harmful unless they hold significant proportions of *respirable* (<1 μm) particles *and* are inhaled over long periods of time, or contain unusually high concentrations of biologically deleterious water-soluble inorganic salts (Note B). In the former case, they would eventually cause serious lung disease (pneumoconiosis; see Note C), and in the latter, accumulation of coal dust in streams or lakes could conceivably render the water unsuitable as an aquatic habitat. Whether, as is sometimes asserted, coal dusts are also inimical to plants when depositing on their leaves is very doubtful.

Rather greater potential hazards could, however, be posed by *inorganic particulate matter*, which can be identified with fly ash emitted from suspension-fired combustion systems (see Section 10.4) or gasifiers (see Sections 12.2 and 12.3). The bulk of such matter consists of partly fused (and therefore relatively inert) mixed oxides of iron, calcium, aluminum, and silicon; and this threatens the environment in the main only through the inordinately large amounts of solid material that, unless otherwise controlled, would be loaded into the atmosphere (Note D). But more serious dangers to ecosystems and public health could develop if certain trace constituents of fly ash (which are generally complexed with organic matter in the coal and only isolated by combustion) are set free to enter a biological cycle. Particular anxieties are being expressed over mercury (which is potentially the most toxic substance in coal ash), as well as over arsenic, lead, cadmium, selenium, and fluorine. Details of the ecological pathways that these elements follow are, to some extent, still uncertain, and much remains also to be learned about how their oxides are converted into more destructive forms. However, the toxicity of arsenic and lead in mammals is well known; cadmium, selenium, and fluorine have been shown to be potent plant poisons; and the lethal effects of organic mercury compounds (notably methyl mercury, which can apparently accumulate in several species of fish and shellfish) have recently been underscored by the so-called Minamata disease (Note E).

(b)   Among pollutant *gases* or vapors emitted by coal plants, environmentally the most dangerous are acid gases, i.e., SO_2, SO_3, and NO_x from coal combustion and H_2S from carbonization, gasification, and liquefaction of coal. Aside from their pungent, nauseating odors (and consequent impact on ambient air quality), all these gases are extremely toxic; and, over time, they can severely stunt plant growth or impair respiratory functions in animals even when diluted to concentrations deemed "safe" for short exposures. Oxides of nitrogen and sulfur have also been identified as major factors in smog formation (Note F) and can cause particularly widespread ecological damage, sometimes far from their source, when precipitating as "acid rains" or forming weak solutions of nitric and sulfurous/sulfuric acids with atmospheric moisture at ground levels.[1] As such, they can do extensive damage to man-made structures as well as to biosystems (Note G).

Hydrocarbons and related oxygenated compounds generated by carbonization, liquefaction, and some types of gasification (e.g., Lurgi gasification; see Section 12.2) are, strictly speaking, by-products rather than process wastes, but become potent pollutants if accidentally discharged. Especially persistent and deleterious are phenols and polycyclic aromatics, all of which tend to be much less susceptible to microbial or abiotic degradation than aliphatic or monocyclic compounds. Because of their mobility, these substances can readily pass into rivers and lakes and then enter plants and animals by ingestion of the contaminated waters. Except in massive doses, absorption of hydrocarbons (or their derivatives) by plants appears to have only transitory ill effects, but in mammals, many of such coal tar chemicals are carcinogenic or can cause various respiratory and metabolic diseases.

(c)   With respect to *liquid effluents*, environmental concerns center mostly on acid mine waters and surface runoffs from open coal storage piles. Such effluents develop where inorganic (pyritic) sulfur in exposed coal is oxidized by air to SO_2 and a variety of iron sulfates and there is water to carry them further afield as aqueous solutions.[2] The ecological effects of these pollutants are then very similar to the effects of acid rains, but always more localized and therefore even more destructive.

Some attention must, however, also be paid to safe disposal of less well-defined waste water streams from coal processing. Coming from plants in which coal is cleaned, carbonized, gasified, or liquefied, etc., such streams may contain entrained coal fines, mineral matter (or ash), and coal tar hydrocarbons, as well as a variety of spent organic or inorganic chemicals

---

[1] Once discharged into the atmosphere, SO_2 can remain free for up to seven days and migrate with air currents over hundreds of kilometers. Acid rains in Scandinavia have thus been traced to SO_2 emitted from generating plants in southern England and continental Europe.

[2] It is important to note that organically bound sulfur in coal does not participate in these oxidation processes, and that coals containing little pyrite consequently pose no environmental hazards from acid mine waters or runoffs even if their total sulfur contents are substantial.

used at various points in the processing sequence to condition cooling water, promote catalysts, improve product separation, or otherwise facilitate the conduct of the particular operation. Because the compositions of the result-ant effluents are process- and sometimes even site-dependent, little is known about their environmental effects; but for practical purposes it may be assumed that they are more likely to be deleterious than beneficial.

Except for coal dusts, the pollutant wastes emitted from coal plants are, of course, not unique to such installations. Airborne inorganic particulate matter resembling fly ash in composition and properties is produced in many mineral-quarrying and -processing operations. SO_x and NO_x form and are released by combustion of fuel oils or natural gas, as well as by com-bustion of coal. H_2S and SO_2 are major waste products of natural gas and petroleum processing. And inadvertent or unavoidable hydrocarbon emis-sions from automobiles, refineries, oil tankers, etc., certainly far surpass present (or any reasonably projected) levels of hydrocarbon emission from coal plants.

Nor are *all* potential ecological and public health hazards thought to be posed by coal pollution likely to become real problems. Past experience suggests that much of the waste material released from industrial operations tends to be harmlessly dissipated or eliminated by chemical breakdown and/or fixation in innocuous forms before it can accumulate and do serious damage. However, the severe air and water quality deterioration already experienced in most industrialized countries makes it obvious that current pollutant loads, i.e., the quantities of deleterious wastes discharged into the natural environment, already exceed the levels that the environ-ment can accept without suffering long-term degradation; and concern must also be expressed over the fact that recent studies have revealed several previously unsuspected ill effects from pollution. There are, for example, indications that hazards may on occasion be created by synergy between pollutants that individually are relatively harmless.

In efforts to remedy earlier, still persistent, ecological damage and ensure a cleaner environment despite continued industrial growth, most countries have, since the late 1950s, enacted legislation which requires that concentra-tions of noxious effluents do not exceed specific limits set for each such pollutant (Note H), and industry compliance with these requirements has made emission control measures integral components of modern coal operations.

15.2   Pollution Abatement Technologies

In most cases it is technically possible to capture or destroy a particular type or class of pollutant matter by any one of several alternative methods, and which is used in any specific instance depends primarily on costs. How-

ever, since industrial wastes almost always contain a wide variety of eco-
logically harmful constituents, reduction of pollutant concentrations to or
below permitted maximum discharge levels usually requires submitting the
raw effluents to a number of successive or simultaneous treatments before
they are finally released.

(a) *Suspended Solids* (or "*Particulates*"). Solid matter entrained in gas
streams (e.g., fly ash in flue gas) is usually removed by electrostatic precipitation or by capturing it in scrubbers or fabric ("baghouse") filters. Cyclones,
which can at best only extract 80% of gas-entrained particles and which are
therefore insufficient to meet contemporary particulate emission standards,
are only used in conjunction with other devices (e.g., ahead of a scrubber;
see below).

In large installations (e.g., thermal power stations or large industrial
boilers), the preferred gas cleanup system is an electrostatic precipitator,
which consists of negatively charged wire or plate electrodes between parallel,
positively charged collector plates and is normally operated with a 40–60
kV potential difference. Gas entering such a system will then ionize and
electrically charge entrained solid particles, which consequently move trans-
versely to the collector plates; and accumulated solids are transferred to a
disposal hopper by periodically vibrating (or "rapping") the plates. Mathematical models that describe the working of electrostatic precipitators, and
which can be used to design them and predict their performance, have been
developed by, inter alios, Gooch and Francis [3].

Suitably sized and optimally operated modern "high-efficiency" precipitators can capture over 99.5% of entrained solids from a gas stream without
causing a significant pressure drop in it. Actual performance is, however,
critically affected by

(1) the gas velocity, which must be sufficiently low to permit transverse
migration of particles to the collector plates and which therefore cannot, as
a rule, exceed 4–8 ft/sec (1.2–2.5 m/sec), and

(2) the effective resistivity of the particles, which depends on their temperature, moisture content, and chemical composition.

Where resistivities at conventional operating temperatures (250–300°F $\equiv$
120–150°C) are greater than $\sim 5 \times 10^{10}$ ohm-cm, the precipitators must be
used at a higher temperature (typically, 350°F $\equiv$ 175°C)[3] and/or designed
to present a larger collector plate area.[4]

---

[3] In practice, this is achieved by placing the precipitator *below* rather than, as is more usual,
above the air-heater section of the boiler. However, the higher temperature also expands the
volume of raw gas that must be treated and consequently requires a larger precipitator than
would otherwise be needed.

[4] Because of widely different needs, electrostatic precipitators are designed with plate areas
ranging from 100 ft² to over 900 ft²/Mcf min⁻¹.

Wet scrubbers, which are well-proven means for stripping gaseous constituents (e.g., SO_2 and NO_x; see below) from gas streams, have also been successfully used as dust collectors in some industrial operations. The most effective types employ either a venturi, through which a countercurrent water jet is injected into the gas, or a packed bed of plastic spheres, which are violently agitated by the incoming gas as it moves against a water spray. However, for efficient particle capture, scrubbers require a relatively large pressure drop; and this, coupled with the fact that they tend to compound downstream problems from excessive gas cooling and carry-over of mists, has so far precluded their use in large boiler installations. Nevertheless, scrubbing could emerge as an attractive alternative method of removing particulates from flue gas if it proves feasible to adapt it for simultaneous capture of SO_2, e.g., by adding lime to the scrubber fluids and making use of the reaction

$$CaO + SO_2 + \tfrac{1}{2}O_2 \rightarrow CaSO_4$$

Several electric utility companies and the US Environmental Protection Agency (EPA) are actively investigating this option [4].

More complete and less costly particulates removal from flue gases may also become possible through use of fabric bag filters through which gas flow can be periodically reversed in order to discharge accumulated solids and unclog pores [5]. Usually made from knitted glass fibers, such filters can retain very fine (< 1-μm) particles, which are difficult to capture by electrostatic precipitation but which future, more stringent emission standards will undoubtedly require to be removed before flue gas is released into the atmosphere (Note I). Aside from having to be operated at somewhat higher temperatures than electrostatic precipitators in order to avoid condensation of moisture and attendant clogging, bag filters are generally simple to install, use, and maintain. But there is still little practical experience of their performance on large industrial or utility boilers, and considerable further testing is required before their value can be assessed.

(b)   *Liquid Effluents.*   Because of the great variety of industrial effluent streams and the wide periodic swings in composition of streams that originate from a *single* source (e.g., a power station or coal-processing plant). liquid effluent treatment is, as a rule, specifically designed to meet the requirements of each particular case.

The usual approach involves temporary containment of the raw effluents in a waste treatment pond from which appropriately clarified water is eventually either recycled to the plant or released into a natural water course.

(In the former case water quality is primarily determined by process requirements, and in the latter it is now mostly governed by standards under statutory provisions that limit the maximum concentrations of deleterious contaminants that may be discharged.) Suspended solids in the ponded waste are then allowed to settle (with settling sometimes accelerated by additives that promote aggregation of fine particles), and noxious dissolved matter is rendered harmless by addition of acids or alkalis, by microbial or abiotic oxidation, by solvent extraction, and/or by adsorption (e.g., on active carbon).

Because of its flexibility ponding, followed by appropriate treatment of the confined waste, is technically a very effective pollutant control method. However, in order to avoid problems that could arise from faulty containment, close attention must be given to proper siting, sizing, and construction of the ponds; and in many instances it is also important to exercise great care in the final disposal of pond sludges, which, if leachable, could adversely affect the quality of groundwater. Since it is obviously impractical to completely remove all potentially hazardous substances from a complex waste stream, it is also imperative, before completing the design of the treatment circuit, to determine which components must be removed, how completely they need to be stripped out, and in which sequence the different treatment steps must be carried out.

(c) *Gaseous Pollutants.* Most of the noxious or potentially injurious gases generated in the course of coal combustion or processing form in such small amounts or are so quickly destroyed by atmospheric oxidation that they can be safely vented. But serious environmental damage and health hazards would be caused if the very much larger volumes of SO_2 and NO_x from combustion or of H_2S from carbonization, gasification, and liquefaction were similarly dispersed;[5] and even before imposition of statutory air quality standards that limited the emission of these gases, much attention was given to modified operating procedures that either reduced their formation or more or less completely removed them from the stack gas before it was released into the air.

In a few instances, where this is not precluded by excessive costs, compliance with contemporary SO_2 emission standards is now ensured by using low-sulfur coal; and in the future it is also expected to reduce SO_2 emissions to near zero by substituting desulfurized coal products (such as SRC; see

[5] Attempts to disperse "raw" stack gases more effectively by releasing them through taller chimneys sometimes succeeded in reducing ground-level concentrations of SO_2, etc., to acceptable values in the vicinity of the emitting plant, but then merely transferred environmental problems to other (downwind) areas (see Section 15.1).

Section 14.2) for coal per se. However, as matters stand, by far the most common technique involves capturing SO_2 from the flue gas by

(i)   direct interaction with limestone via

$$CaCO_3 \rightarrow CaO + CO_2$$
$$CaO + SO_2 \rightarrow CaSO_3$$

which is partly followed by

$$CaSO_3 + \tfrac{1}{2}O_2 \rightarrow CaSO_4$$

or

(ii)   transiently fixing it by absorption in an alkaline solution and re-generating this solution by reaction with lime.

An example of such "indirect" trapping of SO_2 in calcium salts is the sodium-based double-alkali process, in which the capture reactions are

$$Na_2SO_3 + SO_2 + H_2O \rightarrow 2NaHSO_3$$
$$2NaOH + SO_2 \rightarrow Na_2SO_3 + H_2O$$

with the latter followed by

$$2Na_2SO_3 + O_2 \rightarrow 2Na_2SO_4$$

and regeneration is accomplished by

$$2NaHSO_3 + Ca(OH)_2 \rightarrow Na_2SO_3 + CaSO_3 + 2H_2O$$
$$Na_2SO_4 + Ca(OH)_2 + 2H_2O \rightarrow 2NaOH + CaSO_4 \cdot 2H_2O$$

Variants of these processes include capture of SO_2 by *ammonia*, with formation of ammonium sulfite, which is subsequently oxidized to the sulfate; by an *aqueous solution of ferric sulfate*, which catalyzes conversion of sulfurous acid, formed by absorption of SO_2 in water, to sulfuric acid; and by *magnesia*, in which case the capture and regeneration reactions are

$$MgO + SO_2 \rightarrow MgSO_3$$
$$MgSO_3 + SO_2 + H_2O \rightarrow Mg(HSO_3)_2$$
$$Mg(HSO_3)_2 + MgO \rightarrow 2MgSO_3 + H_2O$$

with magnesia then being recovered by thermal decomposition, i.e., by

$$MgSO_3 \rightarrow MgO + SO_2$$

and SO_2 thus freed processed to elemental sulfur (see Section 12.6).

Unlike direct or indirect fixation of SO_2 in $CaSO_3/CaSO_4$, which is discarded as waste, these variants sometimes offer an economic advantage by yielding a salable product, i.e., ammonium sulfate (a fertilizer), sulfuric acid, or sulfur, but this advantage is at least partly offset by higher operating costs.

Except for SO_2 capture by lime, which can sometimes be effected by injecting pulverized limestone into the burning fuel bed,[6] removal of SO_2 from flue gas is always accomplished by passing the raw gas through a scrubber where, depending on the particular process, it moves against either an incoming slurry (of CaO or MgO) or an aqueous solution (of $NaOH/Na_2SO_3$, NH_3, or ferric sulfate). The exiting gas is then freed of entrained mists and reheated (in order to prevent condensation of moisture in the stack) before it is discharged into the atmosphere.

The simplified flow diagrams shown in Fig. 15.2.1 illustrate typical modes of operation by which 70–90% of SO_2 in flue gas can be stripped out.

NO_x emissions can be controlled by analogous methods, and in some operations (see below) flue gases are in fact *simultaneously* stripped of SO_2 and NO_x. More commonly, however, equally if not more effective reductions to current NO_x emission standards (Note J) are brought about by relatively simple modifications of the combustion process itself.

Since, at sufficiently high temperatures, NO_x (a mixture of NO and NO_2) forms by interaction of nitrogen and oxygen in combustion air, as well as from nitrogen in the coal, and oxidation of atmospheric nitrogen may contribute up to 80% of the total, it has been found practical to suppress formation of NO_x by

(i)   reducing the volumes of excess air admitted through the burners,
(ii) recycling part of the flue gas ($\sim 15\%$) into the combustion air [6], or
(iii) resorting to "staged" combustion, i.e., supplying less than the stoichiometric combustion air to the burners and admitting additional air through separate ports above them [7].

These measures inhibit formation of NO_x by lowering flame temperatures and reducing the volume of oxygen available for $N_2 + O_2 \rightarrow 2NO$. Their efficiencies do, however, depend on the firing method (see Section 10.2). Inherent advantages, due to relatively low flame temperatures, are offered by fluidized coal combustion systems (which typically generate only 60–180 ppm NO_x, as compared with 400–800 ppm formed in conventional pulverized fuel boilers); and in pulverized fuel boilers, tangential firing is most effective for suppressing formation of NO_x.

---

[6] Limestone injection has, however, only proved attractive in fluidized bed combustion systems (see Section 10.5). In conventional pulverized fuel boilers (see Section 10.4), it has generally not been satisfactory, and in such installations, wet scrubbing of flue gas is preferred.

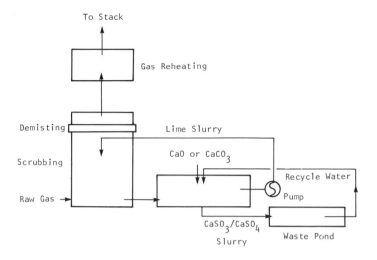

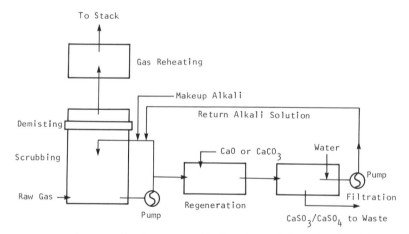

Fig. 15.2.1 Flow sheets for direct (top) and indirect (bottom) desulfurization of flue gas with limestone.

Of the scrubbing methods, the most successful to date is one in which NO_x is reduced to nitrogen by ammonia. The reaction can be formally represented by

$$NO + NO_2 + 2NH_3 \rightarrow 2N_2 + 3H_2O$$

and is advantageously conducted over a transition metal (e.g., copper, chromium, manganese, or vanadium) supported on alumina.

Provided that the raw flue gas is first carefully freed of entrained particulates (which would otherwise tend to plug the catalyst), as much as 80 to 90% NO_x can be abstracted in this manner; and where SO_2 must also be removed, satisfactory capture of both contaminants can be accomplished by using metal oxides (such as MgO) which absorb SO_2 as well as catalyze reduction of NO_x.

Another promising scrubbing technique, which also allows simultaneous removal of SO_2 and NO_x, is based on absorption and resembles the SO_2 absorption procedures outlined above. In one version, in which ammonia is used, the reactions are

$$(NH_4)_2SO_3 + SO_2 + H_2O \rightarrow 2NH_4HSO_3$$
$$6NH_4HSO_3 + 2NO \rightarrow 2NH(NH_4SO_3)_2 + (NH_4)_2S_2O_6 + 2H_2O$$

and the sodium imidodisulfonate is then hydrolyzed to yield ammonium sulfate via

$$NH(NH_4SO_3)_2 + 2H_2O \rightarrow (NH_4)_2SO_4 + NH_4HSO_4$$
$$NH_4HSO_4 + NH_3 \rightarrow (NH_4)_2SO_4$$

Simultaneous decomposition of the ammonium dithionate via

$$(NH_4)_2S_2O_6 \rightarrow (NH_4)_2SO_4 + SO_2$$

yields SO_2, which, where so desired, can be converted to elemental sulfur.

In an alternative version, use is made of

$$Na_2SO_3 + SO_2 + H_2O \rightarrow 2NaHSO_3$$
$$6NaHSO_3 + 2NO \rightarrow 2NH(NaSO_3)_2 + Na_2S_2O_6 + 2H_2O$$

and the sodium imidodisulfonate is decomposed by treatment with calcium nitrite, calcium oxide, and sulfuric acid to yield calcium sulfate, sodium sulfate, and nitrogen. The dithionate is decomposed as above.

Methods for stripping H_2S from process gas streams and for converting H_2S and/or SO_2 to elemental sulfur have been summarized in Section 12.6.

Notes

A. Environmental concerns over coal mining per se relate primarily to the nature and extent of land disturbance caused by operation of open (surface) pits and to the feasibility of acceptable land reclamation after mining ceases. In some cases, concerns are also expressed over destruction of near-surface aquifers. For a discussion of these topics, see Doyle and/or Downs and Stocks [1].

B. It should be emphasized that coal itself, far from being harmful to plants and/or animals, may actually be directly or indirectly beneficial to them. After atmospheric oxidation, which enriches coal in humic acids (see Section 5.1), it becomes a useful soil amendment and organic fertilizer; and because of its high porosity (see Section 4.1), coal can also sorb noxious odor- and color-forming substances (including bacteria) from water. For this latter reason, coal is in fact frequently advocated as a low-cost substitute for active carbons in waste water treatment [2].

C. Pneumoconiosis, sometimes also referred to as black lung disease, was for centuries an inescapable fate of coal miners but has now been largely eradicated by systematic dust control programs in mines. Permissible dust concentrations in coal mines are now set by standards under various industrial health and safety ordinances and enforced by government inspectors.

D. Since pulverized coal combustion systems emit up to 90% of the total ash contents of the coals they burn as fly ash, the flue gas from such installations would contain $10–14 \ gm/m^3$, and a modern 1000-MW generating plant burning bituminous coal with, say, 15% ash would emit some 1200 tons/day. If subbituminous coal with 15% ash is burned, the daily emission would be some 1800 tons.

E. In the 1950s, several fishermen and their families on Minamata Bay, Japan, died or were permanently disabled by paralysis of their central nervous systems after ingesting mercury-polluted fish. This episode prompted studies of mercury levels in North American and European water courses and bans on fishing in lakes in which possibly dangerous Hg levels were indicated.

F. In November 1952, a dense fog in London, England, which was attributed to air pollution by coal combustion and which lasted several days, was the direct cause of death of over 4000 persons, mostly through respiratory or heart failure. It was later found that elderly persons and children are particularly prone to disablement by such smogs.

G. Especially vulnerable to attack by such acids are buildings faced with limestone or limestone aggregates, which are corroded via, e.g., $CaCO_3 + H_2SO_4 \rightarrow CaSO_4 + CO_2 + H_2O$. Similar damage can, however, also be done to metal sheathings, which, additionally, are liable to corrosion by acid sulfates. Classic examples of such attack can be seen in almost all large North American and European cities where coal has been a traditional home-heating and industrial fuel. Enforcement of modern environmental legislation, which among other matters limits the sulfur contents of permitted fuels, has gone far toward eliminating this type of destruction.

H. In some instances, presently permitted maximum emission limits are scheduled to be progressively lowered over the next few years. However, the need to maintain economically viable industry clearly demands some environmental trade-offs, and introduction of increasingly stringent standards, even where otherwise justified, will therefore depend on progress in the development of more refined control technology than is now available.

I. Particles smaller than $\sim 1 \ \mu m$ usually account for less than 0.5 wt % of a fly ash, but are potentially the most hazardous because they are respirable. Inhaled over extended periods of time, they will induce serious lung disease.

J. Current upper NO_x emission limits in the United States range from 0.7 lb per million Btu (set by EPA in 1971) to 0.15 lb (demanded by the State of New York in *new* plants). In this connection, it is however worth noting that pollution by NO_x is almost always a lesser problem than pollution by SO_2: A 1000-MW power station burning a typical bituminous coal with 2% sulfur generates ~ 350 tons/day SO_2, but only ~ 100 tons/day NO_x.

References

1. W. S. Doyle, Strip Mining of Coal: Environmental Solutions, 1976. Noyes Data Corp., New Jersey, 1976; C. G. Downs and J. Stocks, "Environmental Impact of Mining." Halstead Press, Somerset, Britain, 1977.

2. U. S. Dept. Interior, Office of Coal Research, Contr. No. 1401-0001-483, Rep. No. 55 (1971); A. E. Perrotti and C. A. Rodman, *Chem. Eng. Progr.* **69** (11), 63 (1973); Biospheric Research Inc., Washington, D.C., Rep. PB-18455 (1969); A. J. Kehoe, U. S. Patent No. 3,300,403 (1967); C. O. Bunn, U. S. Patent No. 3,798,158 (1974); *Chem. Abstr.* **80,** 809 405 (1974).

3. J. P. Gooch and N. L. Francis, *Symp. Elec. Precip. Contr. Fine Particles, U. S.* EPA-650/2-75-016 (1975).

4. EPA Alkali Scrubbing Test Facility, Limestone Wet Scrubbing Test Results, EPA-650/2-74-010 (1974).

5. *EPA Symp. Fabric Filters Contr. Sub-Micron Particles* EPA-650/2-74-043 (1974).

6. R. E. Thompson, M. W. McElroy, and R. C. Carr, *Proc. Seminar NO_x Contr. Technol.* Electric Power Research Institute, California, SR-39 (1976).

7. W. W. Habelt and B. M. Howell, *Proc. Seminar NO_x Contr. Technol.* Electric Power Research Institute, California, SR-39 (1976); G. A. Hollinden, N. P. Moore, and R. L. Zielke, *ibid.*

INDEX